Management Science and Decision Technology

Jeffrey D. Camm
University of Cincinnati

James R. Evans
University of Cincinnati

South-Western College Publishing
Thomson Learning™

Australia • Canada • Denmark • Japan • Mexico • New Zealand • Philippines
Puerto Rico • Singapore • South Africa • Spain • United Kingdom • United States

Management Science and Decision Technology, 1e, by Jeffrey D. Camm & James R. Evans

Publisher: Dave Shaut
Senior Acquisitions Editor: Charles McCormick, Jr.
Senior Developmental Editor: Alice C. Denny
Senior Marketing Manager: Joseph A. Sabatino
Production Editor: Elizabeth A. Shipp
Media Production Editor: Robin K. Browning
Manufacturing Coordinator: Sandee Milewski
Internal Design: Joe Devine
Cover Design: Tin Box Studio, Inc.
Cover Photographer: © 1999, Photonica
Production House: Pre-Press Company, Inc.
Printer: Westgroup

Printed in the United States of America
2 3 4 5 02 01 00

For more information contact South-Western College Publishing, 5101 Madison Road, Cincinnati, Ohio, 45227 or find us on the Internet at http://www.swcollege.com

For permission to use material from this text or product, contact us by
• telephone: **1-800-730-2214**
• fax: **1-800-730-2215**
• web: **http://www.thomsonrights.com**

Library of Congress Cataloging-in-Publication Data

Camm, Jeffrey D.
    Management science and decision technology / Jeffrey D. Camm and
James R. Evans.
       p. cm.
    Includes bibliographical references and index.
    ISBN 0-324-00715-9
    1. Management science.   2. Industrial management—Mathematical
models.   3. Industrial management—Decision making.   4. Decision
making—Mathematical models.   5. Risk management.   I. Evans, James
R. (James Robert).   II. Title.
    HD30.25.C359   1999
    658.4'03—dc21

                                      99-38723

This book is printed on acid-free paper.

Contents

Preface ix

PART 1 MODELS, DATA MANAGEMENT, AND ANALYSIS

1. Modeling for Decisions and Insight 1

APPLICATIONS AND BENEFITS OF MANAGEMENT SCIENCE 3
Strategic Business Decisions 3   Scheduling Airline Crews 3   Poultry Production 3
Search for Sunken Treasure 4   Financial Strategies 4

DECISION TECHNOLOGY 4
Spreadsheets 5   Spreadsheet Add-Ins 7   Stand-Alone Software 7

DEVELOPING MODELS 8
Descriptive Models 8   *Management Science Practice: Labor Scheduling at Taco Bell 9*
Optimization Models 13   Models Involving Uncertainty 16   The Importance of Good
Data 16

ANALYZING AND SOLVING MODELS 16
Model Analysis 17   *Management Science Practice: Model Analysis of HIV Transmission through
Needle Sharing 17*   Tornado and Spider Charts 22   Solving Optimization Models 24

INTERPRETING AND USING MODEL RESULTS 26

A CASE STUDY OF MODELING AND ANALYSIS: INVENTORY MANAGEMENT 27
Managing Inventory Systems 27   Model Development 29   Model Analysis and Solution 32   Interpreting Results 35   Applying Inventory Theory to Cash Management 37

PROBLEMS 38

NOTES 43

BIBLIOGRAPHY 43

APPENDIX: BASIC SPREADSHEET SKILLS 44
Functions 44   Auditing Your Spreadsheets 46   Creating Tornado and Spider Charts 46

2. Data Management and Analysis 49

APPLICATIONS OF DATA MANAGEMENT AND ANALYSIS 50
Law Enforcement 50   Health Care 50   Retail Operations and Strategy 51   Hotel
Location 51

DATA STORAGE AND RETRIEVAL 51
Management Science Practice: Allders, International 52   Sorting and Filtering Data 55   Using
AutoFilter to Check Data Accuracy 59   Using Data Retrieval Methods in Optimization
Modeling 59

DATA VISUALIZATION 60
Charts 61   Geographic Data Maps 63

DATA ANALYSIS 66
Descriptive Statistics 67   Pivot Tables 68   Probability Distributions and Data Fitting
with Crystal Ball 71   Covariance and Correlation 74   Applying Covariance in Financial
Portfolio Optimization Models 77

REGRESSION ANALYSIS 78
Simple Linear Regression 79   Statistical Issues in Regression 82   Multiple Linear
Regression 86   Building Regression Models 90   Using Regression in Optimization
Models 90

TECHNOLOGY FOR STATISTICAL ANALYSIS 91

PROBLEMS 92

NOTES 94

BIBLIOGRAPHY 95

3. Forecasting 96

APPLICATIONS OF FORECASTING 97
Plant Closure at Allied-Signal 97   Call-Center Demand at L. L. Bean 98   Inventory
Management at IBM 98

TIME SERIES COMPONENTS 98

MODELS FOR TIME SERIES WITH ONLY AN IRREGULAR COMPONENT 102
Single Moving Average 102   Measuring Forecast Accuracy 102   Single Exponential
Smoothing 105

MODELS FOR TIME SERIES WITH A TREND COMPONENT 107
Double Moving Average 107   Double Exponential Smoothing 109

MODELS FOR TIME SERIES WITH A SEASONAL COMPONENT 111
Additive Seasonality 111   Multiplicative Seasonality 113

MODELS FOR TIME SERIES WITH TREND AND SEASONAL COMPONENTS 115
Holt-Winters Additive Model 115   Holt-Winters Multiplicative Model 116

FORECASTING WITH REGRESSION MODELS 117
Management Science Practice: A Forcasting Model for Supermarket Checkout Services 118
Linear Regression for Time Series Forecasting 121   Using Indicator Variables in Regression Models 123

FORECASTING TECHNOLOGY 125

CHOOSING THE BEST FORECASTING METHOD 126

FORECASTING WITH CB PREDICTOR 126
Interpreting CB Predictor Outputs 133

APPLICATIONS OF FORECASTING TO MODEL BUILDING 133

PROBLEMS 136

NOTE 139

BIBLIOGRAPHY 139

PART 2 OPTIMIZATION

4. Linear and Multiobjective Optimization Models 140

THE STRUCTURE OF LINEAR AND MULTIOBJECTIVE
OPTIMIZATION MODELS 141

APPLICATIONS OF LINEAR AND MULTIOBJECTIVE
PROGRAMMING MODELS 143
Land Management at the National Forest Service 143   Credit Collection 144   Research and Development Funding 144   Site Selection 144

MODELING OPTIMIZATION PROBLEMS IN EXCEL 145

BUILDING LINEAR PROGRAMMING MODELS 146
Dimensionality Checks 148   LP Modeling Examples 148   Standard Form 154   *Management Science Practice: Using Transportation Models for Procter & Gamble's Product Sourcing Decisions 162*

SOLVING LINEAR PROGRAMMING MODELS 164
Intuition and LP Solutions 164   When Intuition Can Fail 165   Using Excel Solver 166

INTERPRETING SOLVER REPORTS AND SENSITIVITY ANALYSIS 169
Answer and Limit Reports 170   Sensitivity Report 171   Scenario Analysis 174

SOLVING MULTIOBJECTIVE MODELS 177
Weighted Objective Approach 177   Absolute Priorities Approach 180   Goal-Programming Approach 181   *Management Science Practice: Goal Programming for Site Location at Truck Transport Corporation 183*

TECHNOLOGY FOR LINEAR OPTIMIZATION 184

PROBLEMS 185

NOTES 196

BIBLIOGRAPHY 196

APPENDIX: USING PREMIUM SOLVER TO SOLVE LINEAR PROGRAMS 197

5. Integer and Nonlinear Optimization Models 199

APPLICATIONS OF INTEGER AND NONLINEAR MODELS 200
Project Selection at the National Cancer Institute 200   Crew Scheduling at American
Airlines 200   Mortgage Valuation Models at Prudential Securities 201   Paper Produc-
tion 201

BUILDING INTEGER-PROGRAMMING MODELS 201
Management Science Practice: Sales Staffing at Qantas 204

SOLVING INTEGER-PROGRAMMING MODELS 212
Using Solver for Integer Programs 212   Sensitivity Analysis 214

BUILDING NONLINEAR OPTIMIZATION MODELS 218
Management Science Practice: Kanban Sizing at Whirlpool Corporation 225

SOLVING NONLINEAR PROGRAMMING PROBLEMS 226
Using Excel Solver 226   Sensitivity Analysis 228   The Problems of Nonlinearity 229
Using Premium Solver's Evolutionary Algorithm for Difficult Problems 231

TECHNOLOGY FOR INTEGER AND NONLINEAR OPTIMIZATION 233

PROBLEMS 234

NOTES 241

BIBLIOGRAPHY 241

APPENDIX: USING PREMIUM SOLVER TO SOLVE INTEGER AND
NONLINEAR PROGRAMS 242
Solving Linear Integer Programs 242   Solving Nonlinear Programs 243

PART 3 DECISION, RISK, AND SIMULATION

6. Decision and Risk Analysis 246

APPLICATIONS OF DECISION AND RISK ANALYSIS 248
Environmental Impact Assessment 248   Assessment of Catastrophic Risk 248   Sports
Strategies 249   Risk Assessment for the Space Shuttle 249

STRUCTURING DECISION PROBLEMS 249
Generating Alternatives 249   Defining Outcomes 250   Decision Criteria 250   Deci-
sion Trees 251   *Management Science Practice: Collegiate Athletic Drug Testing 252*   Decision
Strategies 254

UNDERSTANDING RISK IN MAKING DECISIONS 254
Average Payoff Strategy 255   Aggressive Strategy 255   Conservative Strategy 256
Opportunity Loss Strategy 256   Quantifying Risk—Insights from Finance 257   An Ap-
plication of Decision and Risk Analysis: Evaluating Put and Call Options 258

EXPECTED VALUE DECISION MAKING 260

Management Science Practice: The Overbooking Problem at American Airlines 261 An Application of Expected Value Analysis: The "Newsvendor" Problem 263 Expected Value of Perfect Information 265

OPTIMAL EXPECTED VALUE DECISION STRATEGIES 266

Sensitivity Analysis of Decision Strategies 268

TECHNOLOGY FOR DECISION ANALYSIS 268

RISK TRADE-OFFS AND MULTIOBJECTIVE DECISIONS 269

UTILITY AND DECISION MAKING 272

Exponential Utility Functions 275

PROBLEMS 277

NOTES 283

BIBLIOGRAPHY 284

7. Monte Carlo Simulation 285

APPLICATIONS OF MONTE CARLO SIMULATION 286

New Venture Planning 286 Pharmaceutical Research 287 Project Management 287

BUILDING AND IMPLEMENTING MONTE CARLO SIMULATION MODELS 288

SAMPLING FROM PROBABILITY DISTRIBUTIONS 292

BUILDING SIMULATION MODELS WITH CRYSTAL BALL 294

Interpreting Crystal Ball Output 298

STATISTICAL ISSUES IN MONTE CARLO SIMULATION 302

MONTE CARLO SIMULATION EXAMPLES 303

Newsvendor Problem 304 *Management Science Practice: Simulating a CD Portfolio 306* Pricing Stock Options 308

CRYSTAL BALL *TORNADO CHART EXTENDER* 310

OPTIMIZATION AND SIMULATION 312

PROBLEMS 316

NOTES 321

BIBLIOGRAPHY 321

APPENDIX: ADDITIONAL CRYSTAL BALL FEATURES 321

Correlated Assumptions 321 Freezing Assumptions 323 Overlay Charts 323 Trend Charts 323 Sensitivity Charts 324

8. Systems Modeling and Simulation 327

APPLICATIONS OF DYNAMIC SYSTEM MODELS 328

Toll Booth Improvement 328 Designing Security Checkpoints 329 Technology Evaluation 329 Dental Practice Management 329 Forest Fire Management 330

MODELING AND SIMULATING DYNAMIC SYSTEMS 330

QUEUEING SYSTEMS 335

Customer Characteristics 335   Service Characteristics 336   Queue Characteristics 337
System Configuration 337   *Management Science Practice: One Line or More? 338*

MODELING AND SIMULATING QUEUEING SYSTEMS 339

THE DYNAMICS OF WAITING LINES 342

ANALYTICAL QUEUEING MODELS 346

The M/M/1 Queueing Model 347   Other Queueing Models 348   Little's Law 349
Analytical Models vs. Simulation 350   Sensitivity Analysis 351   *Management Science Practice: Queueing Models for Telemarketing at L.L. Bean 353*

MODELING AND SIMULATING DYNAMIC INVENTORY SYSTEMS 354

Using Simulation to Optimize Inventory Systems 357

SYSTEMS SIMULATION TECHNOLOGY 359

Arena Business Edition 359   *Management Science Practice: Designing an Air Force Repair Center Using Simulation 360*

USING SIMULATION SUCCESSFULLY 370

Verification and Validation 371   Statistical Issues 372

PROBLEMS 372

NOTES 378

BIBLIOGRAPHY 374

Appendix A   Normal Distribution Table 379

Index 381

Preface

Management science—the use of mathematical models and quantitative approaches for decision making—originated some 50 years ago in military applications. The discipline flourished in business and industry during the 1950s and early 1960s, as computing power developed and researchers refined the mathematics underlying the core approaches of mathematical programming, decision analysis, and simulation. During the 1970s and 1980s, however, many in business and industry viewed management science as too theoretical and irrelevant to the real problems they faced. The 1990s brought about a renaissance of management science, fueled by the ubiquitous personal computer and powerful desktop software like Microsoft® Excel and many other user-friendly tools for implementing management science theories. More than any time in history, strategic data—such as scanner data, geodemographic data, and electronic data interchange—are available through corporate information systems. What managers need is the ability to make sense out of these data and use them effectively for making sound decisions. It is with this need in mind that we have written this book.

FEATURES FOR TEACHING AND LEARNING

This book focuses on three fundamental topics of management science in eight chapters—decision and risk analysis, optimization, and simulation. The topical material is presented along with the decision technology support required to solve real problems—data management, data analysis, and forecasting, as implemented with spreadsheets and other special software. Although we could have included many other topics, we felt that the inclusion of too many topics would remove the "forest from the trees." Essential topics such as inventory models and queueing, for example, *are* included, but within a broader context of modeling and analysis. This feature not only makes the book unique, but takes management science back to its roots—not a collection of independent models and algorithms, but a philosophy for modeling and gaining insight into complex decision problems. Also, by condensing the material into a shorter and more focused presentation, this book can be used more easily in shorter courses and modules, which are becoming increasingly prevalent in business schools today. However, the book includes sufficient material to make it suitable for traditional quarter or semester-length courses.

With these perspectives in mind, the key features of the book include:

1. Integration of Microsoft Excel for modeling and analysis, but not in a "cookbook" fashion. The focus is on using Excel to gain insight through good modeling and analysis practice. To this end we exploit Excel capabilities like data tables, charts, and database support wherever possible to help illustrate and explain ideas.
2. A unique chapter on Data Management and Analysis that describes key issues needed to use management science effectively—sorting and filtering data, data visualization, and relevant statistical applications.
3. Excel Exercises throughout each chapter to provide students with hands-on practice. These are almost always extensions of text examples, and are experiential in nature so that students can extend the examples and gain insights about the models or the fundamental theory behind them. Answers or explanatory notes to these exercises are provided on the CD-ROM that accompanies the book.
4. Many examples drawn from finance make the book especially relevant to today's business student.
5. Software add-ins and supplements provided on a CD-ROM that accompanies the book: academic versions of *Crystal Ball Pro* and *CB Predictor, TreePlan, Premium Solver for Excel*, and a fully functional evaluation version of *Arena Business Edition*, a state-of-the-art systems simulation package designed expressly for business process improvement. More details about the software are provided in a separate section of this Preface.
6. "Management Science Practice" examples of real applications in real businesses. In addition, each chapter presents several short application write-ups of practical applications of the tools and techniques.

BRIEF OVERVIEW

The first three chapters deal with basic concepts of modeling and decision technology. Chapter 1, Modeling for Decisions and Insight, focuses on model development and analysis. Excel spreadsheets are introduced as a modeling tool, with a significant focus on using data tables, tornado and spider charts, and other spreadsheet-based capabilities to understand and gain insight from the models. A case study that applies these principles and tools to inventory management is included. An appendix describes some essential spreadsheet skills. Chapter 2 addresses Data Management and Analysis, and includes discussions of sorting and filtering data using Excel and the role of data management in optimization modeling. Topical coverage also includes data visualization with charts and geographic data maps; data analysis, including a basic statistics review, pivot tables, probability distributions, and covariance and correlation; and regression analysis. The discussions of data analysis and regression are focused particularly on their use in management science applications. *Crystal Ball* is introduced as a tool for distribution fitting. Chapter 3 focuses on forecasting time series data. The use of forecasting data in modeling and *CB Predictor* add-in are key features of this chapter.

Chapters 4 and 5 address optimization. Chapter 4, Linear and Multiobjective Optimization Models, focuses on developing and solving linear programs for both the single and multiobjective cases. This chapter develops a variety of applications, cultivates intuition for solving LP models, and shows how intuition can fail and why optimization algorithms are needed. *Excel Solver* is used to obtain optimal solutions, and an extended discussion of sensitivity and scenario analysis is provided. The appendix to Chapter 4 describes the use of *Premium Solver* for linear optimization. Chapter 5, Integer and Nonlinear Optimization Models, describes approaches for building and solving integer and nonlinear models. The chapter parallels Chapter 4, with *Excel Solver* being used to find optimal solutions. In the nonlinear case, the problems of nonlinearity and their impact on Solver are explained from an intuitive point of view, and the use of the Evolutionary Solver option in *Premium Solver* is illustrated for difficult nonlinear optimization problems.

The remaining chapters focus on decisions, risk, and simulation. Chapter 6, Decision and Risk Analysis, discusses the formulation of decision problems and approaches for selecting decisions. A major theme of this chapter is the importance and impact of risk in making decisions, drawing on insights from finance. Expected value decision making is described from a broad perspective, encompassing not only decision trees, but other examples such as the newsvendor problem. This chapter also includes a discussion of risk tradeoffs for multiobjective decisions and utility functions.

Chapter 7, Monte Carlo Simulation, focuses on the use of simulation for analyzing risk in decision models. This chapter includes general concepts of Monte Carlo sampling, the use of *Crystal Ball* for implementing Monte Carlo simulations, interpretation of output statistics, and the use of the *Crystal Ball* Tornado Chart extender for understanding model sensitivity. An example of how simulation can be integrated with optimization using *OptQuest* is also included.

Chapter 8, Systems Modeling and Simulation, deals with dynamic system models, principally queueing and inventory. A brief discussion of analytical queueing models is included, but the principal focus is on the use of simulation for modeling and analyzing these systems. Excel spreadsheets are used to develop fundamental concepts and gain intuition about the dynamics of queuing and inventory models. *Arena Business Edition* is introduced as a practical means of modeling and solving dynamic system problems.

SOFTWARE

We are very proud to introduce instructors and students to several exceptional software tools for implementing management science techniques. In completing their coursework, students will gain practical experience with software from Systems Modeling, Decisioneering, Frontline Systems, and Decision Support Services. We now provide a brief description of the software and add-ins found on the CD-ROM that accompanies *Management Science and Decision Technology*.

- **Academic version of Arena® Business Edition, Systems Modeling Corporation, www.sm.com**
  Arena BE is a stand-alone simulation package used for mapping and analyzing business processes. Comparisons of current "as-is" operations with potential "to-be" scenarios allow the right course of change to be charted with confidence.

The academic version of *Arena BE* has all the features and functions of *Arena BE* but limits the size of model that can be created to 50 modules placed in the model and no more than 100 concurrent entities during a run.

- **Crystal Ball® Pro (Student Version), OptQuest™ for Crystal Ball® (Student Version), and CB Predictor™ (Student Version), Decisioneering, Inc., www.decisioneering.com**
  Crystal Ball Pro is a suite of forecasting and risk analysis tools that enhance spreadsheet modeling. Crystal Ball Pro users can immediately use the assumptions CB Predictor creates in a spreadsheet model and move right to optimization with OptQuest.

  Crystal Ball is a Microsoft® Excel add-in that provides the ability to perform Monte Carlo analysis (a technique for simulating real-world situations involving elements of uncertainty) on spreadsheet models. Users can make better-informed decisions based on the true probability of specific outcomes.

  The student version of the Crystal Ball software allows a maximum of six assumptions and six forecasts to be defined per model. Available distribution types include normal, triangular, uniform, custom, Poisson, and exponential. One pairwise correlation can be created per model, there are no overlay charts, simulations are limited to 1,000 trials, and the best fit function can fit a maximum of 100 points to a curve. There is no technical support, except to answer installation questions.

  OptQuest is a Crystal Ball add-in that offers global optimization for spreadsheets under conditions of uncertainty. It determines the optimal choice for a given business decision based on multiple Crystal Ball simulations. The student version allows two decision variables per model. There is no limitation on the output of OptQuest or the extenders.

  CB Predictor is a forecasting tool designed to work seamlessly with risk analysis and optimization technology. It incorporates advanced methods of time-series forecasting into a Microsoft Excel add-in. These techniques analyze historical data to predict future performance. The student version of the CB Predictor software allows one data series per run and does not include regression.

- **Premium Solver™ Version 3.5 for Education, Frontline Systems, Inc., www.frontsys.com**
  Premium Solver is an upwardly compatible extension of the standard Microsoft® Excel Solver, with the capacity to solve much larger problems at speeds much faster than the standard Solver.

  The Premium Solver for Education V3.5 is a full-featured version of Frontline's commercial Premium Solver product, but its capacity (200 variables maximum) and speed are about the same as the standard Excel Solver. Moreover, it starts up in standard Solver mode, where it looks and acts just like the standard Excel 97 Solver, except for a "Premium" button in the main Solver Parameters dialog. When pressed, the Premium Solver functionality appears. This includes the dropdown list for selecting the Simplex Solver, GRG Solver, or Frontline's new Evolutionary Solver, plus revised Options dialogs for all of the extra Premium Solver functionality. Operating in "Premium Solver mode," this Solver offers a range of new functionality that makes it easier to teach optimization and to get students excited about the possibilities of management science and operations research.

- **Academic version of TreePlan™ for Microsoft® Excel, Decision Support Services, www.treeplan.com**
  TreePlan is an Excel add-in that allows the user to build decision trees in worksheets using menu commands and dialog boxes. The decision trees include formulas for summing cash flows to obtain values and for calculating rollback values for determining optimal strategy.

- **Student resource files**
  The CD-ROM accompanying the text also includes answers and notes to the Excel Exercises and Excel spreadsheet files for the examples, excercises, and certain problems in the book.

ANCILLARIES

All instructor ancillaries will be available on a single CD-ROM. The *Instructor's Resource CD* (ISBN: 0-324-00716-7) includes problem solutions, test bank questions and solutions, and PowerPoint slides.

ACKNOWLEDGMENTS

We are extremely grateful to the following reviewers who provided excellent feedback and suggestions for improvement on our original proposal, and on the manuscript of this book:

Sharad Chitgopekar,
Illinois State University

Damodar Golhar,
Western Michigan University

V. Daniel R. Guide, Jr.,
Suffolk University

William Kasperski,
University of Nebraska–Omaha

Iihyung Kim
Purdue University

Ruth A. Maurer,
Walden University

Satish Mehra,
University of Memphis

Bradley Meyer,
Drake University

Taeho Park,
San Jose State University

Finally, we also thank senior acquisitions editor Charles McCormick, Jr., senior developmental editor Alice Denny, production editor Libby Shipp, and others at South-Western College Publishing for the editorial support during the preparation of this text.

Jeffrey D. Camm
James R. Evans

DEDICATION

To
Karen, Jennifer, Stephanie, and Allison
and
Beverly, Kristin, and Lauren

Modeling for Decisions and Insight

O U T L I N E

Applications and Benefits of Management Science
Strategic Business Decisions
Scheduling Airline Crews
Poultry Production
Search for Sunken Treasure
Financial Strategies

Decision Technology
Spreadsheets
Spreadsheet Add-Ins
Stand-Alone Software

Developing Models
Descriptive Models
Management Science Practice: Labor Scheduling at Taco Bell
Optimization Models
Models Involving Uncertainty
The Importance of Good Data

Analyzing and Solving Models
Model Analysis
Management Science Practice: Model Analysis of HIV Transmission through Needle Sharing
Tornado and Spider Charts
Solving Optimization Models

Interpreting and Using Model Results

A Case Study of Modeling and Analysis: Inventory Management
Managing Inventory Systems
Model Development
Model Analysis and Solution
Interpreting Results
Applying Inventory Theory to Cash Management

Problems

Appendix: Basic Spreadsheet Skills
Functions
Auditing Your Spreadsheets
Creating Tornado and Spider Charts

M anagers at all organizational levels face a variety of problems. Consider, for example, a fictitious company, Hanover Incorporated, a producer of printed circuit boards used in custom computer assembly.[1] Hanover produces three types of boards, which we will refer to as boards A, B, and C. Demand for Hanover's products has been steadily increasing, and management recognizes that total demand for its boards will soon exceed production capacity. The company has production facilities in Paris, France, and Austin, Texas, and sells to companies in Brazil, China, France, Malaysia, and the United States. Some of the types of decisions that Hanover faces might be

- forecasting future demand,
- determining how much capacity to have in each plant,
- whether to build another facility or close an existing one,
- how to best ship products from plants to markets,
- where and when to invest cash reserves,
- making pricing decisions,
- determining the mix of employee skills to meet corporate objectives,
- developing advertising plans,
- scheduling production of its three major products to meet demand,
- assigning workers to tasks,

and many others. Developing effective strategies to deal with these types of problems can be a daunting task.

Management science is the scientific discipline devoted to the analysis and solution of complex decision problems. The process of using management science generally consists of three major phases:

1. *Developing a model* of the problem to be solved or the decision situation to be studied. Models help us focus on the important aspects of a problem.
2. *Solving and/or analyzing the model* (usually with the help of a computer). The solution or analysis provides insights that we probably could not obtain without the help of the model.
3. *Interpreting the results.* Model results provide information to make good decisions.

Management science is aided by a diverse collection of computer-based methods and tools for building, manipulating, and solving models, which we refer to as **decision technology.** These methods and tools include spreadsheets, data management, data analysis, special software for implementing management science approaches, and communication tools for linking these together. Spreadsheets, in particular, provide a convenient means to manage data, formulas, and graphics simultaneously, using intuitive representations instead of abstract mathematical notation. Although the early applications of spreadsheets were primarily in accounting and finance, spreadsheets have developed into powerful general-purpose managerial tools for applying techniques of management science. The marriage of management science and decision technology allows managers to better convert data into information and provide powerful insights for decision making. This is perhaps best stated by noted business consultants Michael Hammer and James Champy, who said, "When accessible data is combined with easy-to-use analysis and modeling tools, frontline workers—when properly trained—suddenly have sophisticated decision-making capabilities."[2]

In this chapter we introduce you to the basic elements of management science and decision technology through various examples. Throughout this book, as appropriate, we will describe real applications of management science in business to further support the concepts.

APPLICATIONS AND BENEFITS OF MANAGEMENT SCIENCE

Management science developed as a formal discipline from efforts to improve military operations prior to and during World War II. Applications included the effective deployment of radar and sizing of convoys for transporting materiel across the Atlantic. After the war, scientists recognized that the tools and techniques developed for the military could be applied successfully to many types of business problems. The growth of the field was aided greatly by developments in computing during subsequent decades, and is experiencing a renaissance because of the personal computer.

Management science can be applied to a wide variety of problems, a sampling of which we present next. Complete descriptions of these applications can be found in the references listed at the end of this chapter, and we encourage you to read them.

Strategic Business Decisions

Federal Express has used management science models to help make its major business decisions since its overnight package delivery operations began in 1973. In the early stages of the company's growth, they developed a model to select cities and decide on a route system. This model led to a planning system that has served as the foundation for the over $8 billion corporation that it is today. Management science has been applied to many other aspects of the company: finances, personnel assignments, engine use, and operations (Mason et al., 1997).

Scheduling Airline Crews

The monthly scheduling of airline crews to flights is enormously complex, due to union and FAA work rules and the sheer overall size of the problem. American Airlines, for instance, has over 16,000 flight attendants, over 2,300 flight segments per day, and over 500 aircraft. Crews reside in 12 different cities, called crew bases; a crew must therefore be assigned to a sequence of flights (typically lasting three days) that starts and ends at the same crew base. Total crew costs exceed $1.3 billion annually; a 1 percent improvement in crew utilization therefore translates into a $13 million savings each year. American Airlines has developed and refined a scheduling system called TRIP—Trip Reevaluation and Improvement Program—using a variety of management science techniques (Anbil et al., 1991).

Poultry Production

Sadia Concordia SA is the largest poultry producer in Brazil, processing over 300 million chickens and 11 million turkeys a year. The company has used management science since 1990 to improve decision making, resulting in better conversion of feed to live bird weight, improved utilization of birds, almost 100 percent fulfillment of daily production plans with increased output of higher value products, greater flexibility, and reduced lead time in meeting market demand (Taube-Netto, 1996).

Search for Sunken Treasure

Management science techniques were used to devise a search plan to find a ship, the *Central America*, which sank in 1857 off the coast of South Carolina with three tons of gold on board. The search plan incorporated historical accounts of the storm and the shipwreck, mathematical analysis of drift due to ocean currents and winds, and estimates of the navigational instruments of the period. The plan was used to convince investors that their investment had a reasonable chance of success and led to the discovery of the ship in 1989 (Stone, 1992).

Financial Strategies

The secondary mortgage market is a market created for issuing and trading securities built from portfolios of mortgages, primarily single-family mortgages. The market has grown to a size comparable to the corporate bond market. Prudential Securities, one of the top three firms in this market, uses a full range of management science methods to predict the prepayment of mortgages, to estimate the value of mortgage-backed securities and adjustable-rate mortgages under various interest-rate scenarios, and to structure the best portfolios of mortgage-backed securities (Ben-Dov et al., 1992).

The benefits of using management science and decision technology can be impressive. Consider, for example, the following results achieved by several major companies:

- Making the right decisions about overbooking airline flights, determining the number of discount fares to offer on a flight, and controlling reservations to account for connecting flights over its entire system has saved American Airlines over $500 million per year (Smith et al., 1992).
- Procter & Gamble makes and markets over 300 brands of consumer goods around the world. The company used management science and decision technology in reengineering its North American product-sourcing and distribution system, resulting in pretax savings of over $200 million per year and creating a new Center for Expertise in Analytics within the corporation (Camm et al., 1997).
- A spreadsheet model developed by an internal consulting group at Du Pont helped the company to evaluate new strategies for one of its principal businesses. A key element of this model was the assessment of uncertainties such as competitors' strategies, market size, market share, and prices. The strategy that was ultimately chosen was expected to increase the value of the business by $175 million (Krumm and Rolle, 1992).

Do not think that management science applies solely to multimillion-dollar corporate projects. All businesses, whether large or small, can use management science effectively. A list of additional applications of management science is given in Table 1.1.

DECISION TECHNOLOGY

Decision technology provides the supporting tools for taking management science from the realm of abstraction to reality. While specialists will always be needed for large, complex projects, the technological advances in decision technology that

Finance	Production	TABLE 1.1
Pension fund mangement and investment Cash management Portfolio management Financial planning and control	Inventory planning and control Facility layout Product mix analysis	Applications of management science
Human Resources Management	**Education**	
Work shift scheduling Labor-management negotiation Personnel evaluation and selection Recruitment and promotion	Library management Teaching assignment scheduling Classroom assignment Selection of MBA students	
International Business	**Health Care**	
Global financing and capital structure International marketing channels Global manufacturing and plant location	Nurse scheduling Financial planning Diagnosis and therapy Blood distribution	
Marketing and Transportation	**Natural Resources**	
Advertising budget allocation New product sales analysis Market mix analysis Media planning Retail promotion strategy Distribution planning Fleet configuration Airline operations planning	Hydroelectric system management Mining project evaluation Water pollution control	
	Miscellaneous	
	Police beat design Ski-area design Crisis management Energy planning	

Source: H. B. Eom and S. M. Lee, "A Survey of Decision Support System Applications (1971–April 1988)." *Interfaces*, Vol. 20, No. 3, May–June 1990, 65–79.

have paralleled the growth of personal computing have brought management science to the desktops of every manager. In this book we will use three basic types of decision technology: spreadsheets; spreadsheet add-ins; and special, stand-alone software.

Spreadsheets[3]

You undoubtedly are already familiar with spreadsheets for accounting, financial analysis, or statistical applications. Throughout this book, we will use Microsoft Excel for illustrating the analysis and solution of management science problems. We do so for several reasons. First, spreadsheets are ubiquitous—everybody uses them and they will be available in nearly every business and nonprofit organization. While more powerful computer packages exist at universities and larger corporations, each has its own language and idiosyncracies, and they may not be available after you leave the university or change jobs. Second, spreadsheets provide a creative, flexible tool for users of various skill levels to create, analyze, and solve complex management science models with relative ease. Finally, spreadsheets are a wonderful learning tool for fundamental concepts—the real objective of this book—because of the integration of graphics to provide visualization of difficult concepts, their ability to rapidly update calculations as data change, and the wide array of statistical, mathematical, and database functions needed to manipulate data. We do not purport to make you an expert; however, once you know the basic concepts, learning to use more sophisticated software will be much easier. Finally, we note that this is not intended to be a cookbook on using Excel—our main purpose is to gain insight and understanding about the underlying management science models and techniques. However, many times we find it necessary to provide detailed instructions for entering data or selecting the right options to enable you to apply the tools effectively.

The appendix to this chapter provides a review of some basic Excel spreadsheet skills with which you should be familiar. A folder with the example files used in this book is included on the accompanying CD-ROM. You will need these to do the Excel Exercises that are suggested throughout the book to develop your skills and provide further insights about the concepts we discuss. Solutions to these exercises are included on the enclosed CD-ROM.

USING SPREADSHEETS EFFECTIVELY Well-designed spreadsheets should have several characteristics. First, the spreadsheet must provide accurate results. Because the formulas and functions are more susceptible to undetected errors than standard computer programs, the user must be careful. Good design and documentation can make an application more understandable and less vulnerable to incorrect use. Cross-checking, batch totals (such as row and column sums), and breaking down lengthy formulas into smaller steps can help to reduce errors. Careful and complete testing should help to detect errors and verify results. Second, spreadsheets need to be understood by both their creators and their users. Clear and complete descriptions and a consistent approach to the layout greatly contribute toward comprehensibility. Third, spreadsheets often need to be adapted to changes in a problem or application. In designing a spreadsheet you should consider its possible future uses or variations.

Some tips that will enhance the spreadsheet models you develop are given next.

- *Planning saves time and trouble.* Although there are many ways to correctly design problems with spreadsheet models, much time will be saved if you lay out a tentative structure for the model on paper before turning on the computer.
- *Use help files.* When you have a question, don't hesitate to examine the on-line help files. Spreadsheet software have many capabilities of which few users are aware.
- *Document your worksheet.* You should design your model so that six months from now you can remember its purpose and how it works. First, choose a file name for your model that is both descriptive and simple. You may wish to put a short written description of what the model does on the worksheet and a list of assumptions, definitions, and formulas. Where appropriate, include important instructions and explanatory notes on the face of the worksheet.

NOTE: In the example models provided with this book, we have used the *Comment* feature of Excel to provide explanatory notes and summaries of key formulas. Whenever you see a cell with a small red triangle in the upper right corner, position your cursor over that cell and a comment text box will appear on the screen.

- *Use an effective layout.* A common strategy for laying out a spreadsheet is to separate areas of different use. In general, a good layout should be vertical rather than horizontal. Every spreadsheet should begin with an identification that includes a descriptive title, and often the author's name and date of creation. For many applications, users will enter data directly; thus, a *data section area*, which will hold these data, should be separate from the work area. A separate *model* or *output area* should contain calculated results. Some input data may need to be repeated to enhance understanding and readability. Such input data should be referenced only with cell references or range names from the data section so that if the data change (and this should be done *only* in the data section), calculations will be updated automatically. If new information is provided, changes are made to the data section only; no formulas need to be revised. We will use this structure in the examples in this book.

- *Pay attention to alignment and format.* Put labels over all columns. Use the currency and comma formats where appropriate. Use an integer format (no decimal places) where decimal accuracy is not needed. Use uppercase and lowercase letters just as you would in a handwritten or typed report. Arrange all worksheet data in either columns or rows, not a combination of both.
- *Write clear formulas.* Clear formulas speed up calculations by eliminating unnecessary computation, reducing errors, and making it easier to maintain your spreadsheet. Complex calculations should be divided into several cells. Explanatory comments should be placed next to formula cells if appropriate.
- *Use sample data during development.* Developing a blank spreadsheet model with correct formulas and format is quite difficult. In general, if you develop a model with a data section, you should enter some sample data to determine if your answers make sense. If not, you should carefully check the formulas and assumptions you made.

Spreadsheet Add-Ins

Many add-ins for applying specific management science approaches either are included with software such as Microsoft Excel or are available commercially. Excel includes several add-ins, which are essentially Visual Basic modules that provide additional commands and functions. A typical installation of Excel using the Excel Setup program installs several add-ins, including the Analysis Toolpak, which is a collection of data analysis tools that we will use in this book. However, one important add-in that we will use, Solver, is *not* installed unless you specifically select it. To do so, select the "Custom installation" option and ensure that Solver is included. You may check to see what add-ins are installed by selecting *Add-ins* from the *Tools* menu, and ensuring that the correct add-ins are checked. If the add-in does not appear in the list of available add-ins, click the *Browse* button to find the missing .xla files in the Library folder.

Other add-ins are commercially available from third parties. In this book we will use four other add-ins—the student version of Crystal Ball Pro, a suite of applications for risk analysis; CB Predictor, an add-in for forecasting; Premium Solver, an extended version of the Excel Solver; and TreePlan, an add-in for decision tree analysis. All of these are available on the CD-ROM accompanying this book and should be installed on your computer.

Stand-Alone Software

Many excellent commercial packages for specific management science applications and decision technology support are available. We will discuss some of these throughout the book as appropriate to provide you with some knowledge as to their capabilities, but cannot delve into many details. However, in Chapter 8 we will introduce one such package—*Arena Business Edition*—a limited, but fully functional version of a commercial simulation package. You may wish to install this program on your computer at this time also.

DEVELOPING MODELS

A **model** is a representation or abstraction of the key factors in the problem or decision situation and the relationships among them. A model can be a simple picture, a spreadsheet, or a set of mathematical relationships. Models capture the most important features of a problem and present them in a form that is easy to interpret. Models complement decision makers' intuition and often provide insights that intuition cannot. For example, one early application of management science in marketing involved a study of sales operations (Brown et al., 1956). Sales representatives had to divide their time between large and small customers and between acquiring new customers and keeping old ones. The problem was to determine how the representatives should best allocate their time. Intuition suggested that they should concentrate on large customers and that it was much harder to acquire a new customer than to keep an old one. However, intuition could not tell whether they should concentrate on the 500 largest or the 5,000 largest customers or how much effort to spend on acquiring customers. Models of sales force effectiveness and customer response patterns provided the insight to make these decisions.

People use models to make decisions constantly, and most of the time, are not aware they are using them! For example, a naïve investor who wants to choose between a stock fund and a bond fund uses some "mental model" that considers expected returns, current and future cash requirements, and risks. A sophisticated investor might use a more formal spreadsheet model to project cash flows, while an analyst at a brokerage firm would have access to sophisticated mathematical and computer models. The major difference is the formality and complexity of the models used; formal models provide structure to data and information, and allow you to incorporate many more factors and relationships than can mental models.

We classify models into three major categories: (1) descriptive models, (2) predictive models, and (3) optimization models. Descriptive models simply tell "what is." They describe relationships and provide information for evaluation. An important type of descriptive model is a simulation model. Simulation models replicate the behavior of business situations or systems that have many interacting components and sources of uncertainty to understand the impacts of policy or design decisions. Predictive models, such as forecasting models, characterize relationships among quantitative historical data in order to estimate future outcomes. Optimization models recommend the best decisions to satisfy an objective, such as maximizing profit or minimizing cost. In a highly competitive world, where one percentage point can mean a difference of hundreds of thousands of dollars or more, knowing the best solution can mean the difference between success and failure. Taco Bell uses all three types of models in its labor scheduling system (see Management Science Practice box).

Descriptive Models

A simple, yet powerful, descriptive model is called an **influence diagram**, a diagram that shows the relationships between various quantities and is a useful approach for conceptualizing the elements of a decision problem and their relationships to one another. The elements of the model are represented by circles called *nodes*. Branches

Management Science Practice:
Labor Scheduling at Taco Bell

Taco Bell has approximately 6,500 company-owned, licensed, and franchised locations around the world, with annual sales of about $4.6 billion. Taco Bell labor costs represent approximately 30 percent of every sales dollar and are among the largest controllable costs. They are also among the most difficult to manage because of the direct link that exists between sales capacity and labor. Because the product must be fresh when sold, it is not possible to produce large amounts during periods of low demand to be warehoused and sold during periods of high demand, as is typical in most manufacturing enterprises. Instead, the product must be prepared virtually when it is ordered. And since demand is highly variable and is concentrated during the meal periods (52 percent of daily sales occur during the three-hour lunch period from 11 a.m. to 2 p.m.), determining how many employees should be scheduled to perform what functions in the store at any given time is a complex and vexing problem.

In developing a new labor-management system, Taco Bell identified several requirements. The system should

- be responsive and economical;
- be able to predict labor requirements that minimize labor cost and meet all corporate standards for hospitality, quality, service, and cleanliness for any existing or planned restaurant configuration;
- serve as an effective and efficient in-store labor-management tool to help the store manager to plan and schedule;
- serve as an effective and efficient tool for providing the labor required to achieve Taco Bell's financial targets;
- provide timely feedback for controlling labor cost at all levels; and
- have the inherent flexibility to evolve as Taco Bell evolves.

Although work such as cleaning the store and preparing and restocking ingredients can be scheduled at the discretion of the restaurant manager, there is little or no discretion for scheduling work in response to direct customer needs, such as taking orders, making change, producing a custom product, and delivering orders. Different customers order different items and combinations of items. From 20 to 80 percent of these orders, depending on region, require custom preparation (no hot sauce, extra hot sauce, and so forth). Consequently, few products are inventoried, and those that are can be held for only a few minutes. As a result, customer arrivals drive the work to be performed in the restaurant, and labor capacity must be scheduled accordingly.

The crux of the labor-management problem, then, lies in understanding the quantitative trade-off relationship between labor and speed of service, and speed of service and revenues. To minimize payroll and deliver desired customer service, store managers require a tool that will help them to determine which employee should begin work and which employee should go home at every time interval of the day for every day of the week. At the beginning of this study, this interval was 30 minutes because of restrictions imposed by the

continued

in-store data-storage devices. Today, the company is overcoming this limitation and implementing a 15-minute time increment systemwide. In addition to the number of employees required, such a tool must indicate where they must be positioned and how they are to share the required duties. And before it can determine that, it must know how many customer transactions will take place at the store during every 15-minute interval of every day of the week.

This means that the new labor-management system needed three types of management science models: (1) a *forecasting model* for predicting customer transactions during each 15-minute interval of the restaurant's open time during the next week; (2) a *simulation model* to translate customer transactions to labor requirements, taking into account a wide variety of restaurant configurations, and to generate data for a scheduling model; and (3) an *optimization model* to schedule employees to satisfy labor requirements and minimize payroll. These models must operate in concert while permitting the manager to intervene and make appropriate revisions before implementing the results (Figure 1.1).

Between 1993 and 1996, the number of restaurants using this new system more than tripled, and savings in labor costs went from $3.51 million in 1993 to $16.4 million in 1996. Other benefits, such as reducing variability in the level of service, hospitality, quality, and cleanliness among stores, provides increased levels of customer satisfaction, and hence, higher sales. The senior vice president of technology and quality for Taco Bell stated, "This is not just a labor-management system. It is an incredibly powerful and adaptive decision-making tool to help us reliably predict the potential impact of new, strategic initiatives . . . (and) to assess and avoid risk."

FIGURE 1.1 The structure of Taco Bell's labor-management system

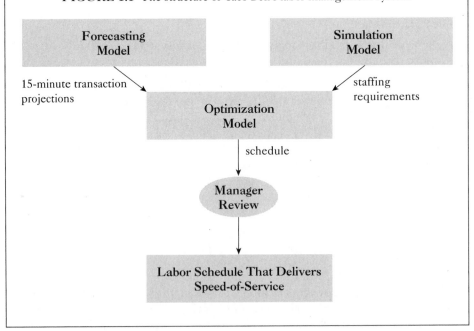

Source: Adapted from Jackie Hueter and William Stuart, "An Integrated Labor-Management System for Taco Bell," *Interfaces*, Vol. 28, No. 1, January–February 1998, 75–91.

connect the nodes and show which elements influence one another. Because of their visual nature, influence diagrams are highly effective means of communication, and they also help in developing more formal models.

Example 1: A Profit Model

To illustrate a simple influence diagram, consider the familiar equation that describes profit:

$$\text{Profit} = \text{Revenue} - \text{Cost}$$

Since revenue and cost influence profit, an influence diagram that captures these relationships is shown in Figure 1.2. We can develop a more detailed model by noting that revenue depends on (is influenced by) the unit price and quantity sold; and that total cost depends on the unit cost, quantity produced, and relevant fixed costs. The enhanced model is shown in Figure 1.3. (An important principle of modeling is to start simply and embellish the model as necessary.)

We may use the influence diagram to guide the development of a more formal model for profit. We must first specify the precise nature of the relationships among the various quantities in Figure 1.3. For example, revenue is computed by multiplying the unit price of the item by the quantity sold. Thus,

$$\text{Profit} = (\text{Unit price})(\text{Quantity sold}) - \text{Cost}$$

Cost is often expressed as fixed cost plus variable cost. Therefore, we have

$$\text{Profit} = (\text{Unit price})(\text{Quantity sold}) - [\text{Fixed cost} + (\text{Unit cost})(\text{Quantity produced})]$$

Note that this model does not tell a manager how much to produce or what price to charge; it simply describes the relationships among these quantities.

In this model we have three types of quantities:

1. **Parameters,** a term used to describe numerical constants; in this case, unit price, unit cost, and fixed cost.
2. **Uncontrollable inputs,** quantities that can change, but cannot be directly controlled by the decision maker. For instance, the demand may vary, but the decision maker cannot directly manipulate this value; thus, demand is an uncontrollable input.
3. **Controllable variables,** often called **decision variables.** These may be selected at the discretion of the decision maker. For example, the quantity produced is a controllable variable.

We may think of models simply as relationships among these three types of quantities.

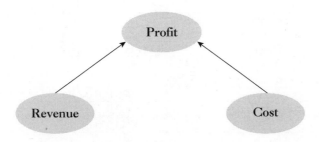

FIGURE 1.2
A simple influence diagram for profit

FIGURE 1.3
An enhanced
influence dia-
gram for profit

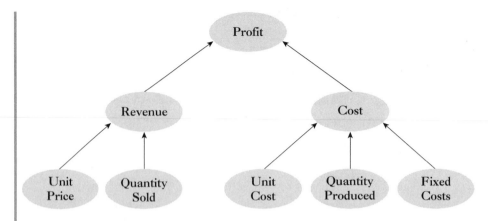

Figure 1.4 shows a spreadsheet for this model. The range A3:B8 is the data section containing the parameters and uncontrollable inputs. The model section is specified in the range A11:C20. The only decision variable, quantity produced, is entered in cell B17. The spreadsheet allows you to enter any value for the quantity produced as well as the uncontrollable input, demand. The spreadsheet shows the output when a value of 40,000 is used for quantity produced and demand is 50,000. Note that the formulas for revenue and cost refer to the cells in the input section. As noted earlier, it is easier to change the input data than to modify the formulas.

FIGURE 1.4
Spreadsheet
model for profit
evaluation

	A	B	C	D	E
1	**Profit Model**		*Key cell formulas*		
2					
3	*Parameters and Uncontrollable Inputs*				
4				Position cursor over this cell to	
5	Unit Price	$40.00		display cell formulas	
6	Unit Cost	$24.00			
7	Fixed Cost	$400,000.00			
8	Demand	50000			
9					
10					
11	*Model*				
12					
13	Unit Price	$40.00			
14	Quantity Sold	40000			
15	Revenue		$1,600,000.00		
16	Unit Cost	$24.00			
17	Quantity Produced	**40000**			
18	Variable Cost		$960,000.00		
19	Fixed Cost		$400,000.00		
20	Profit		$240,000.00		

We can develop a formal mathematical model by defining symbols for the parameters, uncontrollable inputs, and decision variables:

$$P = \text{profit}$$
$$p = \text{unit price}$$
$$c = \text{unit cost}$$
$$F = \text{fixed cost}$$
$$S = \text{quantity sold}$$
$$Q = \text{quantity produced}$$

we have

$$P = pS - [F + cQ] = pS - F - cQ$$

Note the correspondence between the spreadsheet formulas and the mathematical model:

$$\text{Revenue} = pS = \text{B13} * \text{B14} = \text{C15}$$
$$\text{Variable Cost} = cQ = \text{B16} * \text{B17} = \text{C18}$$

If you can write a spreadsheet formula, you can develop a mathematical model!
   This example shows three ways of representing the same model—a picture, a spreadsheet, and a mathematical function. We will focus on spreadsheets and mathematical models in this book, but you should use pictures whenever possible to organize your thoughts.

Excel Exercise 1.1

Modify the profit model spreadsheet (*profit model.xls*) by assuming that demand is related to price by the function $S = 90,000 - 1,000p$. What is the profit for a price of $36 and production quantities of 50,000, 55,000, and 60,000 units?

Optimization Models

An **optimization model** seeks to identify the best values of decision variables to achieve some objective. For example, a grocery-store manager who believes that waiting lines are too long might have dual objectives of maximizing customer service and minimizing costs. Possible decisions might include hiring additional employees (how many?), building larger facilities (what size?), or acquiring new scanner technology (what kind?). Customer service might be measured by the length of waiting lines or the average time that customers wait. Costs might be measured as discounted present value. Notice that the objectives conflict: the decisions that will improve customer service will increase cost. Thus, the manager must make a tradeoff between these objectives. In a similar fashion, a consumer-products company manager who feels that distribution costs are too high might investigate redesigning the distribution system to lower the total delivered cost of the product. This might involve deciding on new locations for manufacturing plants and warehouses (where?), new assignments of products to plants (which ones?), and the amount of each product to ship from different warehouses to customers (how much?).

A third example is one that most financial managers face. Companies have demands for cash: payroll, taxes, accounts payable, and so on. They typically maintain their liquid assets in cash or short-term securities such as treasuries, CDs, and commercial paper, which generally earn higher amounts of interest than checking accounts. Firms incur an opportunity cost in terms of lost interest for holding too much cash, and incur transaction charges for converting securities to cash. The questions these managers face is how much of a cash balance to maintain, and when to convert securities to cash to ensure that demands can be met so as to minimize the opportunity and transaction costs.

Most optimization models have **constraints**—limitations, requirements, or other restrictions that are imposed on any solution, such as "Do not exceed the allowable budget" or "Ensure that all demand is met." For instance, the consumer-products company manager would probably want to ensure that a specified level of customer service is achieved with the redesign of the distribution system. The presence of constraints makes identifying the best solution considerably more difficult.

Example 2: A Product Mix Model

A manufacturer of high-tech radar detectors, Santa Cruz MicroProducts, assembles two models: LaserStop and SpeedBuster. Both models use many of the same electronic components. Two of these components, which have very high quality and reliability requirements, can only be obtained from a single overseas manufacturer. For the next month, the supply of these components is limited to 6,000 of component A and 3,500 of component B. Table 1.2 shows the number of each component required for each product and the profit per unit of product sold. How many of each product should be assembled during the next month to maximize the manufacturer's profit? Assume that the firm can sell all it produces.

To develop an optimization model for this problem, we first determine the decision variables. To simplify the presentation, let us define the decision variables as LS, representing the number of units of LaserStops to assemble, and SB, the number of units of SpeedBusters to make. Next, we model the objective function. Profit is computed as the quantity sold times the profit/unit, added for both products. Therefore, an expression for profit is

$$\text{Profit} = 25LS + 40SB$$

We also observe that we have limited quantities of the two components; these define constraints. Since only 6,000 of component A will be available, the number of component A used in assembling both products cannot exceed 6,000. Since each product requires 12 of component A, the number of component A that will be used is

$$\text{Quantity of component A used} = 12LS + 12SB$$

TABLE 1.2
Data for Santa Cruz Micro-Products radar detector assemblies

	Component Requirements/ Unit		
	A	B	Profit/Unit
LaserStop	12	6	$25
SpeedBuster	12	10	$40

This number must be less-than-or-equal-to the number available (6,000), or

$$12LS + 12SB \leq 6,000$$

Similarly, the quantity of component B used ($6LS + 10SB$) cannot exceed the amount available:

$$6LS + 10SB \leq 3,500$$

Mathematically, there is no reason why the values of LS and SB cannot be less than zero. However, in terms of the real situation, this is illogical. Therefore, we must also add constraints restricting these quantities to be at least zero, or

$$LS \geq 0 \text{ and } SB \geq 0$$

Therefore, the complete model is

$$\text{Maximize profit} = 25LS + 40SB$$

subject to the constraints

$$12LS + 12SB \leq 6,000 \text{ (component A limitation)}$$
$$6LS + 10SB \leq 3,500 \text{ (component B limitation)}$$
$$LS \geq 0, SB \geq 0 \text{ (nonnegativity)}$$

The company could use this model to find the best combination of products that satisfy all constraints and maximize profit.

Figure 1.5 shows a spreadsheet model for this problem. To show the correspondence more clearly, we will write the model in terms of the spreadsheet:

$$\text{Maximize profit} = F16 = B6 * B13 + D6 * D13$$

subject to the constraints

$$B7 * B13 + D7 * D13 \leq F7 \text{ (component A limitation)}$$
$$B8 * B13 + D8 * D13 \leq F8 \text{ (component B limitation)}$$
$$B13 \geq 0, D13 \geq 0 \text{ (nonnegativity)}$$

	A	B	C	D	E	F	G
1	**Santa Cruz MicroProducts**					Key cell formulas	
2							
3	*Parameters and Uncontrollable Inputs*						
4							
5	Product	LaserStop		SpeedBuster			
6	Unit Profit	$25.00		$40.00		Availability	
7	Component A	12		12		6000	
8	Component B	6		10		3500	
9							
10	*Model*						
11							
12		LaserStop		SpeedBuster		Total	Unused
13	Quantity Produced	375		125		500	
14	Component A Used	4500		1500		6000	0
15	Component B Used	2250		1250		3500	0
16	Profit	$9,375.00		$5,000.00		$14,375.00	

FIGURE 1.5
Spreadsheet model for Santa Cruz Micro-Products

Excel Exercise 1.2

Modify the *Santa Cruz MicroProducts.xls* spreadsheet to include a third major component resulting from a new product design. Each LaserStop requires 3 units of component C, while each SpeedBuster requires 1 unit. Twelve hundred units of component C are available. Save your model for use in a later exercise.

Models Involving Uncertainty

In the examples we have seen thus far, all the data—particularly the uncontrollable inputs—are assumed to be known and constant. In many situations, this assumption may be far from reality, as uncontrollable inputs often exhibit random behavior. Some examples would be customer demand, arrivals at ATMs, and returns on investments. We often assume such variables to be constant in order to simplify the model and the analysis. However, many situations dictate that randomness be explicitly incorporated into our models. This is usually done by specifying probability distributions for the appropriate uncontrollable inputs. For instance, in the profit model of Example 1, we might assume that demand is normally distributed with a mean of 50,000 and a standard deviation of 10,000 units. Models that include randomness are called *probabilistic*, or *stochastic*, models. We will see many examples of probabilistic models and techniques to address them throughout this book.

The Importance of Good Data

Models require data, and an important part of modeling is obtaining good data. Managers must collect, "clean," and organize data in a useful form. For many models, managers must estimate or forecast important inputs, a process that requires good support from the information systems function in an organization. For example, in the Santa Cruz MicroProducts example, data on component availability and profit margins might be easy to obtain from inventory records or the accounting department. Other models require data that must be estimated using statistical analysis. Care must be taken when working with data, and every effort should be made to ensure that data are sufficiently accurate. In one application the authors have encountered, a distribution system design model relied on data obtained from the corporate finance department. Transportation costs were determined using a formula based on the latitude and longitude of the locations of plants and customers. But when the solution was represented on a geographic information system (GIS) mapping program, one of the customers was in the Atlantic Ocean! Thus, good data management and analysis capabilities are essential to applying management science. We discuss these issues in the next several chapters.

ANALYZING AND SOLVING MODELS

A model helps managers to gain insight into the nature of the relationships among components of a problem, aids intuition, and provides a vehicle for communication. We might be interested in evaluating and analyzing risks associated with decision alternatives, determining the best solution to an optimization model, or studying the impact of changes in assumptions on model solutions. Management scientists have spent decades developing and refining a variety of approaches to address different types of problems. Much of this book is devoted to helping you understand these techniques and gain a basic facility in using them.

Model Analysis

Models allow you to easily evaluate different **scenarios**—a specific combination of model inputs that reflect key model assumptions (see Management Science Practice box on HIV transmission models). Spreadsheets, in particular, facilitate **sensitivity**, or **what-if, analysis**—the process of changing key model inputs to determine their effect on the outputs. This is one of the most important and valuable approaches to gaining insight into a problem. Microsoft Excel provides two useful tools for what if analysis: *Data Tables* and the *Scenario Manager*. **Data tables** summarize the impact of one or two inputs. We will illustrate the use of data tables to evaluate the risks of uncertain demand and alternative decisions for the profit model in Example 1.

Management Science Practice: Model Analysis of HIV Transmission through Needle Sharing

By 1986, 17 percent of newly reported AIDS cases in the U.S. were attributable to IV drug use; by 1987 this figure had risen to 20 percent. Intravenous drug users therefore represent a key target group for behavioral change efforts. Researchers in Los Angeles developed a model focusing on the dynamic consequences of HIV transmission through needle sharing. The model is based on the following assumptions:

- Only drug users who inject frequently enough to be considered addicts can become infected. The total addicted population in a given area may be considered a collection of small cliques—groups of individuals who mutually share needles. Needles from one clique may be shared with other cliques.
- The nonsharing population is fixed in size and the sharing population is initially constant.
- Needle sharing is the only route of HIV transmission (i.e., sexual transmission is not considered). All members of a particular clique share the same HIV infection status. One implication of this assumption is that all sharers may be at risk for infection while nonsharers face no such risk.
- An uninfected sharer risks infection with a fixed probability whenever an infectious needle is used or not effectively cleaned.
- A stock of used needles is maintained in proportion to the sharer population. These are disposable and discarded after an average number of uses.

The model is a rather complex set of mathematical equations whose simultaneous solution provides predictions of the behavior of key variables in the system over time. The model starts at time zero with an initial IV drug user

continued

population and an initial sharer fraction of that population. As time progresses, the total user population changes only when the population of sharers changes, while the nonsharer population remains fixed. The population of sharers is subdivided into two stocks: uninfected sharers and infected sharers, initialized so that the stock of infected sharers is only a tiny fraction (0.01 percent) of all sharers at time zero.

The model was used to predict behavior patterns and to analyze intervention strategies. Figures 1.6 to 1.8 show the model predictions of key population categories and the annual incidence of new infections, population fractions and the infectious fraction of uninfected sharers' used needles, and HIV deaths over a 16-year time horizon. These graphs show how the key variables reach an equilibrium.

The model can assist policy makers in studying the potential benefits of campaigns to increase the effective cleaning of shared needles. The following questions were addressed:

- How important is the magnitude of such a campaign?
- How important is the timeliness of the campaign?
- How might the openness of needle sharing affect the impacts of the campaign's magnitude and timeliness?
- If the campaign results in a greater attraction to needle sharing, to what extent might its benefits be undermined?

FIGURE 1.6 Baseline model behavior pattern

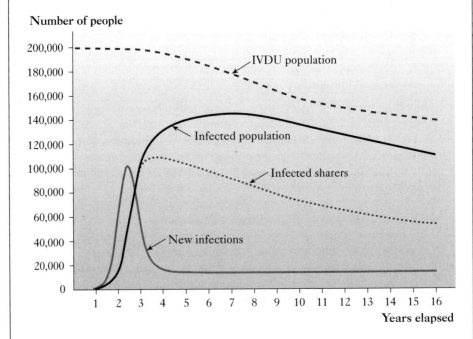

continued

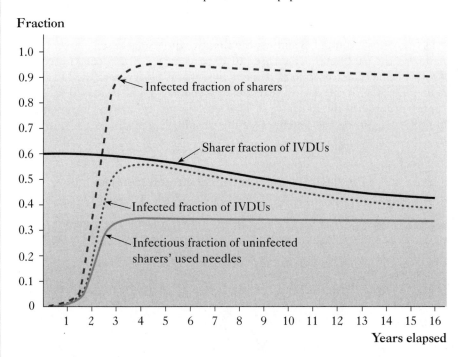

FIGURE 1.7 Model predictions for population fractions

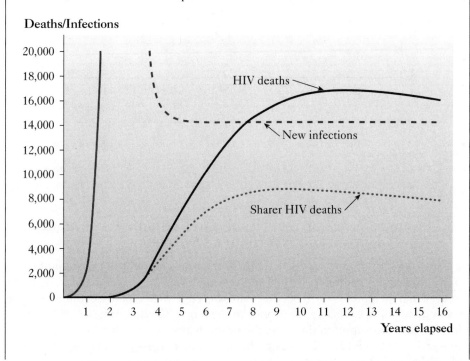

FIGURE 1.8 Model predictions for HIV deaths and new infections

continued

Figure 1.9 shows a summary of several policy scenarios in which a key parameter—the effective cleaning fraction of shared uses—is changed at different rates and compared to the baseline with an initial value of 10 percent (scenario # 1). Scenarios 2 through 5 address the first two questions raised above. In scenario # 2, the parameter is increased to 30 percent (for all 16 years) in scenario # 3, this value is increased to 50 percent; in scenario # 4, it is increased to 50 percent, but not until the second year; and in scenario # 5, it is increased to 50 percent, but not until the fourth year. The results show the critical importance of both the magnitude and timeliness of the cleaning campaign.

FIGURE 1.9 HIV deaths under alternative scenarios

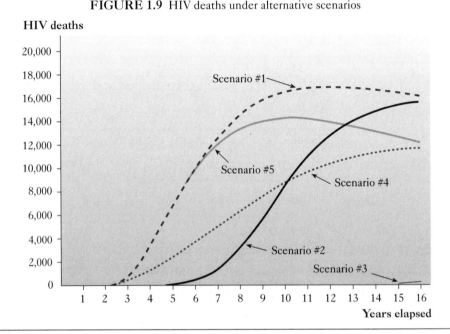

Source: Adapted from Jack B. Homer and Christian L. St. Clair, "A Model of HIV Transmission through Needle Sharing," *Interfaces*, Vol. 21, No. 3, 1991, pp. 26–49.

Example 3: Using Data Tables to Evaluate Uncertainty and Alternative Decisions

Figure 1.10 shows a one-way data table for the profit model to evaluate profit for a fixed production quantity of 40,000 and varying levels of sales demand from 20,000 to 60,000 in increments of 5,000. To create this table, first create a column of inputs (sales demand) to evaluate; this is done in the range B26:B34. In cell C25, enter the formula for the output; in this case, the profit, which is the value in cell C20 (from Figure 1.4, on page 12). Next, select the data table range—the smallest rectangular block that includes the formula and all the values in the input range (B25:C34). Select *Table* from the *Data* menu, and specify the location of the input cell in your

FIGURE 1.10
One-way
data table

	B	C	D
23	Data Table		
24	Demand		
25		$240,000.00	◄—[=C20]
26	20000	$ (560,000.00)	
27	25000	$ (360,000.00)	
28	30000	$ (160,000.00)	
29	35000	$ 40,000.00	
30	40000	$ 240,000.00	
31	45000	$ 240,000.00	
32	50000	$ 240,000.00	
33	55000	$ 240,000.00	
34	60000	$ 240,000.00	

FIGURE 1.11
Data table
dialog box

model (cell B8, which contains the input value of demand) in the dialog box, as shown in Figure 1.11. Use the column input cell because the data table is arranged as a column (if it were in a row, you would place this cell location in the *Row input cell* box). Click OK, and Excel evaluates the profit for each value in the data table.

The data table shows that demand must be at least approximately 35,000 in order to achieve a positive profit. This analysis also shows the risk of losses if demand falls below 30,000. As this spreadsheet is designed, you could change the production quantity in the model and easily evaluate the associated risks.

A logical extension would be to evaluate the profit for varying levels of sales demand and production quantity simultaneously. This can be done with a *two-way data table*. For example, suppose that we wish to evaluate the profit for demands and production quantities between 20,000 and 60,000. The table, shown in Figure 1.12, is an extension of the one-way data table in which production quantities are added in the range F25:N25. The profit formula is written in cell E25 and the range E25:N34 is selected as the table range. After selecting *Table* from the *Data* menu, both the row and column input cells must be specified. The row input cell is B17, corresponding to quantity produced, and the column input cell is B8, corresponding to sales demand (see Figure 1.4). The results show, for instance, that any production quantity runs a risk of incurring a loss, and that this loss increases as the production quantity increases. A minimum of 30,000 units must be produced to achieve any profit, although both the profit potential and loss risk increase for higher production quantities. Another modification would be to incorporate probabilistic assumptions about the levels of demand. We will address this issue in a later chapter.

FIGURE 1.12
Two-way
data table

	E	F	G	H	I	J	K	L	M	N
23	Two-Way Data Table									
24	Demand	Production Quantity								
25	$240,000.00	20000	25000	30000	35000	40000	45000	50000	55000	60000
26	20000	$ (80,000)	$ (200,000)	$ (320,000)	$ (440,000)	$ (560,000)	$ (680,000)	$ (800,000)	$ (920,000)	$ (1,040,000)
27	25000	$ (80,000)	$ -	$ (120,000)	$ (240,000)	$ (360,000)	$ (480,000)	$ (600,000)	$ (720,000)	$ (840,000)
28	30000	$ (80,000)	$ -	$ 80,000	$ (40,000)	$ (160,000)	$ (280,000)	$ (400,000)	$ (520,000)	$ (640,000)
29	35000	$ (80,000)	$ -	$ 80,000	$ 160,000	$ 40,000	$ (80,000)	$ (200,000)	$ (320,000)	$ (440,000)
30	40000	$ (80,000)	$ -	$ 80,000	$ 160,000	$ 240,000	$ 120,000	$ -	$ (120,000)	$ (240,000)
31	45000	$ (80,000)	$ -	$ 80,000	$ 160,000	$ 240,000	$ 320,000	$ 200,000	$ 80,000	$ (40,000)
32	50000	$ (80,000)	$ -	$ 80,000	$ 160,000	$ 240,000	$ 320,000	$ 400,000	$ 280,000	$ 160,000
33	55000	$ (80,000)	$ -	$ 80,000	$ 160,000	$ 240,000	$ 320,000	$ 400,000	$ 480,000	$ 360,000
34	60000	$ (80,000)	$ -	$ 80,000	$ 160,000	$ 240,000	$ 320,000	$ 400,000	$ 480,000	$ 560,000

Excel Exercise 1.3

Construct a one-way data table for the profit model that shows the impact of varying the unit cost between $20 and $30. Also, construct a two-way data table in which the unit price also varies between $37 and $43.

For situations involving more than two variables, the Excel Scenario Manager provides a means of evaluating changes in input value assumptions. We will not discuss this, but encourage you to learn how to use it and experiment with it.

Tornado and Spider Charts

Charts, graphs, and other visual aids play an important part in model analysis. Two important tools are tornado charts and spider charts. A **tornado chart** graphically shows the impact that *variation in one model input* has on some output while holding all other inputs constant. A **spider chart** shows the sensitivity that *percentage changes in inputs* have on a model output. Both charts help you to identify and focus on critical factors that influence model results. The charts show which inputs are the most influential on the output. If these inputs are uncertain, then you would probably want to study them further to reduce uncertainty and its effect on the output. If the effects are small, you might ignore any uncertainty or eliminate them from the model. They are also useful in helping you select the inputs that you would want to analyze further with data tables or scenarios.

Excel does not have an automatic routine for creating tornado and spider charts; however, they can be constructed relatively easily using data tables and Excel's charting capabilities. The appendix to this chapter also provides step-by-step guidelines on how to do this, and you might want to try them after reading the next example.

Example 4: A Tornado and Spider Chart for the Profit Model

Figure 1.13 shows the information needed to construct a tornado chart for the profit model. For each of the input variables, we assume a lower limit, base case value, and upper limit (B21:E26). These would be values that you assume represent the range of uncertainty of uncontrollable inputs, or in the case of the unit price, a range that you might consider for evaluating model results. The output results, shown in the range B28:F33, represent the total profit when an input variable is set to one of its assumptions, holding all other variables constant. These were computed using the data tables on the right. The last column computes the range between the lower and upper limits and is used to sort the rows in descending order of sensitivity. The tornado chart shows that changes in the unit price will have the most effect on profits, and that uncertainty about the fixed cost will have the least effect. The step-by-step process for creating this tornado chart is found in the appendix to this chapter.

To create a spider chart, we vary each input by some percentage increment from its base case value and use data tables to evaluate the impact on the output. Then we simply display these results on the same line graph. Figure 1.14 shows the calculations and the spider chart for profit as the unit price, unit cost, fixed cost, and sales demand change up to ±80 percent of their base case values. The slopes of the lines indicate whether a positive change in the input has a positive or negative effect on the output. The steeper the slope, the more that variable affects the profit. In this case, we see that the unit price is the most sensitive while the fixed cost is the least. Sales demand is highly sensitive for values below a 20 percent reduction from the base case, although it has no effect above this value (can you explain why?).

	B	C	D	E	F	G	H	I
21	Input	Lower	Base	Upper		Data Table Calculations		
22	Assumptions	Limit	Case	Limit		Unit price	$ 240,000	
23	Unit price $	30	$ 40	$ 50		$ 30	$ (160,000)	
24	Unit cost $	20	$ 24	$ 30		$ 40	$ 240,000	
25	Fixed Cost $	300,000	$ 400,000	$ 500,000		$ 50	$ 640,000	
26	Sales Demand	30000	50000	70000				
27						Unit cost	$ 240,000	
28	Output	Lower	Base	Upper		$ 20	$ 400,000	
29	Results	Limit	Case	Limit	Range	$ 24	$ 240,000	
30	Unit price $	(160,000)	$ 240,000	$ 640,000	$ 800,000	$ 30	$ -	
31	Sales Demand $	(160,000)	$ 240,000	$ 240,000	$ 400,000			
32	Unit cost $	400,000	$ 240,000	$ -	$ 400,000	Fixed Cost	$ 240,000	
33	Fixed Cost $	340,000	$ 240,000	$ 140,000	$ 200,000	$ 300,000	$ 340,000	
34						$ 400,000	$ 240,000	
35						$ 500,000	$ 140,000	
36								
37						Sales Demand	$ 240,000	
38						30000	$ (160,000)	
39						50000	$ 240,000	
40						70000	$ 240,000	

Profit Model Tornado Chart

FIGURE 1.13
Tornado chart

FIGURE 1.14
Spider chart

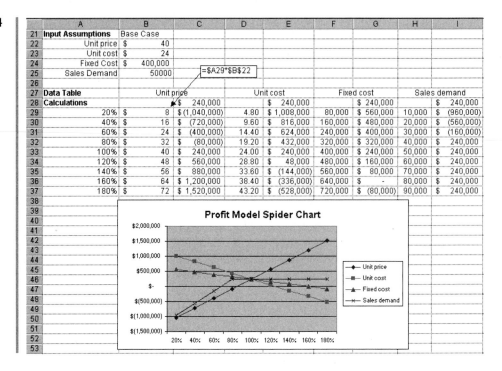

	A	B	C	D	E	F	G	H	I
21	**Input Assumptions**	Base Case							
22	Unit price	$ 40							
23	Unit cost	$ 24							
24	Fixed Cost	$ 400,000							
25	Sales Demand	50000	=$A29*$B$22						
26									
27	**Data Table**		Unit price		Unit cost		Fixed cost		Sales demand
28	**Calculations**		$ 240,000		$ 240,000		$ 240,000		$ 240,000
29	20%	$ 8	$(1,040,000)	4.80	$1,008,000	80,000	$ 560,000	10,000	$ (960,000)
30	40%	$ 16	$ (720,000)	9.60	$ 816,000	160,000	$ 480,000	20,000	$ (560,000)
31	60%	$ 24	$ (400,000)	14.40	$ 624,000	240,000	$ 400,000	30,000	$ (160,000)
32	80%	$ 32	$ (80,000)	19.20	$ 432,000	320,000	$ 320,000	40,000	$ 240,000
33	100%	$ 40	$ 240,000	24.00	$ 240,000	400,000	$ 240,000	50,000	$ 240,000
34	120%	$ 48	$ 560,000	28.80	$ 48,000	480,000	$ 160,000	60,000	$ 240,000
35	140%	$ 56	$ 880,000	33.60	$ (144,000)	560,000	$ 80,000	70,000	$ 240,000
36	160%	$ 64	$1,200,000	38.40	$ (336,000)	640,000	$ -	80,000	$ 240,000
37	180%	$ 72	$1,520,000	43.20	$ (528,000)	720,000	$ (80,000)	90,000	$ 240,000

Profit Model Spider Chart

Legend: Unit price · Unit cost · Fixed cost · Sales demand

Solving Optimization Models

Finding solutions to optimization models is where management science theory and research have proven extremely valuable. For some models, **analytical solutions**—closed-form mathematical formulas—can provide an answer. In most cases, however, an algorithm provides the solution. An **algorithm** is a systematic procedure that finds a feasible or an optimal solution to a problem. A **feasible solution** is one that satisfies all constraints of a model. A feasible solution that minimizes or maximizes the objective is called an **optimal solution**. Researchers in management science have developed algorithms to solve many types of optimization problems, and we will discuss some of these in this book. However, we will not be concerned with the detailed mechanics of these algorithms, since they are widely available on personal computers; our focus will be on the use of the algorithms to solve the models we develop.

If possible, we would like to have an algorithm that finds an optimal solution. However, some models are so complex that it is impossible to solve them optimally in a reasonable amount of computer time, because of the extremely large number of computations that may be required. (Some problems may not be possible to solve in your lifetime, even on the fastest available computer!) For many optimization problems, therefore, a manager might be satisfied with a solution that is good but not necessarily optimal. Other reasons why managers may not be concerned with finding an optimal solution to a model include the following:

- Inexact or limited data used to estimate uncontrollable quantities in models may contain more error than that of a nonoptimal solution.
- The assumptions used in a model make it an inaccurate representation of the real problem, making the "best" solution moot.

- Managers may not need the very best solution; anything better than the present one will often suffice, so long as it can be obtained at a reasonable cost and in a reasonable amount of time.

Solution procedures that generally find good solutions without guarantees of finding an optimal solution are called **heuristics**, from a Greek word meaning "to discover." Heuristics are often very intuitive and guided by common sense, as the following example illustrates.

Example 5: A Heuristic Solution to the Product Mix Problem

In Example 2 we developed a spreadsheet for the Santa Cruz MicroProducts model. We could try to find the best solution by trial and error by simply choosing different values for the decision variables.[4] Even though spreadsheets allow us to do this quite easily, it would be better to apply some intuition or heuristics. A reasonable heuristic might be to produce as many units of the product with the largest unit profit as possible. In this case, we would attempt to produce as many Speed-Busters as possible without violating any constraint. Thus, if we set $LS = 0$, the first constraint must satisfy

$$12SB \leq 6,000$$

or

$$SB \leq \frac{6,000}{12} = 500$$

From the second constraint we have

$$10SB \leq 3,500$$

or

$$SB \leq \frac{3,500}{10} = 350$$

We can therefore produce at most 350 SpeedBusters without violating a constraint. The total profit would be $40(350) = \$14,000$. While this looks like a good solution, it is *not* an optimal solution. The optimal solution is to produce 375 Laser Stops and 125 SpeedBusters for a total profit of $14,375. By comparing these solutions, can you explain *why* the heuristic solution is not optimal? Intuitive approaches based on common sense often yield very good solutions; however, they generally cannot guarantee optimal solutions. You will learn how to solve optimization problems of this type using Excel in Chapter 4.

Excel Exercise 1.4

Use the model you developed in Excel Exercise 1.2 for this problem with the additional third component. Is the solution "$LS = 375$ and $SB = 125$" feasible? If not, what is the best solution you can find?

The purpose of models is to provide insight for making decisions; thus, we should not take optimal solutions at face value without exploring other alternatives or investigating the effects of uncertainties or inaccuracies in the model data. As with the profit model, we could perform sensitivity analysis by changing parameters. The difference with optimization models, however, is that a change in a parameter may cause the optimal solution to change—it is not a case of simply re-evaluating the spreadsheet. In the Santa Cruz MicroProducts problem, suppose that an additional 4 units of component B are available. If you change the data in the spreadsheet, then you will find that you can produce an additional SpeedBuster but will have to produce one fewer LaserStop. However, profits will increase by $15 (you gain the additional profit from the SpeedBuster but lose the profit on a LaserStop). Thinking of this as a rate, the one additional unit of component B is worth $15/4 = $3.75 in extra profits. This type of information can enable a manager to determine the value of trying to obtain additional resources. Other questions that might be asked are, How will the optimal product mix and total profit change as the unit profits are changed? How will profits be affected if a contract requires production of 50 LaserStops? We will explore such questions in the optimization chapters.

Excel Exercise 1.5

Use the Santa Cruz spreadsheet to try to determine the value of an additional unit of component A.

INTERPRETING AND USING MODEL RESULTS

Interpreting the results from the analysis phase is crucial in making good decisions. Models cannot capture every detail of the real problem, and managers must understand the limitations of models and their underlying assumptions. **Validity** refers to how well a model represents reality. A "perfect" model corresponds to the real world in every respect; unfortunately, no such model has ever existed, and never will exist in the future, because it is impossible to include every detail of real life in one model. One technique for judging the validity of a model is to identify and examine the assumptions made in a model to see how they agree with the manager's perception of the real world; the closer the agreement, the higher the validity.

Because many important considerations usually cannot be included in formal models, managers should never implement model solutions blindly. Quantitative information must therefore be integrated with qualitative analysis drawn from one's experience and judgment. The more well structured a problem is, the more you can rely on model results. As problems become more fuzzy and strategic in nature, more weight is usually placed on qualitative issues. Good modeling and analysis provide alternatives by which managers can assess impacts of different decisions.

One question that anyone should ask about model results is, What do the results tell you about the problem or decision that must be made? For example, in the Santa Cruz MicroProducts model, producing 375 LaserStops and 125 SpeedBusters uses the entire stock of all components (substitute these values into the constraints or enter them into the spreadsheet). This suggests that we cannot make more of each product and increase our profit without having more component inventory. To produce more SpeedBusters, for example, we would have to produce fewer LaserStops. The better the ability of a decision maker to understand a model and believe in its results, the more likely it is that he or she will accept it and use it.

Sensitivity to political and organizational issues is an important skill that management scientists must possess when looking beyond the technical results. A case in point is a study conducted for the New York City Department of Sanitation that focused on allocating 450 new street cleaners throughout the city. A mathematical model was developed to allocate the cleaners in an optimal fashion based on the marginal improvement that could be achieved with additional cleaners. The model suggested that many of the new cleaners should be assigned to districts that were already quite clean. In the final allocation, however, most of the new street cleaners went to the dirtiest districts, but each district was assigned some as a political compromise. The value of the model was not in providing an optimal solution, but in predicting the effects of alternative decisions.

A CASE STUDY OF MODELING AND ANALYSIS: INVENTORY MANAGEMENT

We now present an example that integrates the three themes we have developed in this chapter that characterize management science and decision technology: *modeling, analysis,* and *interpretation of results*.

Managing Inventory Systems

In managing the inventory of retail goods, managers must make a fundamental decision about how much and when to order. These decisions have economic impacts. Inventory costs fall into four major categories:

1. Purchase costs
2. Order preparation costs
3. Inventory holding costs
4. Shortage costs

The costs of purchasing the items may be constant for any quantity ordered, or discounts might be given as an incentive to order larger quantities. Order preparation costs involve the time spent preparing and placing orders, such as clerical, telephone, receiving, and inspection time. Inventory holding costs include all expenses associated with carrying inventory, such as rent on storage space, utilities, insurance, taxes, and the cost of capital. Shortages can be either a back order in which the customer will wait for an item that is out of stock, or a lost sale in which the customer purchases the item elsewhere. Back orders incur additional costs for shipping, invoicing, and labor; lost sales result in lost profit opportunities and possible future loss of revenues.

These costs are influenced by the amount ordered and the timing of the ordering decision. For example, if many small orders are placed, then the ordering cost will be high, but little inventory will be carried, reducing holding costs. On the other hand, if few large orders are placed, then ordering costs will be low, but inventory holding costs will be high. Similarly, if orders are placed too early, excessive holding will result. If orders are placed too late, the firm risks running out of stock and incurring shortages. Thus, decision makers seek a minimum cost balance among these costs. Establishing such tradeoffs is typical of many management science problems. Let us first describe the typical inventory system and its environment.

Most inventory systems operate on a continuous review policy. In a **continuous review system,** the inventory position—the amount on hand plus any amount on order but not yet received minus back orders—is monitored continuously. Whenever the inventory position falls to or below a level r, called the reorder point, an order for a fixed amount, say Q, units is placed. Figure 1.15 illustrates the operation of a continuous review system from the viewpoint of both stock level and the inventory position. The dark curve is the stock level, the light curve is the inventory position. As we move to the right, the stock level is initially the same as the inventory position. When it reaches the reorder point, an order for Q units is placed, and the inventory position is increased by Q. The amount of time it takes for an order to arrive is called the **lead time**. During the lead time, the stock level continues to decrease as new demands are met. When the order arrives, the stock level increases by Q, and the stock level and inventory position are again equal. If the stock level hits zero and demands continue to arrive, back orders accumulate until an order arrives. When an order arrives, the accumulated back order demand is filled and the remainder becomes the stock level. Notice that the inventory position does not change when an order arrives, because the inventory position was updated when the order was placed. Figure 1.16 is a flowchart that describes the logical operation of such a continuous review system with back orders. For this situation, we might be interested in developing a total cost model to evaluate various inventory policies and strategies; for instance, the impact

FIGURE 1.15
Stock level and inventory position curves for a continuous review system

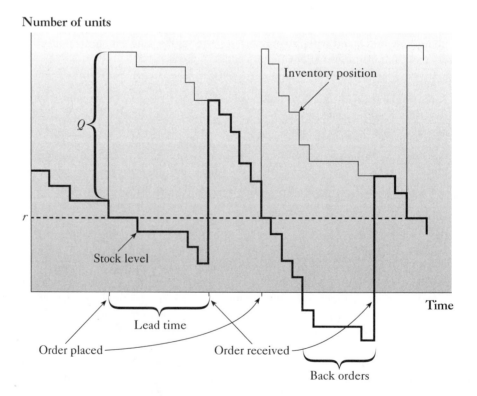

FIGURE 1.16 Flowchart for a continuous review inventory system

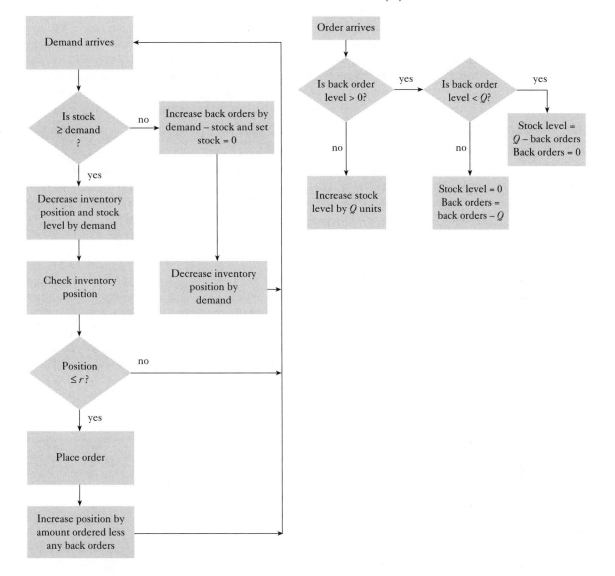

of smaller, more frequent deliveries (just-in-time purchasing) on the cost. Or our goal might be to determine the order quantity that minimizes total cost. We will do both for this example.

Model Development

We need to make several important assumptions:

1. Only a single inventory item is considered.
2. The entire quantity arrives at one time.
3. The demand for the item is constant over time.
4. No shortages are allowed.

For many inventory situations, assumptions 1 and 2 are usually valid. However, assumptions 3 and 4 are critical. If the demand varies considerably or exhibits randomness, then the model would not provide useful results to the decision maker. On the other hand, if demand is not exactly constant, but generally smooth, then assuming constancy is probably appropriate. Assumption 4 may be questionable in practice, but it is made to simplify the model that will be developed. With these assumptions, the stock level and inventory position curves in Figure 1.15 smooth out and take the form shown in Figure 1.17. Notice that with no shortages allowed, the stock level curve cannot fall below zero.

The key data requirements for this model are the annual demand, holding cost per unit, and ordering cost. The annual demand is never known with certainty; it must be based on a forecast. We must be cautious, because forecasts are often wrong and subject to error. The holding cost is normally computed as a percentage of the item's unit price. The percentage used—called the carrying charge rate—is one of the more difficult pieces of data to estimate. Clearly, the holding cost rate should be at least as great as the cost of capital (at which the firm can invest elsewhere), but beyond that, it is virtually impossible to account for all of the costs (none of which will remain constant) associated with holding inventory, particularly on a unit-by-unit basis. The ordering cost likewise is difficult to estimate. Thus, it should be clear that any model that is developed will have inaccurate data at best. For this example, suppose that annual demand is 15,000 units, ordering cost is $200 per order, unit cost is $22, and the carrying charge rate is 20 percent.

An influence diagram for the total inventory cost is shown in Figure 1.18. We break down the total cost into the holding cost term and ordering cost term. Holding cost is a function of the holding cost per unit and the number of units in inventory. As described earlier, the unit holding cost is based on the cost of the item and the carrying charge rate. Ordering cost is a function of the number of orders placed and the cost of each order. Both the number of units in inventory and the number of orders placed depend on the order quantity.

FIGURE 1.17
Stock level and inventory position curves under simplifying assumptions

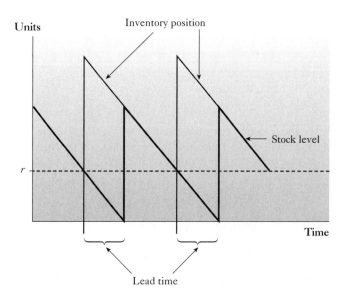

FIGURE 1.18 Influence diagram for total inventory cost

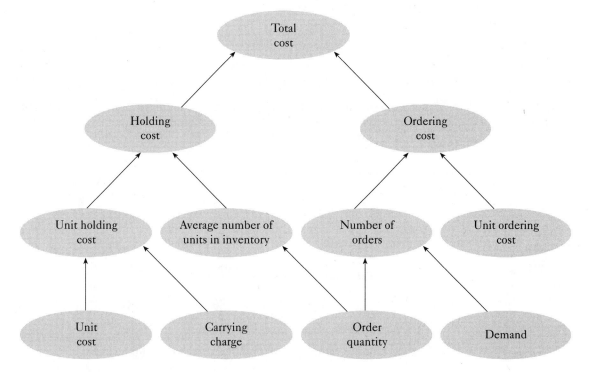

To develop a model, let

Q = order quantity

D = annual demand

C = unit cost of the item

C_0 = cost per order placed

i = inventory carrying charge per unit

We can use the influence diagram in Figure 1.18 for guidance. Let us first deal with the ordering cost. With an annual demand D and order quantity Q, the number of orders that must be placed each year is D/Q. Therefore the total annual ordering cost is

$$\text{Annual ordering cost} = (\text{number of orders per year})(\text{cost per order}) = \left(\frac{D}{Q}\right)C_0$$

Inventory holding cost per unit is computed as a fixed percentage of the unit cost of the item. Thus, if C is the unit cost of the item and i is the annual carrying charge rate (expressed as a decimal fraction), the cost of holding one unit for one year is

$$\text{Holding cost per unit} = iC$$

The assumption of constant demand becomes important in determining an expression for the average number of units in inventory. If Q units are ordered and then received when the inventory level is depleted, then the inventory will decrease to zero at a constant rate until the next order arrives. Since the inventory level starts at Q and ends at zero, the average inventory is

$$\text{Average inventory} = \frac{(Q + 0)}{2} = \frac{Q}{2}$$

Thus, the annual inventory holding cost is

$$\text{Annual inventory holding cost} = (\text{average inventory})(\text{annual holding cost per unit})$$
$$= \left(\frac{Q}{2}\right)iC = \frac{iCQ}{2}$$

We may now express the total annual inventory cost as

$$\text{Total annual cost} = \frac{DC_0}{Q} + \frac{iCQ}{2}$$

Using the data we have for this specific example, we have

$$\text{Total annual cost} = \frac{(15,000)(200)}{Q} + \frac{(.2)(22)Q}{2}$$
$$= \frac{3,000,000}{Q} + 2.2Q$$

We would like to find the value of Q that minimizes the total annual cost.

Model Analysis and Solution

We may set up a simple spreadsheet for the inventory cost model as shown in Figure 1.19. The basic model is given in the top half of the spreadsheet; in the bottom half, we set up a table to examine the effects of different order quantities on the costs. The objective is to find the point at which the total cost is minimized. Figure 1.20 provides more insight into the nature of the solution through a graph generated from the spreadsheet data. It shows that the order cost decreases and the holding cost increases as the order quantity goes up. We see that the minimum total cost occurs in the vicinity of $Q = 1,200$. This may not be the true optimal solution, because we only studied order quantities in increments of 100. Thus, within this level of precision, it appears that the optimal order quantity lies somewhere between 1,100 and 1,300. With the spreadsheet, we could easily construct a table for order quantities within this range and get closer to the optimal solution. From the graph (as well as from the table in the spreadsheet), we clearly see that this model finds a minimum cost tradeoff between the order cost and the holding cost.

Trying to identify an optimal solution in this fashion can be at best time consuming, and at worst, impractical. As we noted earlier in this chapter, analytical solutions and algorithms often provide means of finding optimal solutions. For this inventory model we can find a mathematical solution showing the value of Q that minimizes the total cost, using calculus:

$$Q^* = \sqrt{\frac{2DC_0}{iC}}$$

	A	B	C	D
1	**Economic Order Quantity Model**		Key cell formulas	
2				
3	*Parameters and uncontrollable inputs*			
4				
5	Annual Demand Rate	15000		
6	Ordering Cost	$200.00		
7	Unit Cost	$22.00		
8	Carrying Charge Rate	0.2		
9				
10	*Model*			
11				
12	Order Quantity	500		
13	Annual Demand Rate	15000		
14	Ordering Cost	$200.00		
15	Order cost		$6,000.00	
16	Carrying Charge Rate	0.2		
17	Unit Cost	$22.00		
18	Inventory cost		$1,100.00	
19	Total cost		$7,100.00	
20				
21	*Order Quantity*	*Order cost*	*Inventory cost*	*Total cost*
22	500	$6,000.00	$1,100.00	$7,100.00
23	600	$5,000.00	$1,320.00	$6,320.00
24	700	$4,285.71	$1,540.00	$5,825.71
25	800	$3,750.00	$1,760.00	$5,510.00
26	900	$3,333.33	$1,980.00	$5,313.33
27	1000	$3,000.00	$2,200.00	$5,200.00
28	1100	$2,727.27	$2,420.00	$5,147.27
29	1200	$2,500.00	$2,640.00	$5,140.00
30	1300	$2,307.69	$2,860.00	$5,167.69
31	1400	$2,142.86	$3,080.00	$5,222.86
32	1500	$2,000.00	$3,300.00	$5,300.00

FIGURE 1.19
Spreadsheet for
EOQ model

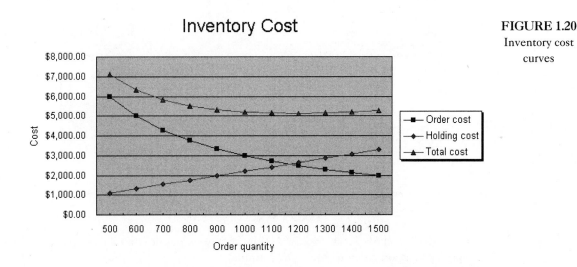

FIGURE 1.20
Inventory cost
curves

This value is called the **economic order quantity (EOQ)** and the model is sometimes called the **economic order quantity model.** Substituting the values of the uncontrollable inputs into this formula, we have

$$Q^* = \sqrt{\frac{2(15,000)(200)}{(0.20)(22)}} = 1,167.75, \text{ or } 1,168$$

This results in a total cost of

$$\text{Total cost } = \frac{DC_0}{Q} + \frac{iCQ}{2}$$

$$= \frac{(15,000)(200)}{1,168} + \frac{0.2(22)(1,168)}{2} = \$5,138$$

However, suppose that you did not know this formula. Fortunately, decision technology comes to the rescue! Excel provides a tool called Solver, which finds optimal solutions to many types of optimization problems. To use Solver, select it from the *Tools* menu in Excel. Excel displays the dialog box in Figure 1.21. The *target cell* is the worksheet cell that contains the objective function that we wish to optimize; in this case, the total cost in cell C19 in Figure 1.19. Next, click on the appropriate radio button to define the objective. Solver allows you to maximize or minimize the target cell, or to find a solution using the *Value of* a specified number. Finally, tell Solver the cell or range that contains the decision variables (Changing Cells); in this case, the order quantity in cell B12. Click the *Solve* button to find the solution; the changing cells and target cells are automatically updated in the spreadsheet. We will use Solver extensively in Chapters 4 and 5 and describe the other features of Solver and the reports it can generate.

FIGURE 1.21
Solver parameters dialog box

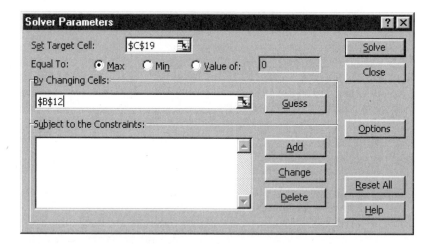

Excel Exercise 1.6

Open the Excel file *Inventory model.xls* and use Solver as described to verify the optimal solution given by the EOQ formula.

The advantages of either an analytical solution procedure or a decision technology tool like Solver are that they are easier to apply than a heuristic procedure and generally guarantee an optimal solution. Therefore, whenever possible, you should use an exact solution method rather than relying on heuristic approaches.

One issue that the model does not address directly is the decision when to order; that is, the reorder point. If we order too early, we will have more stock than we need and incur unnecessary holding costs. If we order too late, we run the risk of running out of stock before the order is received. Ideally, we would like the order to be received at the same time we run out of stock. Therefore, the reorder decision depends on the lead time. We should set the reorder point to be the inventory level that provides enough stock to satisfy all demand during the lead time. If the demand rate, D, is constant as assumed, and the lead time is t, then the reorder point should be equal to Dt. To illustrate this for the numerical example, we have $D = 15,000$ units per year. Suppose that the lead time is one week, or .0192 years. The demand during the lead time is $(15,000)(.0192) = 288$ units. If we place an order when the inventory position reaches 288, then the order will arrive when the stock level falls to zero. This is illustrated in Figure 1.22.

Interpreting Results

To obtain a quick sense of the validity of the EOQ model, let us examine the two terms that constitute the total cost:

1. Ordering cost, $\dfrac{DC_0}{Q}$

2. Holding cost, $\dfrac{iCQ}{2}$

FIGURE 1.22 Computing the reorder point

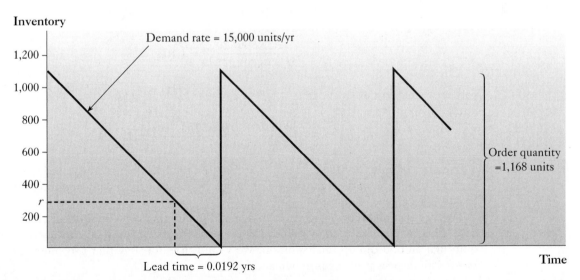

Demand during lead time = Annual rate × Lead time
= (15,000 units/yr) (0.0192 yrs)
= 288 units

As the order quantity Q increases, we can see that the ordering cost decreases, because Q is in the denominator. This makes sense because fewer orders of larger size are being placed. At the same time, however, the holding cost increases (since Q is in the numerator) because larger orders will increase the average inventory. The model therefore appears to reflect what we would observe in practice. These observations were demonstrated in Figure 1.20.

Today, many firms are practicing "just-in-time" purchasing, in which frequent orders for small quantities are made. To make the just-in-time concept economically attractive, the ordering cost must be low. By analyzing the economic order quantity formula:

$$Q^* = \sqrt{\frac{2DC_0}{iC}}$$

we see that as C_0 decreases, so does Q^*, as does the total cost. Here again, the model reflects what we would expect to happen.

We may also examine the sensitivity of the model to variations in the uncontrollable data. To illustrate this, let us consider the numerical instance of the economic order quantity model that we discussed. Recall that $D = 15{,}000$, $C_0 = 200$, $C = 22$, and $i = .20$, resulting in $Q^* = 1{,}168$ and total cost = \$5,138. Now suppose that we err by 10 percent in our estimate of the annual demand, meaning that D can be as low as 13,500 or as high as 16,500. How would this affect the total cost? Note that since we do not realize that our estimate is wrong, we would still use $Q^* = 1{,}168$, but that the actual cost would be computed using the *true* value of D. (If demand is lower, we would order less frequently; if it is higher, we would order more frequently.) Thus, if $D = 13{,}500$, the total cost actually incurred is

$$\text{Total cost} = \frac{(13{,}500)(200)}{1{,}168} + \frac{0.2(22)(1{,}168)}{2} = \$4{,}881$$

The optimal order quantity corresponding to $D = 13{,}500$ would be

$$Q^* = \sqrt{\frac{2(13{,}500)(200)}{(0.20)(22)}} = 1{,}108$$

with an annual cost of

$$\text{Total cost} = \frac{(13{,}500)(200)}{1{,}108} + \frac{0.2(22)(1{,}108)}{2} = \$4{,}874$$

a difference of only \$7. Similarly, if $D = 16{,}500$, the total cost of ordering 1,168 units is \$5,395, while the optimal order quantity would have been 1,125 with a total cost of \$5,389, a difference of only \$6.

We see that relatively large errors in demand forecasting have little effect on the cost. This can also be seen in Figure 1.20 by observing that the total cost curve is quite flat around the optimal solution. We would therefore say that this model is *robust* with respect to changes in demand.

Excel Exercise 1.7

Develop a sensitivity analysis table to compute the cost differences in using $Q = 1168$ versus the optimal order quantity for changes in the carrying charge rate from 0.15 to 0.25 in increments of 0.1

Applying Inventory Theory to Cash Management

In introducing optimization models earlier in this chapter, we posed a situation that financial managers face involving cash management. You might easily see an analogy with the economic order quantity model we developed (in fact, analogy can be a very useful modeling approach!). For example, suppose that a firm has cash requirements of D dollars per year and sells Q dollars of securities at a time whenever the cash balance falls to zero. If C_0 represents a fixed transaction cost to sell the securities, and i represents the opportunity cost of holding the cash instead of securities, we have a direct analogy with the EOQ model. In the finance discipline, this is called the *Baumol model* for cash management.

One of the limitations of the Baumol model is that it assumes that the demand for cash is constant over time. In most cases, cash requirements will occur with some uncertainty. The *Miller-Orr model* was proposed to address this situation. This model assumes that the firm will maintain a minimum cash balance, m, a maximum cash balance, M, and an ideal level, R, called the return point. Cash is managed using a *decision rule*. Whenever the cash balance falls to m, $R - m$ securities are sold to bring the balance up to the return point. When the cash balance rises to M, $M - R$ securities are purchased to reduce the cash balance back to the return point. This is illustrated in Figure 1.23.

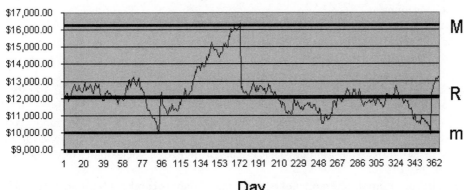

Cash Balance

FIGURE 1.23
Illustration of Miller-Orr decision rule

Using some advanced mathematics, the return point and maximum cash balance levels are shown to be:

$$R = m + Z$$
$$M = R + 2Z$$

where

$$Z = \left(\frac{3C_0\sigma^2}{4r}\right)^{1/3}$$

σ^2 = variance of the daily cash flows, and

r = average daily rate of return corresponding
to the premium associatedwith securities.

For example, if the premium is 4 percent, $r = .04/365$. To apply the model, note that we do not need to know the actual demand for cash, only the daily variance.

Because of the stochastic nature of the problem and the mathematics involved, specifying the actual model is beyond the scope of this book. Essentially, the Miller-Orr model determines the decision rule values to minimize the expected costs of making the cash-security transactions and the expected opportunity costs of maintaining the cash balance based on the variance of the cash requirements. We will use this model in a later chapter to illustrate some issues involving simulation.

PROBLEMS

1. A supermarket has been experiencing long lines during peak periods of the day. The problem is noticeably worse on certain days of the week, and the peak periods are sometimes different according to the day of the week. There are usually enough workers on the job to open all cash registers. The problem is knowing when to call some of the workers stocking shelves up to the front to work the checkout counters. How might management science models help the supermarket? What data would be needed to develop these models?

2. Using knowledge developed from previous courses or interviews with faculty, develop a list of management science applications in each of the following functional disciplines of business:
   a. Accounting
   b. Marketing
   c. Finance
   d. Operations management
   e. Information systems

3. A music club offers new members four free CDs upon signing up. The member need only buy one CD within the next year at half price and then receives three more CDs free. The member may cancel at any time thereafter. How might you develop a model to determine the club's expected profits (at net present value) over, say, a five-year period? What factors influence the decision to offer such an incentive?

4. A (greatly) simplified model of the national economy can be described as follows. The national income is the sum of three components: consumption, investment, and government spending. Consumption is related to the total income of all indi-

viduals and to the taxes they pay on income. Taxes are related to income by the tax rate. Investment is also related to the size of the total income.

   a. Use this information to draw an influence diagram by recognizing that the phrase "A is related to B" implies that B influences A in the model.

   b. If we assume that the phrase "A is related to B" can be translated into mathematical terms as "$A = kB$" where k is some constant, construct a mathematical model for the information provided.

5. Four key marketing decision variables are price (P), advertising (A), transportation (T), and product quality (Q). Consumer demand (D) is influenced by these variables. The simplest model for describing demand in terms of these variables is

$$D = k - pP + aA + tT + qQ,$$

where $k, p, a, t,$ and q are constants. Discuss the assumptions of this model. Specifically, how does each variable affect demand? How do the variables influence each other? What limitations might this model have? How can it be improved?

6. Return on investment (ROI) is computed in the following manner: ROI is equal to turnover multiplied by earnings as a percent of sales. Turnover is sales divided by total investment. Total investment is current assets (inventories, accounts receivable, and cash) plus fixed assets. Earnings equals sales minus the cost of sales. The cost of sales consists of mill cost of sales, selling expenses, freight and delivery, and administrative costs. Construct an influence diagram that relates these variables. Define symbols and develop a mathematical model.

7. Total marketing effort is a term used to describe the critical decision factors that affect demand: price, advertising, distribution, and product quality. Define the variable x to represent total marketing effort. A typical model that is used to predict demand as a function of total marketing effort is

$$D = ax^b$$

Suppose that a is a positive number. Different model forms result from varying the constant b. Sketch the graphs of this model for $b = 0$, $b = 1$, $0 < b < 1$, $b < 0$, and $b > 1$. What does each model tell you about the relationship between demand and marketing effort? What assumptions are implied? Are they reasonable? How would you go about selecting the appropriate model?

8. Automobiles have different fuel economies (mpg), and commuters drive different distances to work or school. Suppose that a state Department of Transportation (DOT) is interested in measuring the average monthly fuel consumption of commuters in a certain city. The DOT might sample a group of commuters and collect information on the number of miles driven per day, the number of driving days per month, and the fuel economy of their cars. Develop a model for the amount of gasoline consumed, using the symbols for the data given below.

G = gallons of fuel consumed per month
m = miles per day driven to and from work or school
d = number of driving days per month
f = fuel economy in miles per gallon

9. A manufacturer of mini-CD players is preparing to set the price on a new model. Demand is thought to depend on the price and is represented by the model

$$D = 2000 - 3p$$

The accounting department estimates that the total costs can be represented by

$$C = 5,000 + 4D$$

   a. Develop a model for the total profit and implement it on a spreadsheet.

   b. Use a one-way data table to estimate the price for which profit is maximized.

10. The demand for airline travel is quite sensitive to price. Typically there is an inverse relationship between demand and price: when price decreases, demand increases and vice versa. One major airline has found that when the price (p) for a round trip between Chicago and Los Angeles is $600, the demand ($D$) is 500 passengers per day. When the price is reduced to $300, demand is 1,200 passengers per day.

   a. Develop an appropriate model that will determine what price to charge in order to maximize the total revenue.

   b. Construct a spreadsheet for this model and use a data table to estimate the price that maximizes total revenue.

11. The Radio Shop sells two popular models of portable sport radios, model A and model B. The sales of these products are not independent of each other (in economics, we call these substitutable products, because if the price of one increases, sales of the other will increase). The store wishes to establish a pricing policy to maximize revenue from these products. A study of price and sales data shows the following relationships between the quantity sold (N) and prices (P) of each model:

$$N_A = 19.5 - 0.6P_A + 0.25P_B$$
$$N_B = 30.1 + 0.08P_A - 0.5P_B$$

   a. Construct a model for the total revenue and implement it on a spreadsheet.

   b. Develop a two-way data table to estimate the optimal prices for each product in order to maximize the total revenue.

12. A forest fire is burning down a narrow valley 3 miles wide at a speed of 40 feet per minute. The fire can be contained by cutting a firebreak through the forest across the valley. It takes 30 seconds for one person to clear one foot of the firebreak. The value of lost timber is $4,000 per square mile. Each person hired is paid $12 per hour, and it costs $30 to transport and supply each person with the appropriate equipment.

   a. Develop a model for determining how many people should be sent to contain the fire and for determining the best location for the firebreak (draw a picture first!).

   b. Implement your model on a spreadsheet and try to find the optimal solution.

13. An investor is considering investing in two stocks: the Quincunx Corporation (QC) and Red Bead Development (RBD). Projected returns are $5 per share for QC, and $3 per share of RBD. The cost per share for QC is $20 while the cost per share of RBD is $15. The investor has $2,000 to invest. How should the money be allocated?

   a. Define symbols and develop a mathematical model for this decision problem.

   b. Implement your model on a spreadsheet, and using trial and error, attempt to find the optimal solution.

14. Marketing managers have various media alternatives, such as radio, TV, maga-zines, etc., in which to advertise. They must determine which to use, the number of insertions in each, and the timing of insertions to maximize advertising effec-tiveness within a limited budget. Suppose that three media options are available: radio, TV, and newspaper. Let R, T, and N be the number of ads placed in each of these media, respectively. The next table provides some information about costs, exposure values, and bounds on the permissible number of ads of each medium desired by the client firm. The exposure value is a measure of the number of peo-ple exposed to the advertisement and is derived from market research studies. A total budget of $40,000 is available. The objective is to maximize the total expo-sure value. Develop a model to determine optimal values of R, T, and N.

Medium	Cost per Ad	Exposure Value	Minimum Units	Maximum Units
Radio	$ 500	2,000	0	15
TV	$2,000	3,500	12	—
Newspaper	$ 200	2,700	6	30

15. A brand manager for a major consumer products company must determine how much time to allocate between radio and television advertising during the next month. Market research has provided estimates of the audience exposure for each minute of advertising in each medium. Costs per minute of advertising are also known, and the manager has a limited budget of $25,000. The manager has de-cided that because television ads have been found to be much more effective than radio ads, at least 70 percent of the time should be allocated to television. Suppose that we have the following data:

Type of Ad	Exposure per Minute	Cost per Minute
Radio	150	$ 400
TV	800	$2000

 a. Develop a model to determine the amount of time spent on radio ads and the amount of time spent on TV ads to maximize total exposure and meet the constraints specified.
 b. Set up a spreadsheet to help you find a good solution.

16. Local Area Networks, or LANs, link several computers together to share common software and hardware, allowing companies to reduce costs by pooling computer resources. One problem that many companies face is determining the best use of printers on a LAN. The company must decide if it should invest in several less expensive, slower printers; in fewer more expensive, high-speed printers; or in some combination. The manager considered two types of printers: a slow model that prints 6 pages per minute (or 1,974.81 bytes per second) and costs $1,400, and a fast model that prints 16 pages per minute (or 5,266.15 bytes per second) but costs $3,250. The manager's decision problem is, "How many printers of each type should be purchased to meet the goal of providing at least 6,500 bytes per second of print capacity at minimum total cost?"
 a. Develop an appropriate model and spreadsheet for analyzing your model.
 b. Using the spreadsheet, determine the best solution that you can.

17. Each year, Kurbe Marketing Research (KMR)[5] conducts hundreds of research studies for a variety of clients. Throughout the year, the demand for telephone re-search varies. KMR conducts most of its research projects in-house and if it cannot

handle all of the work, it must subcontract some of the telephone interviewing at a lower profit margin.

KMR faces conflicting objectives. It needs to have telephone services available when clients need data collected, yet it needs to keep its costs down in order to remain competitive and earn a profit. Costs include the fixed costs of facilities and supervision, general overhead ($125,000 per month), the costs of operating computer-assisted telephone interviewing stations, or CATIs ($3,000 per month per CATI), and the cost of operators ($1,500 per month per operator) . The key decision variable is the number of CATIs to have. If the number of CATIs is high, costs would increase, but the amount of subcontracted work would decrease. If the number is low, subcontracting would increase but costs would decrease. Each CATI can handle $9,000 of research demand per month.

KMR has determined that it receives 15 percent of the revenue for all work subcontracted to an outside vendor. The table below provides the average monthly demands for the past several years.

Month	Demand
January	$940,000
February	$820,000
March	$575,000
April	$860,000
May	$695,000
June	$320,000
July	$840,000
August	$700,000
September	$245,000
October	$770,000
November	$150,000
December	$170,000

   a. Develop a spreadsheet for computing the total annual profit from both in-house and subcontracting work for a variable number of CATIs.

   b. Make a recommendation for the optimal number of CATIs to have.

   c. Examine the sensitivity of your solution if demand estimates vary by 6 percent in either direction. Does this change your recommendation?

18. A firm would like to order its key component by using the EOQ formula. The demand for the component is 6,000 units per year, ordering cost is $75.00 per order, cost per unit is $3.50, and the holding cost rate is 8 percent.

   a. Write out the total cost model and give the EOQ for this component.

   b. Explore the sensitivity of this model to changes in the annual demand by developing a data table for annual demands between 4,000 and 8,000.

   c. Construct a tornado chart for studying the sensitivity of the parameters in the total cost model.

19. The *production lot size model* is similar to the economic order quantity model. However, instead of the ordered quantity of the item being delivered all at once, it is produced at a steady rate for some period of time, say 100 units per day, 500 units per week, and so on. Thus, if D represents the annual demand rate and P represents the annual production rate, during periods of production, inventory increases at the net rate of $P - D$ units. When production is not occurring, however, inventory decreases at the rate of D units per year. Similar to the EOQ model, C_0

represents the ordering cost (sometimes called the *setup cost*), and iC represents the holding cost per unit.

    a. Sketch the inventory level over time using these assumptions and contrast it to Figure 1.17.

    b. Develop a model for the total annual cost in terms of Q, the amount to produce. How does this model differ from the EOQ model? Does it make sense?

20. Apply your results from Problem 19 to the following situation: Sunshine Tanning Products of Honolulu experiences roughly a constant demand of 10,000 bottles of tanning oil per year. The maximum annual rate of production is 50,000 bottles per year, with a setup cost of $400 per run. The inventory holding rate (annual) is 20 percent, and the per-unit cost of production is $1.35 per bottle. Write out the total cost model for this scenario. Using trial and error, try to find the production run size that will minimize cost.

21. A computer peripheral manufacturer needs $12 million in cash next year. It believes that it can earn 8 percent in marketable securities, whereas its checking account pays 2 percent interest. Each transaction to convert securities to cash costs a fixed amount of $300. Determine the optimal amount of securities to convert to cash, and how often transactions are necessary, assuming a constant cash outflow.

22. Hanover, Inc., estimates the standard deviation of its daily cash flows to be $50,000. The cost of either buying or selling marketable securities is $250 per transaction, and the interest rate premium is 8 percent annually. Apply the Miller-Orr model to find the optimal decision rule for managing Hanover's cash balance, assuming a minimum required balance of $5,000.

23. Using the data in Problem 22, construct tornado and spider charts to examine the sensitivity of the parameters of the term Z in the Miller-Orr model. Explain your results in meaningful managerial terms in the context of managing cash flows.

NOTES

1.   This scenario will be used in several other chapters to illustrate various issues in management science.

2.   Michael Hammer and James Champy, *Reengineering the Corporation* (New York: HarperBusiness, 1993), 96.

3.   Adapted from M. C. Thommes, *Proper Spreadsheet Design* (Boston: boyd & fraser Publishing Co., 1992), and G. N. Smith, *LOTUS 1-2-3 Quick*, 2d ed. (Boston: boyd & fraser Publishing Co., 1990).

4.   Excel for Office 97 has a useful new feature called *Conditional Formatting*. This allows you to change fonts or colors depending on the value of the cell. For example, in the Santa Cruz spreadsheet, you could change the color of cells F14 and F15 to red if their values exceed the amount available in cells F7 and F8. Thus, if you were seeking a solution by trial and error, you would know immediately if the solution was not feasible.

5.   We gratefully acknowledge our students Anita Anderson, Ken Brown, Jerry Tepe, and Jeanne Vennemeyer for developing this scenario as part of a class assignment.

BIBLIOGRAPHY

Anbil, Ranga, Eric Gelman, Bruce Patty, and Rajan Tanga. 1991. "Recent Advances in Crew-Pairing Optimization at American Airlines." *Interfaces*, Vol. 21, No. 1, January–February 1991, pp. 62–74.

Ben-Dov, Yosi, Lakhbir Hayre, and Vincent Pica. 1992. "Mortgage Valuation Models at Prudential Securities." *Interfaces*, Vol. 22, No. 1, January–February, pp. 55–71.

Brown, A. A., E. L. Hulswit, and J. D. Detelle. 1956. "A Study of Sales Opportunities." *Operations Research*, Vol. 4, pp. 296–308.

Camm, Jeffrey D., Thomas E. Chorman, Franz A. Dill, James R. Evans, Dennis J. Sweeney, and Glenn W. Wegryn. 1997. "Blending OR/MS, Judgment, and GIS: Restructuring P&G's Supply Chain." *Interfaces*, Vol. 27, No. 1, January–February, pp. 128–142.

Emery, Douglas R., John D. Finnerty, and John D. Stone. 1998. *Principles of Financial Management*. Upper Saddle River, N.J.: Prentice-Hall.

Evans, James R. 1986. "Spreadsheets and Optimization: Complementary Tools for Decision Making." *Production and Inventory Management*, First Quarter, pp. 36–45.

Gallagher, Timothy J., and Joseph D. Andrew, Jr., 1997. *Financial Management Principles and Practice*. Upper Saddle River, N.J.: Prentice-Hall.

Homer, Jack B., and Christian L. St. Clair. 1991. "A Model of HIV Transmission through Needle Sharing." *Interfaces*, Vol. 21, No. 3, pp. 26–49.

Krumm, F. V., and C. F. Rolle. 1993. "Management and Application of Decision and Risk Analysis at DuPont." *Interfaces*, Vol. 23, No. 3, May–June, pp. 84–93.

Mason, Richard O., James L. McKenney, Walter Carlson, and Duncan Copeland. 1997. "Absolutely, Positively Operations Research: The Federal Express Story." *Interfaces*, Vol. 27, No. 2, March–April, pp. 17–36.

Smith, Barry C., John F. Leimkuhler, and Ross M. Darrow. 1992. "Yield Management at American Airlines." *Interfaces*, Vol. 22, No. 1, January–February, pp. 83–88.

Stone, Lawrence D. 1992. "Search for the SS Central America Mathematical Treasure Hunting." *Interfaces*, Vol. 22, No. 1, January–February, pp. 32–54.

Taube-Netto, Miguel, 1996. "Integrated Planning for Poultry Production at Sadia." *Interfaces*, Vol. 26, No. 1, January–February, pp. 38–53.

Woolsey, G. 1988. "A Requiem for the EOQ: An Editorial." *Production and Inventory Management*, Vol. 29, No. 3, Third Quarter, pp. 68–72.

APPENDIX: BASIC SPREADSHEET SKILLS

Spreadsheet software continually evolves to provide users with increasingly sophisticated ways to organize, analyze, and present information. This appendix reviews some of the basic skills you will need to use spreadsheets for management science problems in this book. We have standardized our presentation for Microsoft Excel; differences may exist for other software. We assume that you are familiar with the fundamentals of spreadsheets and have used them for accounting, financial, or other types of business applications. You should be comfortable with basic user skills, including the following:

- Moving around a spreadsheet
- Entering and editing data and formulas
- Copying cell contents
- Performing basic arithmetic calculations
- Relative and absolute addressing
- Formatting data and text
- Retrieving, saving, and printing files
- Creating simple graphs

In addition to these basic skills, you should be familiar with elementary functions.

Functions

Functions are used to perform special calculations in cells. You should consult the help files corresponding to your particular version of software to determine what functions are available. The most common functions used in many management science applications include:

=MIN—finds the smallest value in a list

=MAX—finds the largest value in a list

=SUM—finds the sum of values in a list

=AVERAGE—averages a list of values

=STDEVP—finds the population standard deviation (using n in the denominator)

=STDEV—finds the sample standard deviation (using $n - 1$ in the denominator)

=SQRT—finds the positive square root of a number

=RAND—generates a uniform random number between 0 and 1

Many other specific functions for statistical, financial and other applications exist.

Functions may be linked with other standard arithmetic operations and can also be combined with one another. For instance, we may wish to compute a standard normal value using the formula (where μ = mean, and σ = standard deviation):

$$z = \frac{x - \mu}{\sigma}$$

Suppose that 50 values that define the distribution are stored in the range A3:A53 and that the value of x is in cell B5. If we wish to place the value of z in cell D8, we would write this formula in cell D8:

$$= (B5 - AVERAGE(A3:A53))/STDDEV(A3:A53)$$

Two other functions that deserve special explanation are IF and VLOOKUP.

IF(condition,A,B)

This function allows you to choose one of two values to enter into a cell. If the specified *condition* is true, value A will be put in the cell. If the condition is false, value B will be entered. For example, if cell C2 contains the function

$$=IF(A8=2,7,12),$$

the function states that if the value in cell A8 is 2, the number 7 will be assigned to cell C2; if the value in cell A8 is not 2, the number 12 will be assigned to cell C2. "Conditions" may include

- =   equal to
- >   greater than
- <   less than
- >= greater than or equal to
- <= less than or equal to
- < > not equal to

VLOOKUP(A,X:Y,B)

These functions allow you to look up a value in a table. They are similar in concept to the IF function, but they allow the program to pick an answer from an entire table of values. VLOOKUP stands for a vertical lookup table (a similar function, HLOOKUP, for horizontal lookup table, is also available). The function VLOOKUP uses three arguments:

1. The value to look up (A)
2. The table range (X:Y)
3. The number of the column whose value we want (B)

To illustrate this, suppose we want to place in cell G14 a number 1, 2, or 3, depending on the contents of cell B9. If the value in cell B9 is .55 or less, then G14 should be 1; if it is greater than .55 but .85 or less, then G14 should be 2; and if it is greater than .85, then cell G14 should be 3. We must first put the data in a table in cells. Here we put the data in B4 through C6:

	B	C
4	0	1
5	0.55	2
6	0.85	3

Now consider the formula in cell G14:

$$=\text{VLOOKUP(B9,B4:C6,2)}$$

The VLOOKUP function takes the value in cell B9 and searches for a corresponding value in the first column of the range B4:C6. The search ends when the first value greater than the value in cell B9 is found. The program then returns to the previous row in column B, picks the number found in the cell in the second column of that row in the table range, and enters it in cell G14. Suppose the number in cell B9 is .624. The function would search column B until it finds the first number larger than .624. This is .85 in cell B6. Then it returns to row 5 and picks the number in column C as the value of the function. Thus, the value placed in cell G14 is 2.

Auditing Your Spreadsheets

Excel has a useful auditing function in the *Tools* menu that allows you to examine the structure of formulas graphically and to identify errors in your spreadsheet. An example of an audit trace is shown in Figure 1.24. We encourage you to use this option to check and debug your models.

Creating Tornado and Spider Charts

To create a tornado chart:

1. First, set up a table specifying the lower limit, base case value, and upper limit for each input variable (see the range B21:E26 in Figure 1.13)
2. Next, you will need to compute the value of the output variable when each input variable is changed, while holding all other input variables at their base case values. This can be done easily by creating a data table for each input variable as shown in the range G21:H40 in Figure 1.13. (Hint: use *Edit/Copy* and *Edit/Paste Special* [with *Transpose* option checked] commands to copy the row range C23:E23 to the column range G23:G25.)
3. Copy the data table results to a new table that shows the output results for each input variable change (see the range B28:F33 in Figure 1.13). The last column computes the difference between the largest and smallest value in each row.
4. Sort the rows in this table in descending order of the largest range using the *Data/Sort* command.

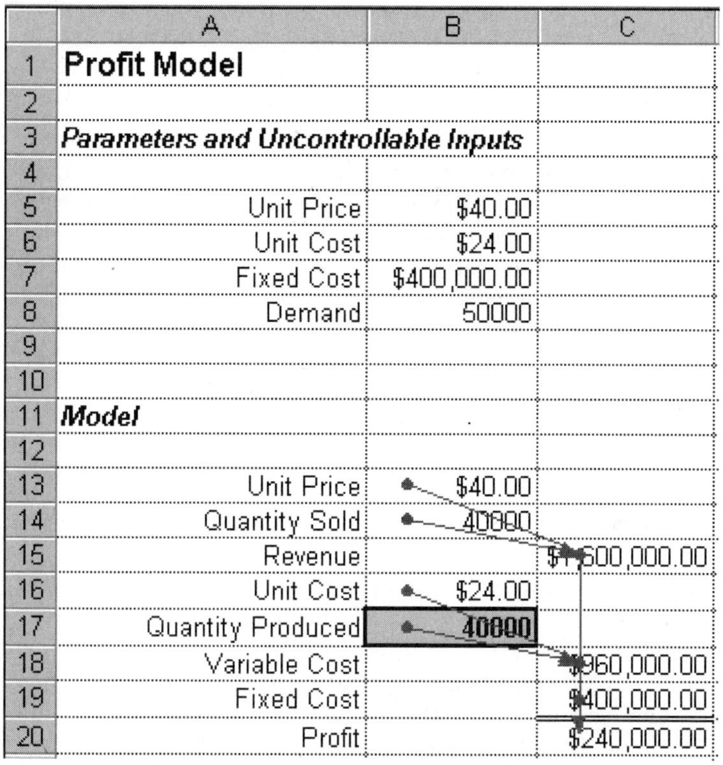

	A	B	C
1	**Profit Model**		
2			
3	***Parameters and Uncontrollable Inputs***		
4			
5	Unit Price	$40.00	
6	Unit Cost	$24.00	
7	Fixed Cost	$400,000.00	
8	Demand	50000	
9			
10			
11	***Model***		
12			
13	Unit Price	$40.00	
14	Quantity Sold	40000	
15	Revenue		$1,600,000.00
16	Unit Cost	$24.00	
17	Quantity Produced	40000	
18	Variable Cost		$960,000.00
19	Fixed Cost		$400,000.00
20	Profit		$240,000.00

FIGURE 1.24
Example of Excel auditing function

5. Create a chart by clicking on the Excel chart icon to run the Chart Wizard. First, select the type of chart as a *Clustered Bar* chart. In Step 2 of the Chart Wizard, the *Data Range* should include the second (lower limit values) and fourth columns (upper limit values) in the output results table created in step 3 (do not include the column headings). You can select noncontiguous column by holding down the *Ctrl* key while highlighting the ranges. In the *Series* tab of the *Source Data* box, set the *Category (X) axis labels* as the first column of the table (the variable names).
6. Customizing the chart is the key to the process. In Step 3 of the Chart Wizard (or select *Options* from the *Chart* menu), perform the following tasks.
   a. In the *Axes* tab, make sure that the check box for *Category (X) axis* is checked, and also click the *Category* button under it. The *Value (Y) axis* check box should also be checked. In the *Gridlines* tab, all check boxes should be unmarked. In the *Legend* tab, unmark the box *Show Legend*. Under *Data Labels*, click on *None*. *Show data table* should be unmarked in the *Data Table* tab. The final step of the Chart Wizard saves the chart object. However, several other features must be customized.
   b. Double click on the vertical (Y) axis. A *Format Axis* dialog box appears. Under the *Patterns* tab, check the following: *Axis—Automatic*, *Major tick mark type—Outside*, *Minor tick mark type—None*, and *Tick mark labels—Low*. Under the *Scale* tab, each input box should have the number 1, and all three check boxes should be marked.
   c. Double click on the horizontal (X) axis. In the *Format Axis* dialog box, under the *Patterns* tab, the *Automatic* button should be marked under *Axis*; *Major tick*

mark type—Outside; Minor tick mark type—None; and *Tick mark labels—Next to axis.* Under the *Scale* tab, enter the base case output value in the box corresponding to *Category (X) axis Crosses at.* (For the text example, you would enter 240,000.)

d. You have charted two data series. Position the pointer on one of the bars in the chart and double click. A *Format Data Series* box will appear. Under the *Patterns* tab, select a custom border (choose your own color) and a color for the area by clicking on one of the colored boxes. The purpose of this is simply to make the bars on the chart the same color so as not to distinguish between the data series, as is normally done in a bar chart. In the *Axis* tab, check *Plot series on primary axis.* Under the *Options* tab, set *Overlap* equal to 100, and *Gap width* arbitrary (the larger the number, the smaller the width of the bars; 150 is a good choice). Repeat this for the other data series if you need to change the bar colors.

To create a spider chart:

1. Set up a table of base case values for the input variables you select (see A21:B25 in Figure 1.14).
2. Determine the percentage changes you want for the inputs (A29:A37). Construct data tables for each input by varying the base case values by the percentage changes (B27:I37).
3. Click on the Chart icon to start the Chart Wizard. For the data ranges, select the outputs from each data table (hold down the *Ctrl* key to select noncontiguous columns). In the *Series* tab of the *Source Data* box, set the *Category (X) axis labels* as the range of percentages (A29:A37).
4. Customize the legend, titles, etc. as appropriate.

Data Management and Analysis

O U T L I N E

Applications of Data Management and Analysis
Law Enforcement
Health Care
Retail Operations and Strategy
Hotel Location

Data Storage and Retrieval
Management Science Practice: Allders, International
Sorting and Filtering Data
Using *AutoFilter* to Check Data Accuracy
Using Data Retrieval Methods in Optimization Modeling

Data Visualization
Charts
Geographic Data Maps

Data Analysis
Descriptive Statistics
Pivot Tables
Probability Distributions and Data Fitting with Crystal Ball
Covariance and Correlation
Applying Covariance in Financial Portfolio Optimization Models

Regression Analysis
Simple Linear Regression
Statistical Issues in Regression
Multiple Linear Regression
Building Regression Models
Using Regression in Optimization Models

Technology for Statistical Analysis
Problems

In Chapter 1 we introduced the basic concepts of management science—building, manipulating, and solving models for decision making. All models require data. For example, in the simple profit model we discussed in Chapter 1, we need values of the unit price, unit cost, fixed cost, and quantity sold in order to compute the profit; likewise, the inventory management model requires estimates of the holding cost, ordering cost, carrying charge, and demand. While such model inputs might magically appear in textbooks, real applications require substantial effort to collect, process, and "clean" (ensure the data are free from errors) them. Data collection, data management, and data analysis often require more effort and time than the actual modeling building and analysis process, and can make an important difference between the success and failure of a management science project.

Both companies and individuals have easy access to vast amounts of data. For instance, scanner data collected automatically in retail outlets allows for the more accurate control of inventory and creates records of customer spending habits, and law enforcement agencies use networked databases to share information about crimes and missing persons. Individuals can easily obtain company reports, current and historical stock prices, interest rates, real estate for sale, and best prices and availability of new and used automobiles, books, and compact disks on the internet. Providing an interface between data and models is a challenging problem. **Data management** involves storing, accessing, transforming, and distributing data and information. **Data analysis** refers to extracting information from data. Both simple visual approaches and sophisticated statistical techniques are important methodologies for data analysis.

In this chapter we discuss these data management and analysis technologies as support tools for management science. Because most models we discuss in this book are implemented using Excel spreadsheets, we focus on the data management, data visualization, and statistical analysis capabilities of Excel.

APPLICATIONS OF DATA MANAGEMENT AND ANALYSIS

The following examples illustrate the benefits of good data collection, accessibility, and analysis for model development and decision making.

Law Enforcement

The Delaware state police consolidate criminal and other data from several police agencies and computers around the state into a single repository of information. The system allows for quick and accurate searches for criminal and victim identification. Identification processes that used to take weeks can now often be performed in hours. Using more accurate and easily accessible data, the state police hope to speed up investigations, analyze crime trends, improve manpower and resource allocation, and better evaluate drug- and gang-control problems (Vijayan, 1997).

Health Care

One of the more difficult jobs confronting hospital administrators is estimating the levels of nurse staffing to meet patient requirements. The further in advance these estimates can be made, the easier it is to plan levels of staffing, vacation time, and training programs. In one short-term, acute-care hospital, data from daily reports by shift, ward, and intensity-of-care levels were used to develop statistical models to predict the number of patients in each level of care and hence, labor requirements. The model supported intuition with more precise information that allowed, for example, hospital management to include part-time nursing schedules in the decision process (Helmer et al., 1980).

Retail Operations and Strategy

Sears, Roebuck and Company recognized that to again become competitive in the retail industry, it needed to provide accurate, easily accessible data to decision makers throughout the firm's retail network. Numerous mainframe databases on sales were consolidated into a single data warehouse, thus eliminating redundant systems and reports. Time previously spent searching and reconciling data from various locations can now be used for analyzing data and making more informed decisions. The company plans to fully integrate other databases including inventory and sales margin data with the sales data warehouse in the near future.

Sears also provided a consulting group with thirteen financial measures, hundreds of thousands of employee satisfaction data points, and millions of data points on customer satisfaction. Using advanced statistical tools, the analysts discovered that employee attitudes about the job and the company are key factors that predict their behavior with customers, which, in turn, predicts the likelihood of customer retention and recommendations—which, in turn, predict financial performance. Sears can now predict that if a store increases its employee satisfaction score by 5 units, customer satisfaction scores will go up by 2 units, and revenue growth will beat the stores' national average by 0.5 percent. Such an analysis can help managers to make decisions, for instance, on improved human resource policies (Cafasso, 1996, and *Fortune*, 1997).

Hotel Location

Location is one of the most important decisions for a lodging firm. All hotel chains search for ideal locations, and in many cases compete against one another for the same sites. La Quinta Motor Inns, a midsized hotel chain that is headquartered in San Antonio, Texas, developed a statistical model that predicted profitability for sites under construction. Input variables were grouped into five categories: competitive factors, demand generators, demographic factors, market awareness, and physical factors. The dependent variable, occupancy, proved to be a very unstable indicator of success; La Quinta settled on forecasting operating margin, which consisted of adding depreciation and interest expenses to profit and dividing by the total revenue. The model was incorporated into a spreadsheet decision model that management uses for selecting sites to minimize the risk of picking an unprofitable site (Kimes and Fitzsimmons, 1990).

DATA STORAGE AND RETRIEVAL

The need to manage and share stored data has created an extensive market for data warehousing software. A **data warehouse** is a large storage facility that stores data from different databases across an organization and allows for the efficient access and dissemination of data. In its simplest form, a data warehouse can be likened to a Microsoft Excel workbook. Each worksheet might represent an individual database; the workbook provides a common storage facility, and various Excel tools provide the means of accessing, manipulating, and reporting the data. For example, it is easy to construct a wide variety of reports and charts, manipulate data interactively, compare data across time periods, and find relationships among variables. This process of digging through data to extract key information and visualize the data in an understandable fashion is often called **on-line analytical processing (OLAP).** OLAP systems allow users to look at data and manipulate them interactively and generate useful reports. Many of the capabilities of Excel can be considered to support OLAP activities.

A data warehouse can provide valuable information for decision making (see the Management Science Practice box on Allders, International). One study of retailers identified five key areas where data warehousing has resulted in major benefits:

- item by item sales analysis
- vendor pricing and performance analysis
- better inventory control through forecasting and management
- strategically planned promotions by location
- targeting individual customers by tracking their purchases[1]

Retail giant Wal-Mart announced in 1997 that it would triple the size of its data warehouse to over 24 trillion bytes of data.[2] Ninety percent of the *Fortune* 1000 companies are believed to have implemented data warehouses.[3] Many companies, such as food cooperative Land O'Lakes, link their sales force with company databases over the Internet, using laptops equipped with analytical software.

Management Science Practice: Allders, International

Allders International specializes in duty-free operations with 82 tax-free retail outlets throughout Europe, including shops in airports and seaports, and on cross-channel ferries. The company employs 1,500 people and had annual sales (1995) of £450 million.

Like most retail outlets, Allders International must track masses of point-of-sale data to assist in inventory and line composition decisions. Which items (known as stock-keeping units, or SKUs) and how many of each SKU to stock at each of its outlets are fundamental decisions that impact the profitability of the company. Allders International implemented a data warehouse to manage this process. After one month, the data warehouse had already saved the firm an estimated £60,000.

Prior to the implementation of a data warehouse for collecting the point-of-sale data, staff had had to analyze large quantities of paper-based data. This manual process was so overwhelming and time consuming that the analyses were often too late to provide information useful for decision making. Through a central data warehouse, the data are now collected and stored electronically

continued

and may be analyzed more quickly. Simple queries for subsets of the data for analysis—for example, the performance of a particular SKU across all outlets, or the overall performance of a particular outlet—are now easy to obtain. Hence the data warehouse achieved significant savings based simply on increased productivity of the staff required to analyze the data.

The data warehouse is even more valuable when one considers the additional benefits from more accurate and timely information provided by these analyses. Managers are now able to quickly identify which SKUs and/or which shops are underperforming. For example, Allders has found that 20 percent of its product lines (groups of similar SKUs) generate 80 percent of its profits. This has allowed the company to selectively eliminate some of the other 80 percent. Eliminating unprofitable SKUs can cut inventory costs, free up shelf space for more profitable SKUs, and reduce the number of suppliers needed.

Finally, the data warehouse has had the side benefit of greatly increasing staff morale. The staff, now freed from the tedious manual approach to the data, is now free to do more creative analyses.

Source: Adapted from Stephen Pass, "Discovering Value in a Mountain of Data," *OR/MS Today*, Vol. 24, No. 5, pp. 24–28, October 1997.

Data mining, which refers to the process of uncovering hidden facts and obscure patterns in data from large databases, is being used increasingly in business. Data mining tools such as neural nets and statistical analysis can sift through large amounts of customer, marketing, production, and financial data to identify meaningful relationships for a business, thus turning data into knowledge.[4] For example, MCI Communications (now MCI Worldcom) used data-mining techniques to evaluate marketing data on 140 million households on as many as 10,000 attributes, such as income, lifestyle, and past calling habits, to identify statistical profiles related to customer retention. Data mining is also used to identify subtle patterns in customer transactions to identify credit card fraud and illegally cloned cell-phone ID numbers. U.S. West designed a program in cooperation with NCR and Sabre Decision Technologies to correlate data about home locations, U.S. West's trunk lines, switch capacity, and other variables to identify prospective customers. The response rate from a low-cost direct mail campaign equaled that of a multimillion-dollar broadcasting campaign. Other applications of data mining include identifying profitable customers and their characteristics, predicting buying behavior, prioritizing credit risk, estimating product demands, and forecasting demand for better inventory management.[5]

Data base management systems (DBMS) are tools that aid in the storage and retrieval of large amounts of data. Some widely used database management systems are SAS, Oracle, and Microsoft Access. For simplicity, we will use Excel workbooks to store and manipulate data in this text. As we shall see, Excel is actually quite powerful in its data manipulation capabilities.

A well-designed database in the form of an Excel workbook should keep similar information in separate worksheets.[6] Each column should contain the same kind of information, and the top one or two rows should contain labels that describe the contents of the column beneath it. The list should not have any blank rows or columns.

To illustrate some of the features of data management, we will describe a database from our fictitious printed circuit board producer, Hanover, Inc., which we introduced in Chapter 1. Because of increasing demand for its products, the company is currently facing a decision of how to increase capacity. Hanover has two existing plants, in Austin and Paris, and seven locations are being considered for a new plant. These consist of two cities in the United States (Charleston, S.C., and Mobile, Ala.) and five other locations simply defined by country (Australia, India, Malaysia, South Africa, and Spain). Although the company has many individual customers, we have aggregated them to eight regions (Malaysia, China, France, Brazil, U.S. Northeast, U.S. Southeast, U.S. Midwest, and U.S. West) to simplify modeling and analysis.

The decision to increase capacity by building a new plant will affect how products are sourced for manufacturing and shipped to customers. We will develop models for evaluating and optimizing this capacity expansion problem in later chapters. Here we focus on the data that are important to this decision. These data are provided in an Excel workbook, *Hanover.xls*, a portion of which is shown in Figure 2.1. Each worksheet in the workbook, which can be accessed by selecting one of the worksheet tabs (at the bottom of Figure 2.1), provides a database of key information. The Hanover database contains the following worksheets:

- *Duty*—the duty rate charged from each plant to each customer. This rate is multiplied by the selling price to get the cost per unit paid in duty to ship into one country from another.
- *Fixed Cost*—the fixed cost of plant operation, including capital costs, insurance, management, etc. These costs depend on the capacity level of the plant (given in thousands of units).
- *Forecast*—predicted demand for each product by customer. Volume is in thousands of units.
- *F&WH Cost*—freight and warehousing cost per unit from each plant to each customer (independent of product type).
- *Price*—the selling price of each product for each customer.
- *Variable Cost*—the variable cost of each product at each plant.

We will use this database, as well as others, for examples, exercises, and problems in this and subsequent chapters.

The level of detail with which data are collected and the manner in which data are stored is a function of how the data will be used to support decision making. For example, collecting and storing sales volumes by product by year will not be of much value if decisions need to be made on a monthly basis. Likewise, how data are labeled and collected impacts how easily different databases can be merged and used together. It would seem that the best thing to do would be to always collect and store data at the most detailed level. However, this could be quite costly to maintain.

Excel Exercise 2.1

Open the workbook *Hanover.xls* and familiarize yourself with the worksheets. Check to see if all data have been collected and recorded at the same level of detail.

	A	B	C	D	E	F
1	**Plant**	**Customer**	**Duty Rate**			
2	Paris	Malaysia	0.22			
3	Austin	Malaysia	0.22			
4	Australia	Malaysia	0.22			
5	Charleston	Malaysia	0.22			
6	India	Malaysia	0.22			
7	Malaysia	Malaysia	0.00			
8	Mobile	Malaysia	0.22			
9	S Africa	Malaysia	0.22			
10	Spain	Malaysia	0.22			
11	Paris	China	0.30			
12	Austin	China	0.30			
13	Australia	China	0.30			
14	Charleston	China	0.30			
15	India	China	0.30			
16	Malaysia	China	0.30			
17	Mobile	China	0.30			
18	S Africa	China	0.15			
19	Spain	China	0.30			
20	Paris	France	0.00			
21	Austin	France	0.09			
22	Australia	France	0.09			
23	Charleston	France	0.09			
24	India	France	0.06			
25	Malaysia	France	0.06			
26	Mobile	France	0.09			
27	S Africa	France	0.06			
28	Spain	France	0.00			
29	Paris	Brazil	0.12			
30	Austin	Brazil	0.12			
31	Australia	Brazil	0.12			

Price \ **Duty** / Forecast / F&WH Cost / Variable Cost / Fixed Cost /

FIGURE 2.1

Hanover.xls
workbook

Sorting and Filtering Data

The most common data manipulation task is sorting. This is often necessary to rank alternatives, particularly when multiple criteria are involved. Excel provides many ways to sort lists by rows or columns, in ascending or descending order, and using custom sorting schemes. Simply highlight the range of data you wish to sort, select the *Sort* option in the *Data* menu, and identify the row or column on which to sort and in what order. For example, Hanover must pay a duty fee when crossing country borders. This fee is expressed as a percentage of the selling price of the good and is dependent on where the good is produced and its destination. This information is contained in the worksheet *Duty* in the *Hanover.xls* workbook, shown in Figure 2.1. If we wanted to access the duty rates from Malaysia to all its customers, we could sort the data by plant to find and extract these data.

Excel Exercise 2.2

Open the worksheet *Duty* in the Hanover database. Sort the data in alphabetical order by plant, customer, and duty rate, in that order.

For large databases, finding a particular subset of records by sorting can be tedious. A more powerful feature in any database management system is the ability to retrieve only the subsets of the data having certain characteristics. Excel has several options that allow data to be "sliced and diced" in almost any way to support model building and decision making. We will demonstrate some of these capabilities.

THE EXCEL AUTOFILTER To **filter** a list means to extract all rows that meet specified criteria. Excel provides two filtering tools: *AutoFilter* for simple criteria, and *Advanced Filter* for more complex criteria. We will use several of the data sets contained in the *Hanover.xls* workbook to illustrate these features.

Example 1: Using the *AutoFilter*

Suppose we wish to access the duty rates from only Malaysia to all customers. First, select the entire list by highlighting cells A1 through C51. Then, from the Excel menu, select *Data*, *Filter*, and *AutoFilter*. Clicking on one of the small down arrow boxes in the lower right-hand corner of each column header will display a drop-down box, as shown in Figure 2.2. These are the options for filtering on that column of data. The options include (*All*), (*Top 10*), (*Custom*), and then any data values that exist in that column. The (*All*) option restores all of the records. The (*Top 10*) option allows you to see the top (or bottom) *n* items or the top or bottom *n* percent of all items (for numeric data). The (*Custom*) option is particularly useful, since it allows you to filter based on logic (or/and), and will be discussed in the next section. In Figure 2.2, we have selected Malaysia from the drop-down box in the *Plant* column. This causes a data query for all data records (rows) for which Plant = Malaysia. The result is shown in Figure 2.3. These records can now be copied into a separate worksheet to create a database for duty rates from Malaysia to the customers. We might use this, for instance, to find the average duty rate from Malaysia by weighting the duty rates by the amount shipped to each customer.

FIGURE 2.2
Using the
Excel
AutoFilter

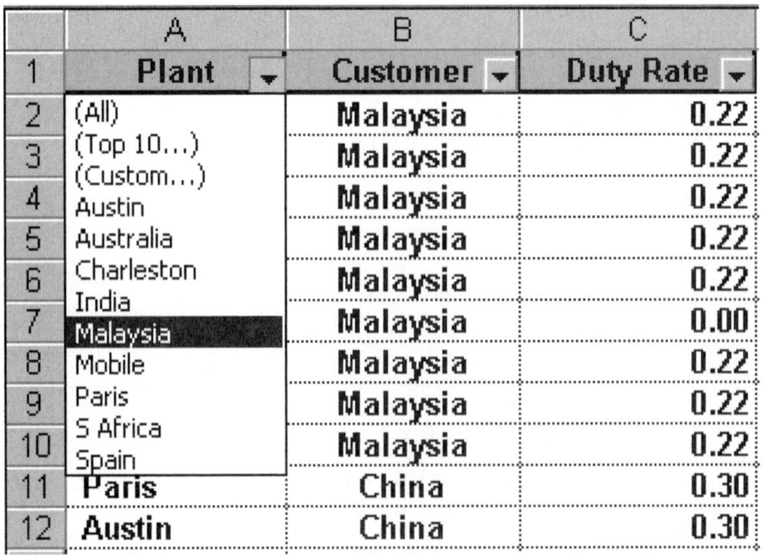

	A	B	C
1	**Plant** ▼	**Customer** ▼	**Duty Rate** ▼
2	(All)	Malaysia	0.22
3	(Top 10...) (Custom...)	Malaysia	0.22
4	Austin	Malaysia	0.22
5	Australia	Malaysia	0.22
6	Charleston India	Malaysia	0.22
7	Malaysia	Malaysia	0.00
8	Mobile	Malaysia	0.22
9	Paris	Malaysia	0.22
10	S Africa Spain	Malaysia	0.22
11	**Paris**	China	0.30
12	**Austin**	China	0.30

	A	B	C
1	Plant	Customer	Duty Rate
7	Malaysia	Malaysia	0.00
16	Malaysia	China	0.30
25	Malaysia	France	0.06
34	Malaysia	Brazil	0.12
43	Malaysia	United States	0.04
47			

FIGURE 2.3
Results of "Plant = Malaysia" query

The *AutoFilter* can be used sequentially to "drill down" the data. For example, in Figure 2.3, we could now select the *Customer* column and select from the list of customers, for example, United States. In this case this would result in a single record (the Malaysia–US duty rate). The *AutoFilter* is limited to a single selection for each column unless the *Custom* option is chosen. To remove the filter from the original data set, simply choose *Data*, *Filter*, and *AutoFilter*.

THE CUSTOM AUTOFILTER The *Custom* option in *AutoFilter* allows you to specify two criteria for the chosen column. The criteria options include *equals*, *does not equal*, *greater than*, *greater than or equal to*, *less than*, *less than or equal to*, *begins with*, *does not begin with*, *ends with*, *does not end with*, *contains*, and *does not contain*. The two criteria may be connected by logical operators "and" or "or." "And" implies that only records meeting both criteria will be listed, whereas "or" lists those records that satisfy either or both of the criteria. The *Custom* option can be used with either numerical or character data.

Example 2: Using the *Custom AutoFilter*

We will illustrate the *Custom AutoFilter* using the *F&WH Cost* worksheet in the Hanover database. Suppose we wish to find all records for which freight and warehousing cost is between $1.25 and $1.30 (inclusive). We choose the F&WH worksheet, highlight the data (cells A1 through C73), select *Data* and *AutoFilter*. In the dialog box, shown in Figure 2.4, enter the criteria in the left input boxes by selecting from the drop-down boxes, and the criteria values in the right boxes. Click the radio button for "and" followed by the OK button. The *AutoFilter* will retrieve all records that satisfy these criteria.

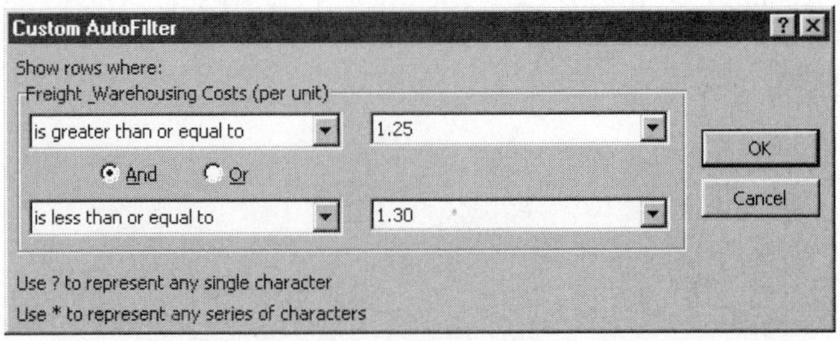

FIGURE 2.4
Custom AutoFilter dialog box

Excel Exercise 2.3

Apply the *Custom AutoFilter* option to find the records satisfying the criteria in Example 2. How many records do you retrieve?

THE ADVANCED FILTER The *Advanced Filter* (also found in the *Data* menu) is a more complicated version of the *AutoFilter*. It allows for specifying criteria for filtering for two or more columns simultaneously and to specify three or more restrictions for a given column. The filtering criteria are entered in a separate section of a worksheet and the cell range of these criteria is an input in the *Advanced Filter* dialog box, shown in Figure 2.5. The location of the list under consideration (including headers) is placed in the *List range* section. The cell range of the criteria for filtering (including headers) is placed in the *Criteria range* section. The *Criteria range* must have at least two rows (column headings that match the list, and filtering criteria). If the *Copy to another location* option is selected (rather than *Filter the list, in place*), the desired location for filtered data is placed in the *Copy to* section.

FIGURE 2.5
Advanced Filter dialog box

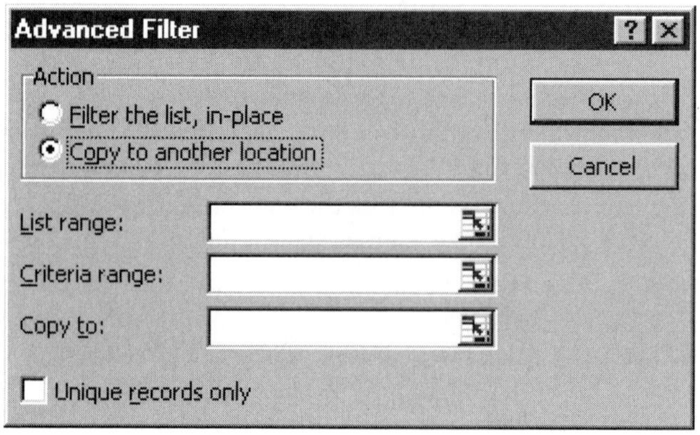

Example 3: Using the *Advanced AutoFilter*

For the *F&WH Cost* worksheet in the Hanover database, suppose we wish to find all customers that have an F&WH cost of less than or equal to $1.30 from the Austin plant. This is easily accomplished by using the *Advanced Filter*. Figure 2.6 shows the *F&WH Cost* worksheet. In the *Advanced Filter* dialog box in Figure 2.5 we would enter the range of the original data (A1:C73) in the *List range*. The *Criteria range* must be specified by the user; in this case, the criteria are listed in the range E7:F8. Finally, the location of the results, entered in the *Copy to* box, is E11:G11. The results are shown in Figure 2.6.

FIGURE 2.6 F&WH cost sheet with *Advanced AutoFilter* criteria and results

	A	B	C	D	E	F	G	H	I	J
1	**Plant**	**Customer**	**Freight & Warehousing Costs (per unit)**							
2	Paris	Malaysia	$1.41							
3	Austin	Malaysia	$1.39							
4	Australia	Malaysia	$1.43							
5	Charleston	Malaysia	$1.37							
6	India	Malaysia	$1.38							
7	Malaysia	Malaysia	$1.33		**Plant**	**Freight & Warehousing Costs (per unit)**				
8	Mobile	Malaysia	$1.37		Austin	<=1.45				
9	S Africa	Malaysia	$1.58							
10	Spain	Malaysia	$1.41							
11	Paris	China	$1.38		**Plant**	**Customer**	**Freight & Warehousing Costs (per unit)**			
12	Austin	China	$1.51		Austin	Malaysia	$1.39			
13	Australia	China	$1.44		Austin	US Northeast	$1.36			
14	Charleston	China	$1.49		Austin	US Southeast	$1.32			
15	India	China	$1.47		Austin	US Midwest	$1.31			
16	Malaysia	China	$1.38		Austin	US West	$1.32			
17	Mobile	China	$1.49							

Using *AutoFilter* to Check Data Accuracy

Well-formulated models and analysis with inaccurate data can lead to disastrous decisions. Bad data caused one corporation to bill only 96 percent of its customers. The $2 billion dollar corporation forfeited $80 million in revenues![7] Thus, it is important to ensure the accuracy of data.

The *AutoFilter* listing of entries in a given column is a quick approach to checking for data errors. By invoking the *AutoFilter* and selecting the drop-down menu for a column, a listing of the values of that column appears. For example, in Figure 2.2, a listing of all plants in the duty worksheet is shown. By scanning this list, misspellings (or in the case of numerical data, numbers out of the normal range) can be easily spotted.

Excel Exercise 2.4

Open the Hanover database and select the Freight & Warehousing Cost worksheet (*F&WH Cost*). Use the *AutoFilter* listing to search for data list errors. Correct any obvious mistakes you find. There is one!

Using Data Retrieval Methods in Optimization Modeling

In Chapter 4, we will discuss the formulation and solution of an optimization model to evaluate the impact of adding a plant in Spain for the capacity expansion decision that Hanover, Inc., faces. The model we will develop is known as the *transportation model*. The transportation model addresses the following problem. Given a set of plants with known capacities, a set of customers with known demands, and the unit cost to supply each customer from each plant, find a transportation plan that minimizes total cost while ensuring that each customer's demand is satisfied and no plant capacities are violated. In its simplest form, the transportation model considers only a single product.

Data needed to construct this optimization model (using product A as an illustration) are:

1. Annual production capacities at the existing Austin and Paris plants, and the proposed site in Spain.
2. Annual demand of the product for each customer.
3. Unit cost to supply the product from each plant to each customer. This is the sum of the variable production cost at the plant, freight and warehousing cost from plant to customer, and duty paid per unit from the plant to the customer. Duty cost per unit is the (duty rate) $\times$ (selling price per unit). For example, using the data in the Hanover database, the cost to source 1 unit of product A from Paris to Malaysia would be $1.26 + $1.41 + .22($4.19) = $3.59.

Let us focus on obtaining the total unit cost data. The variable cost and freight and warehousing costs are found in worksheets in the *Hanover.xls* Excel workbook. Using the *AutoFilter*, we can retrieve the relevant costs (from Paris, Austin, and Spain), and cut and paste them into a new worksheet where we may perform further calculations. Doing this for each plant and each cost component allows us to construct the data we need for the model. These data may be found in the Excel file *transportation.xls*.

Excel Exercise 2.5

Using the approach described in the previous paragraph, compute the unit costs from each plant to each customer for product A and compare your results to the data in the Excel file *transportation.xls*.

DATA VISUALIZATION

The old adage "A picture is worth a thousand words" is probably truer in today's information-rich environment than ever before. Managers need to understand the significance of data in as concise and meaningful way as possible. Data visualization not only can provide excellent communication vehicles, but can also reveal previously surprising patterns and relationships. Through the use of multiple axes and color, higher dimensional data can be simplified for easier interpretation.

Data visualization is also important for management science modeling, both for the construction of analytical models and for effective interpretation of output from these models. Visualizing data is the important first step in analytical model construction. For example, ascertaining the forms of the relationships between sales and advertising, incentives and performance, and resources and output is key to building effective models involving these factors. As we shall see, whether or not the relationship between factors is linear or nonlinear will dictate the type of model to be constructed. Furthermore, complex analytical models often yield complex results. Data visualization is an important tool for expressing these results in an intuitive way for management.

Microsoft Excel 97 includes a variety of tools for effective data visualization and graphical display. These include charts and geographic data maps.

Charts

Charts (graphs and plots) are the most basic way of representing quantitative data. A variety of charts is available in Excel. These include vertical and horizontal bar charts, line charts, pie charts, area charts, scatter plots, three-dimensional surface charts, radar charts, bubble charts, and other special types of charts. The Excel *Chart Wizard* provides an easy way to create charts within a spreadsheet.

Example 4: Creating a Chart

We will illustrate how to construct a pie chart of the forecasted demand for product A in *Hanover.xls*. First, select the *Forecast* worksheet, invoke *AutoFilter*, and select product A from the drop-down list. This results in a list of demand forecasts by customer, as shown in Figure 2.7. To create a chart, click on the *Chart Wizard* icon (a picture of a 3-D bar chart) from the standard tool bar or from the *Insert...Chart...* menu selection. Select the chart type from the list in the dialog box that is displayed (e.g., *Pie*) and then click on the specific chart subtype option. Click *Next* to continue. A second dialog box asks you to define the data to chart. You may enter the data range directly or highlight it in the worksheet. You also need to define whether the data are stored by rows or columns. (Note: If the data to be plotted are not stored in contiguous columns, hold down the *Ctrl* (control) key while selecting each block of data; then start the *Chart Wizard.*) For this example, use the mouse to select cells A1 through A9, then hold the *Ctrl* key while selecting cells C1 through C9. The *Series* tab allows you to check and modify the names and values of the data series in your chart. The third dialog box allows you to specify details to customize the chart and make it easy to read and understand. You may specify titles for the chart, all axis labels, style of gridlines, placement of the legend to describe the data series, data labels, and even a data table of values from which the chart is derived. Finally, the last dialog box allows you to specify whether to place the chart as an object in an existing worksheet or as a new sheet in the workbook. Figure 2.8 shows the result.

	A	B	C
1	Customer ▼	Product ▼	Volume (000 units) ▼
2	Malaysia	A	399.0
3	China	A	3158.3
4	France	A	1406.0
5	Brazil	A	163.5
6	US Northeast	A	68.7
7	US Southeast	A	999.7
8	US Midwest	A	544.9
9	US West	A	1804.0

FIGURE 2.7
Demand forecast for product A by customer

FIGURE 2.8
Pie chart of
product A fore-
casted volume

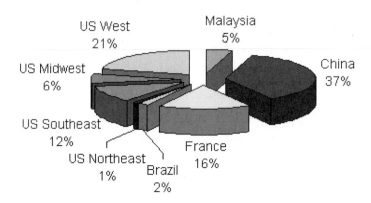

Product A Forecasted Volume (000 units)

MULTIDIMENSIONAL DATA Using bar, line, and pie charts is quite routine in conveying business data such as sales and financial data. Many models used for decision making involve multidimensional data. By collapsing information visually to lower dimensions, we are able to better grasp and synthesize multidimensional data. Special charts, such as *bubble charts* and *radar charts*, provide useful means of doing this.

Example 5: Vendor Selection with a Bubble Chart

Suppose we are trying to decide to which of four vendors we should award a contract. Data on the percentage of defective parts and the percentage of late orders from past contracts are available, and we have also received a bid (price per unit). These data (the price increase is the increase over the current price) are shown in Table 2.1.

TABLE 2.1
Vendor data

Vendor	Percent Acceptable	Percent On Time	Price Increase
Vendor 1	90.00%	80.00%	$0.80
Vendor 2	95.00%	90.00%	$0.50
Vendor 3	91.00%	85.00%	$0.10
Vendor 4	90.00%	87.00%	$0.50

Figure 2.9 is a bubble chart of these data. The center of each bubble is plotted on a grid corresponding to the percent of acceptable parts and on-time deliveries. The size of the bubble shows the relative percentage price increase. From this chart it is clear that Vendor 4 dominates Vendor 1, since Vendor 4 has a higher on-time percentage, the same percent acceptance, and less of a price increase. Likewise, Vendor 4 is in turn dominated by Vendor 2; that is, Vendor 2 has the same price, but higher percent acceptance and higher on-time percentage. Thus, we can easily see that Vendors 2 and 3 are the best candidates. Which of these two vendors to select depends on the importance placed on each of the three factors.

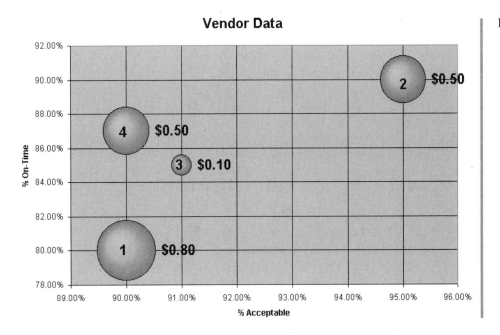

FIGURE 2.9

Bubble chart for vendor selection problem

Although we cannot illustrate it in this text, color is extremely useful in visual data representation, by allowing you to further increase the dimensionality of what is illustrated. For example, adding a different color to a bubble chart increases its capability to four dimensions: x-axis, y-axis, bubble size, and bubble color.

Excel Exercise 2.6

Consider the vendor data in Table 2.1. These data are in *vendor.xls*. Convert the data to % rejected, % late, and % price increase (the current price is $14.00 per unit). Create a radar chart of these data (note: be sure to select *Series In:* rows in the radar dialog box). How can you tell that one vendor dominates another?

Geographic Data Maps

Many applications of management science involve geographic data. For example, geographic data are key aspects of such problems as finding the most cost-effective location of production and distribution facilities, the most equitable sales region assignments, the cost-effective transportation of raw materials and finished goods, and efficient vehicle routing. In geographic-based problems such as these, data mapping can help in a variety of ways. First, the mapping of data is often an easy way to spot data errors (the authors of this text were involved in a project in which a customer appeared in the middle of the Pacific Ocean after the data were mapped!). Second, seeing solutions to problems on a map usually gives some intuition into why the solution makes sense and in this regard, increases the likelihood of managerial acceptance.

Indeed, the use of optimization models in combination with data mapping was instrumental in the success of Procter & Gamble Company's North American Supply Chain Study, which saved the company more than $200 million dollars per year.[8] This project is discussed in a Management Science Practice box in Chapter 4.

A relatively new feature to Excel is *Microsoft Map*, a geographic data-mapping tool that allows you to map sales by region, population by state, units sold by country, etc. To use the mapping tool in Excel, your worksheet must contain a column listing state names, country names, postal codes, counties, or codes used in Microsoft Map. For example, Figure 2.10 shows the western regional sales forecasts for Hanover, Inc. To display these data on a map, highlight the range of the data (A5:B13), then select *Map* from the *Insert* menu, or click on the globe icon on the main toolbar. Next, click on the spreadsheet and size an area by dragging the mouse. If multiple maps exist, a dialog box will appear, allowing you to select the map you would like to show. Excel comes with a variety of standard maps; for example, North America, countries around the world, etc. (Other maps may be added by providing a geocode file. See the *Help* files for instructions on adding new maps. Often, these must be purchased from a third-party vendor.) Excel then displays the *Map Control* dialog box shown in Figure 2.11 and the map in Figure 2.12. The *Map Control* dialog box allows for different representations of the data. The six different formats for representing data appear as icons to the left of the dialog box. The columns of data appear in the upper part of the dialog box. Data are put into the map in a chosen format by clicking and dragging the appropriate format and columns into the white section of the dialog box. In this case, we have selected population and value shading (shaded by population ranges). The other options for data representation include category shading, dot density, graduated symbol (similar to the bubble chart), pie chart, and column chart. Depending on the data, some options are better than others. In this example, value shading is probably the best option. Figure 2.13 shows the dot density option for comparison. Various other formatting options are available. The mapping tool is an excellent way to present geographically related data in a concise and intuitive format.

FIGURE 2.10
Sales forecast
data

	A	B	C
1	**Western Regional Sales Forecast**		
2			
3		**Sales Volume**	
4	**State**	**(000 units)**	
5	Washington	321	
6	Oregon	109	
7	California	421	
8	Idaho	17	
9	Utah	273	
10	Arizona	432	
11	Nevada	54	
12	Montana	12	
13	Colorado	165	

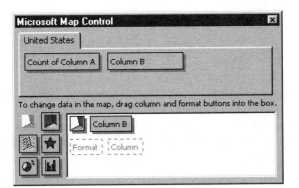

FIGURE 2.11
Map Control
dialog box

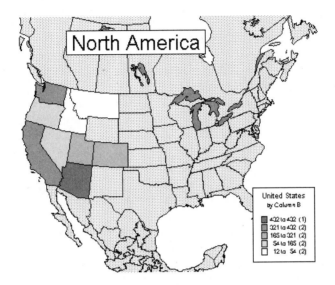

FIGURE 2.12
Value shading of
sales forecasts

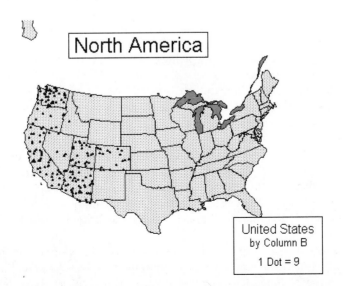

FIGURE 2.13
Density option
for sales
forecasts

Often when using the chart and mapping tools, you will find that the default settings are not satisfactory for your purposes. For example, a 3-D graph might be too compact, an axis on a graph might not be scaled to your satisfaction, or the size of the chart or map might not be adequate. In general, everything can be controlled by double clicking on the object and then on the characteristic you would like to change. This typically invokes a dialog box with various options on how to change the characteristic in question. In some instances, a fair amount of work is required to get graphics into a form suitable for your particular purposes.

Excel Exercise 2.7

Open *Hanover.xls* and select the *Forecast* worksheet. Use the *AutoFilter* to show only the forecasted demand for product A. Copy and paste this information to a new worksheet. Combine the different U.S. territories into one entry with the total demand over the entire United States. This will give you a data set of five customers with forecasted sales of product A. Use the Excel data map tool to map forecasted sales of A by customer, using category shading.

DATA ANALYSIS

Although visualizing data graphically can provide many important insights, quantitative descriptions of data are more useful for modeling and decision making. In this section we discuss several techniques and approaches necessary for effective modeling and analysis:

1. *Descriptive statistics.* For almost any model, costs, revenues, times, production rates, and other significant inputs need to be quantified. Often the only way to do this is to collect data from the system being studied and then estimate population parameters or develop prediction equations using statistical analysis techniques. Statistics play an equally important role in the analysis of the *output* of many models, for instance, simulation, which we will discuss in Chapters 7 and 8.
2. *Probability distributions and distribution fitting.* Many models require that key inputs be expressed as probability distributions. For models to be useful, the distributions chosen should accurately represent the source of the real data. Understanding the properties of probability distributions and how to best fit them to data is crucial.
3. *Covariance and correlation.* Key factors in decision problems often move together in a systematic fashion. Knowing this will often allow one of the factors to be reliably predicted by observations of the other. Also, factor interaction is sometimes quite important in decision models. For example, the relationship between changes in individual stock prices is of fundamental importance to portfolio theory in finance.
4. *Regression analysis.* These techniques can help in developing the structure and the specific functional relationships used in models. For example, knowing the relationship between dollars spent on advertising and sales volume by product would be very important information for developing a model for optimal allocation of advertising budgets.

We assume that the reader has a basic working knowledge of statistics and statistical terminology. We will concentrate on the capabilities of Excel to support statistical analysis for management science applications.

Descriptive Statistics

Excel supports a variety of statistical tools in the functions available and with an add-in called the Analysis Toolpak. A list of statistical functions that can be used alone or incorporated in worksheet formulas can be found by clicking on a blank cell and then the *Paste Function* button [f_x] on the toolbar. The Analysis Toolpak must be installed as an add-in even though it is packaged with Excel. To determine if it is installed, make sure that Analysis Toolpak is checked in the *Tools/Add-Ins* menu option. To invoke the Analysis Toolpak, select *Data Analysis* from the *Tools* menu. You may select one of many different statistical tools from the list of those available.

A basic statistical summary of data may be obtained by selecting the *Descriptive Statistics* tool. This tool provides the following summary statistics:

- *Mean*—the average of the data values
- *Median*—the middle value of the data set (for an even number the average of the two middle values is used).
- *Mode*—the most frequently occurring value in the data set
- *Sample variance*—a measure of the dispersion of the data
- *Standard deviation*—the positive square root of the sample variance
- *Standard error*—a measure of the dispersion of the sampling distribution of the mean, calculated as the (standard deviation/$\sqrt{n}$)
- *Kurtosis*—a measure of the relative flatness or peakedness of the data (distributions with a kurtosis value of less than 3 are flat, and those with a value greater than 3 are peaked)
- *Skewness*—a measure of the asymmetry of a distribution about its mean (a skewness value greater than 1 or less than -1 indicates a highly skewed distribution, and a value between $-.5$ and $.5$ is fairly symmetric)
- *Range*—the largest value minus the smallest value in the data set
- *Minimum*—the smallest value in the data set
- *Maximum*—the largest value in the data set
- *Sum*—the sum of the data points
- *Count*—the number of data points

Although not part of the standard summary statistics, options are available for generating the kth largest and kth smallest values of the data set and a confidence interval for the mean.

Example 6: Descriptive Statistics for F&WH Cost

We will illustrate the *Descriptive Statistics* tool for the freight and warehousing cost data in the F&WH Cost worksheet in the Hanover database. From the *Tools* menu, select *Data Analysis*, then *Descriptive Statistics*. A dialog box will prompt you to define the input range of the data (in the worksheet this is C1:C73). The data must be grouped in a single column and may include a label in the first row (if so, be sure the *Labels in first row* box is checked). If the data are not in a single column, the tool treats each column as a separate data set and provides summary statistics for each. The output is shown in Figure 2.14.

FIGURE 2.14
Results of
descriptive
statistics tool

	A	B
1	*Freight & Warehousing Costs (per unit)*	
2		
3	Mean	1.470789
4	Standard Error	0.014474
5	Median	1.463714
6	Mode	1.365731
7	Standard Deviation	0.122819
8	Sample Variance	0.015084
9	Kurtosis	-0.30866
10	Skewness	0.502511
11	Range	0.452901
12	Minimum	1.274729
13	Maximum	1.727631
14	Sum	105.8968
15	Count	72
16	Largest(1)	1.727631
17	Smallest(1)	1.274729
18	Confidence Level(95.0%)	0.028861

Pivot Tables

A pivot table is a powerful tool for summarizing statistical information for selected data contained in lists or tables within an Excel worksheet. In essence, pivot tables extend the *AutoFilter* capability by incorporating statistical summaries. We illustrate the creation of a one-way pivot table in the following example.

Example 7: Using Pivot Tables to Summarize Data

An important factor in determining a good site for Hanover's new plant will be the ability to ship cost effectively from the new location to its customers. Suppose that we want to examine the average freight and warehousing cost for each site under consideration in the Freight & Warehousing Cost worksheet (*F&WH Cost*) in the Hanover database.

From the *Data* option, select *Pivot Table Report*. This invokes the *Pivot Table Wizard*, which provides step-by-step instructions for constructing the table. These steps are illustrated in Figure 2.15(a)–(e). In Step 1 we specify the source of the data (in this case an Excel list). In Step 2 we specify the location of the data, including headers. Step 3 allows us to specify the table we want. In Figure 2.15(c), the headers from our list—*Plant, Customer,* and *Freight & Warehousing Cost*—are listed to the right.

Using the mouse, click and drag these to where we want them in the table. In this case, we want a table of average cost by plant, so we place *Plant* in the row section and *Freight & Warehousing Cost* in the body of the table, as shown in Figure 2.15(d). Note that the table says average of Freight & Warehousing Cost. Other statistics (for example, max, min, sum, count, etc.) can be obtained by double-clicking on the table entry and selecting from the dialog box shown in Figure 2.15(e). The results of this pivot table are shown in Figure 2.16. Based on these averages, Austin, Charleston, Mobile, and Paris seem to have the most cost-effective freight and warehousing.

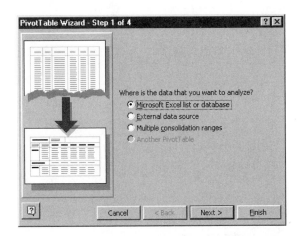

FIGURE 2.15
Steps to creating a pivot table

(a) *Pivot Table Wizard*—Step 1

(b) *Pivot Table Wizard*—Step 2

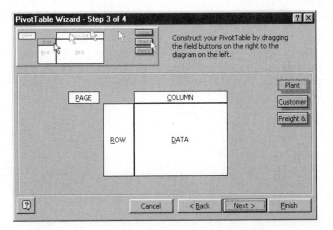

(c) *Pivot Table Wizard*—Step 3

FIGURE 2.15
Steps to creating
a pivot table,
continued

(d) *Pivot Table Wizard*—Step 3, continued

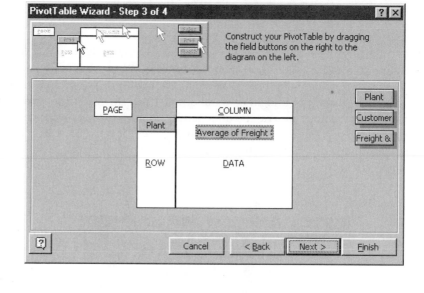

(e) *Pivot Table Field* dialog box

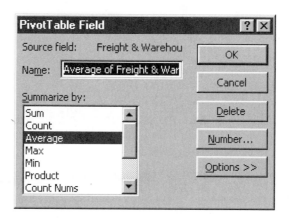

FIGURE 2.16
Pivot Table results for average freight and warehousing costs by plant[9]

	A	B
1	Average of Freight & Warehousing Costs (per unit)	
2	Plant	Total
3	Austin	1.404421841
4	Australia	1.589783887
5	Charleston	1.382848255
6	India	1.517152815
7	Malaysia	1.456926555
8	Mobil	1.470114009
9	Mobile	1.370381719
10	Paris	1.41712329
11	S Africa	1.612705822
12	Spain	1.473286525
13	Grand Total	1.470788583

One-way pivot tables summarize information for one factor—in this case, plant location. Two- and three-way pivot tables can be generated by placing headers in the column and page sections of the pivot table dialog box (Figure 2.15[c]). For example, if we had a database that had several years of data on sales by quarter, by product, and by region, we could generate a three-way table to provide average sales of each product in each quarter in each region. We could accomplish this by placing sales in the body of the table, product in the page section, region in the row section, and quarter in the column section.

Excel Exercise 2.8

Open the Hanover database and select the *Forecast* worksheet. Construct a pivot table that will give total sales by customer. Save the results in a new file.

Probability Distributions and Data Fitting With Crystal Ball

Probability distributions are important components of many models, particularly those that involve uncertainty and risk. A probability distribution is a description of the possible values that a random variable may assume along with the probability of assuming these values. Discrete distributions describe random variables whose outcomes are finite or countable. Continuous distributions describe random variables having an infinite number of outcomes over some range. We summarize the salient properties of common probability distributions next.[10]

- *Uniform distribution.* In the uniform distribution, all values between a fixed minimum and maximum value occur with equal likelihood. The uniform distribution is often used when little information is known about a random variable and only its range can be estimated.
- *Normal distribution.* The familiar bell-shaped normal distribution describes many natural phenomena, such as people's IQs, uncertain inflation rates, or errors in a manufacturing process. The distribution is symmetric about the mean, and the value of the variable is more likely to be close to the mean than far away.
- *Triangular distribution.* The triangular distribution is characterized by three parameters: minimum, maximum, and most likely value that falls between the minimum and maximum. This distribution is often used when no historical data are available and the parameters can be defined judgmentally.
- *Poisson distribution.* The Poisson distribution describes the number of times an event occurs in a specified interval, such as the number of telephone calls arriving at a call center or number of defects per inch of a silicon wafer. The number of possible occurrences in any unit of measurement is unlimited, the occurrences are independent, and the average number of occurrences remains constant.
- *Exponential distribution.* The exponential distribution is widely used to describe events recurring at random times, such as the time between failures of machines or the time between arrivals at a service process. The distribution is not affected by previous events; that is, the future life of a given object has the same distribution, regardless of how long it has existed.

- *Binomial distribution.* The binomial distribution describes the number of occurrences of an event in a fixed number of trials, where only two outcomes can occur for each trial. Trials are independent, and the probability of occurrence remains the same from trial to trial.
- *Lognormal distribution.* The lognormal distribution is used in situations where values are positively skewed. Examples include stock prices (which cannot fall below zero but can increase to any price without limit) and real estate prices. The natural logarithm of a lognormal random variable is normally distributed.

Figure 2.17 shows examples of the form of these distributions. Other distributions are used in different applications. These include the geometric, Weibull, beta, hypergeometric, gamma, logistic, Pareto, extreme value, and negative binomial. We encourage you to consult more advanced texts for information about these distributions.

FIGURE 2.17
Illustrations of probability distributions from the Crystal Ball *Distribution Gallery*

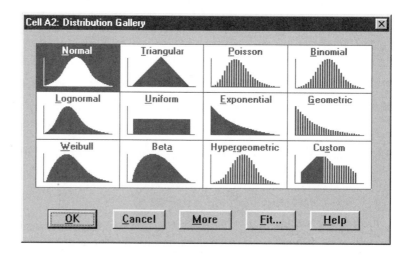

Plotting a histogram of empirical data and comparing the chart to the shape of a common probability distribution is one way of selecting a distribution to include in a model. Knowing the basic properties of distributions can help validate the choice. For example, normal data are symmetric, with a peak in the middle, and exponential data tail off to the right. Summary statistics can provide additional clues about the type of distribution. For example, normally distributed data tend to have a fairly low coefficient of variation, have approximately the same median and mean, a skewness value close to zero, and a kurtosis value of about 3. For exponentially distributed data, however, the median will be smaller than the mean, the mean is about equal to the standard deviation, and the skewness value will be a large positive number.

Visual comparison of data to a theoretical distribution can provide a hypothesis, but it does not provide formal verification. *Goodness-of-fit tests* provide statistical evidence to test hypotheses about the nature of the distribution. The most commonly used goodness-of-fit test is the Chi-square (χ^2) test. The Chi-square goodness-of-fit test tests the hypothesis

H_0: *The sample data come from a specified distribution (e.g., normal)*
against the alternative hypothesis

H_1: *The sample data do not come from the specified distribution*

With this test you can disprove the fit of a particular distribution, but cannot statistically *prove* that data come from the distribution. Other tests are available, such as the Kolmogorov-Smirnov test and the Anderson-Darling test—which are similar except that the Anderson-Darling test places more weight on the tail observations. These tests are used by the Crystal Ball add-in to fit distributions to empirical data using Excel. The Crystal Ball Help option provides some technical details of these tests.

Crystal Ball is an Excel add-in that performs Monte Carlo simulation for risk analysis applications; however, it also includes a very powerful data-fitting capability. We will describe Crystal Ball more completely in Chapter 7. After loading Crystal Ball, two new menu items appear on the main menu bar: *Cell* and *Run*. (A new button bar also provides short-cuts to invoke Crystal Ball menu commands.)

As an example of how to use the Crystal Ball data-fitting option, suppose we have collected data on the time between arrivals at an automated teller machine (ATM) as part of a study to evaluate customer waiting times. These data can be found in the Excel worksheet *ATM.xls*. This worksheet contains the time (measured in hours) between arrivals of 100 customers. To develop models of customer waiting behavior at the ATM, it is useful to know the probability distribution of arrival times; thus, we might wish to fit a theoretical distribution to these data.

To invoke Crystal Ball's distribution-fitting option, select the first cell containing data you would like to fit in the worksheet *ATM.xls*, click on *Cell* from the main menu, and select *Define Assumption...* The Crystal Ball *Distribution Gallery* (Figure 2.17) is displayed. Click on *Fit*. The first of two dialog boxes, shown in Figure 2.18, is displayed. If the data are in the front-most spreadsheet, select *Active Worksheet* (Crystal Ball also allows fitting data from a separate text file), and enter the range of the data in the box to the right. Clicking on *Next* displays the second dialog box, shown in Figure 2.18. You may select all continuous distributions in the distribution gallery, any subset, or just the normal distribution. You must also select the type of test; in this example, we selected the Chi-square test. Check the box "Show Comparison Chart and Goodness-of-Fit Statistics" to display comparative results.

Crystal Ball then fits each of the eleven continuous distributions available in the distribution gallery to the data set. It then rank-orders them, using the test statistic selected. The best-fitting distribution is then displayed in a *Comparison Chart* window with distribution parameters and goodness-of-fit scores, as shown in Figure 2.19. By pressing the *Next Distribution* button, the comparison charts for other distributions can

FIGURE 2.18 Crystal Ball *Fit Distribution* dialog boxes

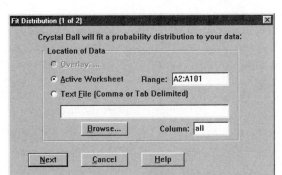

FIGURE 2.19
Comparison Chart
for
Crystal Ball
distribution
fitting

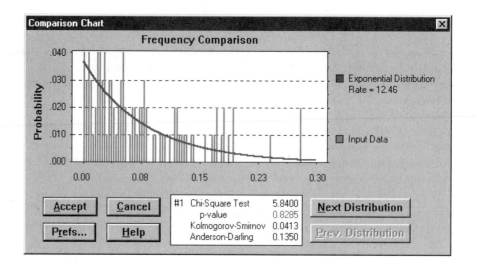

be displayed. In Figure 2.19 we see that the Chi-square p-value is .8285. Generally, a p-value greater than .5 is considered good fit (see Help for information on the other goodness-of-fit tests). Hence, an exponential distribution with a rate of 12.46 arrivals per hour appears to be a good fit.

Excel Exercise 2.9

Make sure Crystal Ball has been loaded into your version of Excel (the student version is on the CD included with this text). Open file *ATM.xls*. Repeat the process just described for fitting a distribution for these data. You should find that the exponential distribution provides the best fit. From the *Comparison Chart* dialog box showing the exponential distribution fit, click on the *Next Distribution* button to find the second and third best fits. What are the Chi-square p-values for these?

Covariance and Correlation

Two statistics that are used to quantify the relationship that exists between pairs of variables are *covariance* and *correlation*. Sometimes, statistical relationships exist between variables even when it is difficult to justify a causal relationship. For example, the *New York Times* reported a strong relationship between the golf handicaps of corporate CEOs and their companies' stock market performance over three years. CEOs who were better than average golfers were likely to deliver above-average returns to shareholders.[11] Thus, you must be cautious in drawing inferences about causal relationships based solely on statistical relationships. On the other hand, you might want to hit the driving range.

Sample covariance is a measure of how two variables move together. For a sample of n pairs of data (x_i, y_i), the sample covariance is defined as:

$$s_{xy} = \frac{\sum_{i=1}^{n}(x_i - \bar{x})(y_i - \bar{y})}{n - 1}$$

where $\bar{x}$ and $\bar{y}$ are the sample means. Note that if both variables are either larger or smaller than their respective means, the product of the terms in the numerator will be positive, as will be s_{xy}. This indicates that as one variable increases, the other does also (see Figure 2.20[a]). On the other hand, if one variable is above the mean and the other below the mean (Figure 2.20[b]), the terms in the expression will be negative and the covariance will be negative. As one variable increases, the other decreases. If no relationship exists, as shown in Figure 2.20(c), then s_{xy} will be close to zero.

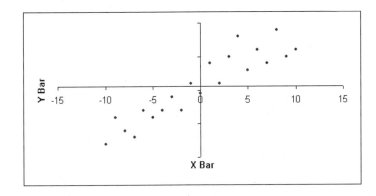

FIGURE 2.20

Covariance examples

(a) Positive covariance

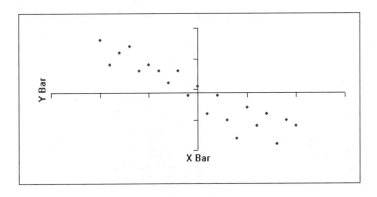

(b) Negative covariance

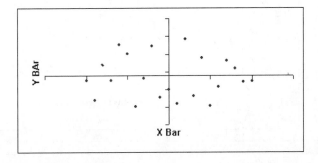

(c) Covariance close to zero

One difficulty in using covariance as a measure of the strength of a linear relationship is that s_{xy} is in the units of measure of x and y. Suppose, for example, that we are interested in the relationship between stock price and dividends. Both could be measured in dollars, cents, or hundreds of dollars. Clearly the measure impacts the value of s_{xy} (although not its sign). A measure of the relationship between x and y that is independent of the metric used is the *sample correlation coefficient*:

$$r_{xy} = \frac{s_{xy}}{s_x s_y}$$

where s_{xy} is the sample covariance and s_x and s_y are the sample standard deviations of x and y. The sample correlation coefficient varies between -1 and $+1$; the closer r_{xy} is to the extremes, the stronger the relationship. Values close to 0 indicate no relationship. The square of the correlation coefficient is called the *coefficient of determination*, R^2, and gives the proportion of the variation explained by the relationship; this will be explained more fully in a later section.

Sample covariance and sample correlation coefficient are both options within Excel's Analysis Toolpak.

Example 8: Studying the Effects of Advertising

A company is trying to determine how sales are related to advertising for one of its products. Data are available for the last twenty quarters. The amount spent on advertising and the resulting sales figures (both in discounted dollars) are shown in Figure 2.21. As the scatter chart shows, the relationship appears to be linear, at least over the range of advertising from $0 to $4,000. Using the Analysis Toolpak, we find that the sample covariance is approximately 6.383 and the sample correlation is 0.97. This supports our intuition that a strong positive linear relationship exists.

FIGURE 2.21 Advertising and sales data

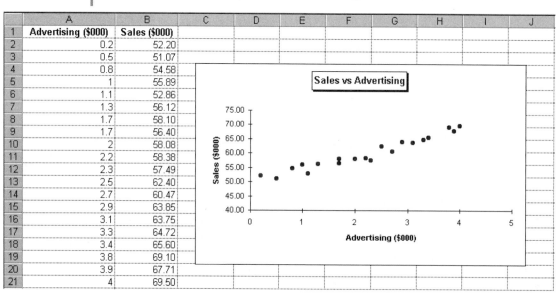

	A	B
1	Advertising ($000)	Sales ($000)
2	0.2	52.20
3	0.5	51.07
4	0.8	54.58
5	1	55.89
6	1.1	52.86
7	1.3	56.12
8	1.7	58.10
9	1.7	56.40
10	2	58.08
11	2.2	58.38
12	2.3	57.49
13	2.5	62.40
14	2.7	60.47
15	2.9	63.85
16	3.1	63.75
17	3.3	64.72
18	3.4	65.60
19	3.8	69.10
20	3.9	67.71
21	4	69.50

Applying Covariance in Financial Portfolio Optimization Models

Suppose we are interested in owning three stocks, and wish to determine the best percentage of our total investment budget to allocate to each stock. Investing in stocks is a risky proposition. As we will discuss in Chapter 6, we can measure risk by the standard deviation (or variance) in stock return. If we select stocks that are highly correlated (such as Internet technology stocks), the total risk of the portfolio is easily magnified.

The **Markowitz portfolio model**[12] is an optimization model that seeks to choose a portfolio of stocks to minimize risk (that is, the variance of portfolio return), subject to a constraint that the expected return of the portfolio meets or exceeds a certain target, and a constraint that ensures we invest all of our budget. Let x_j = fraction of the portfolio to invest in stock j, then for our three-stock example the variance of the portfolio is:

$$\text{Portfolio Variance} = s_1^2 x_1^2 + s_2^2 x_2^2 + s_3^2 x_3^2 + 2s_{12}x_1 x_2 + 2s_{13}x_1 x_3 + 2s_{23}x_2 x_3,$$

where s_j^2 = the sample variance in the return of stock j, and s_{ij} = the sample covariance between stocks i and j. (Recall that if random variables are *not* independent, the variance of their sum must include covariance terms.) The expected return of the portfolio is simply the weighted sum of the expected returns for each stock:

$$\text{Expected Return} = r_1 x_1 + r_2 x_2 + r_3 x_3,$$

where r_j = the expected return of stock j (expressed as a percentage). If the target percent portfolio return is R, the Markowitz optimization model would be:

Minimize $\quad s_1^2 x_1^2 + s_2^2 x_2^2 + s_3^2 x_3^2 + 2s_{12}x_1 x_2 + 2s_{13}x_1 x_3 + 2s_{23}x_2 x_3,$

subject to

$$r_1 x_1 + r_2 x_2 + r_3 x_3 \geq R \quad \text{(meet or exceed return target)}$$
$$x_1 + x_2 + x_3 = 1 \quad \text{(100 percent of budget invested)}$$
$$x_1, x_2, x_3 \geq 0 \quad \text{(cannot invest a negative amount)}$$

Because the variance is a nonlinear function of the x_j variables, the Markowitz model is a *nonlinear optimization model*. These types of models are discussed in more detail in Chapter 5.

Example 9: A Portfolio Optimization Model

The next table shows the expected returns and standard deviations for three stocks.

Stock	1	2	3
Expected Return	15%	20%	10%
Standard Deviation	15%	10%	5%

Stock 1 is a computer manufacturer, stock 2 is a computer component supplier, and stock 3 is a consumer products company. We would expect stocks 1 and 2 to have a high positive correlation with each other, and perhaps a low or even a negative correlation with stock 3. Suppose the covariance values are $s_{12} = 1.35\%$, $s_{13} = -0.15\%$, and $s_{23} = 0.50\%$. If the investor seeks a return of at least 14%, the Markowitz model would be:

Minimize $(15)^2x_1^2 + (10)^2x_2^2 + (5)^2x_3^2 + 2(1.35)x_1x_2 + 2(-.15)x_1x_3 + 2(0.50)x_2x_3,$

subject to

$$15x_1 + 20x_2 + 10x_3 \geq 14 \quad \text{(meet or exceed return target)}$$
$$x_1 + x_2 + x_3 = 1$$
$$x_1, x_2, x_3 \geq 0$$

Excel Exercise 2.10

Construct a spreadsheet for the model in Example 9. Try to find an allocation of stocks that minimizes the total risk while meeting the target return value.

REGRESSION ANALYSIS

In Example 8, although we know that advertising and sales are strongly correlated, we do not know what sales response we might expect for a given level of advertising. **Regression analysis** allows us to develop models that describe the relationships between a dependent variable and one or more independent variables to use for estimation or prediction, as well as to test the significance of the relationship statistically. For example, we might find that sales = 5000 + 3.5(advertising). This model suggests that as advertising increases, sales increases proportionately. A different regression model might be 5000 + 3.5(advertising) + 1.2(advertising)^2. This model suggests that sales would increase more rapidly with higher levels of advertising. Such regression models might be used as objective functions in optimization problems to determine the best level of advertising with budgetary and other restrictions.

Regression models can be linear or nonlinear. **Linear regression** involves models for which the relationship between the dependent variable and independent variable is linear *in the model parameters* (the coefficients of the model variables). While the first sales model is linear in the variable advertising, the second model is nonlinear. However, *both* models are linear in the parameters. In fact, any model for which each term is the product of a constant (model parameter) and some function of the model variables satisfies the assumptions for linear regression.

Even models that appear nonlinear in the parameters can often be transformed into linear models. For example, learning curve data typically follow the form: $y = ax^b$, where y is the time to produce the xth unit, a is the first unit cost, and b is called the learning curve slope. Although this is a nonlinear relationship, it can be linearized by taking the natural log of each side: $ln(y) = ln(a) + bln(x)$. This new model may be written as $z = A + Bw$ where A and B are the model parameters and $z = ln(y)$ and $w = ln(x)$. Hence, we can use linear regression on the transformed data, with z as the dependent variable and w as the independent variable.

In this book we will only discuss linear models; more advanced books on statistics describe regression with nonlinear models. **Simple linear regression** deals with a single independent variable, whereas **multiple linear regression** involves several independent variables.

Simple Linear Regression

A simple linear regression model has the following form:

$$\hat{y} = b_0 + b_1 x$$

where b_0 = the estimated y-intercept

b_1 = the estimated slope of the regression line

$\hat{y}$ = the estimated value of the dependent variable

The values of b_0 and b_1 are calculated to provide the "best-fitting line" to the data points using the *least squares* measure of fit. That is, the values of b_0 and b_1 minimize the sum of squared errors, where the error for each of the data points is defined as the observed value of the dependent variable minus the fitted value of the dependent variable.

Suppose we have n data points; that is, we have observed pairs of independent and dependent variable values (x_i, y_i), $i = 1 , 2 , ..., n$. The simple linear regression model is $\hat{y}_i = b_0 + b_1 x_i$, so that the least squares problem with n observations is

$$\text{Min} \sum_{i=1}^{n}(y_i - \hat{y}_i)^2$$

or

$$\text{Min} \sum_{i=1}^{n}(y_i - b_0 - b_1 x_i)^2$$

The term $y_i - \hat{y}_i = y_i - b_0 - b_1 x_i$ is referred to as the *residual* of the ith observation. The solution to this problem (the optimal values of b_0 and b_1) can be obtained very easily using calculus and are included in two Excel procedures that we will illustrate.

Example 10: Estimating the Effects of Advertising—Continued

Let us continue with the earlier example of the effects of advertising on sales (Example 8). Based on the scatter diagram and the sample correlation coefficient, we believe that a strong linear relationship exists between sales and advertising. For the simple linear regression model

$$\hat{y} = b_0 + b_1 x$$

b_0 = the estimated y-intercept (expected sales with no advertising), b_1 = change in sales per dollar spent on advertising, x = dollar amount (in thousands) spent on advertising, and $\hat{y}$ = estimated sales (in thousands) for x dollars (in thousands) of advertising.

A simple linear regression can be generated directly from a scatter chart of the data by clicking on the chart, placing the cursor over any data point, and right-clicking the mouse. Choose the option *Add Trendline*, and the regression line will be added to the graph (in the *Options* tab, check the boxes to display the equation and R^2 value on the chart). This regression line is shown in Figure 2.22. Here the fitted line has estimates b_0 = 49.49 and b_1 = 4.69. This model allows us to estimate the sales for a given amount of advertising. For example, suppose we allocate \$1,900 for advertising. The estimated sales will be $\hat{y}$ = 49.492 + 4.6945(1.9) = 58.41155. Hence, for an advertising budget of \$1,900, expected sales will be \$58,411.55.

FIGURE 2.22
Trend line for
advertising and
sales data

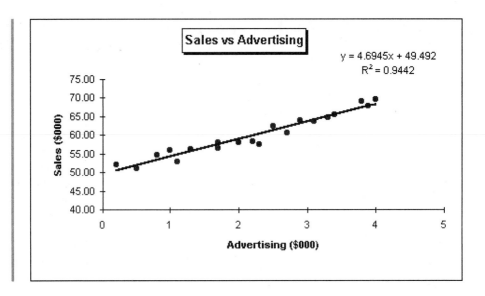

Regression analysis can also be performed in Excel using the *Regression* tool in the Analysis Toolpak. The advantage of using this tool over the quick right-click trend-line approach is that it provides rigorous statistical information to evaluate the model.

Example 11: Using the Regression Tool

We will illustrate regression analysis in Excel using the advertising and sales data in Example 8, which are in the worksheet *Sales.xls* (see Figure 2.21). In the *Regression* tool dialog box, shown in Figure 2.23, enter the range of the dependent variable, sales (Y), as B1:B21, and the independent variable, advertising (x), as A1:A21. We have checked all options with the exception of the "Constant is Zero." Checking this option forces $b_0 = 0$. Forcing $b_0 = 0$ makes sense, for example, if you are modeling variable cost as a function of total output. In this case, the regression line should go through the origin (0,0), since if nothing is produced, there is no variable cost. In the advertising and sales example, however, we would likely still have some sales even with no advertising, so we did not choose this option. Clicking OK in the dialog box performs the regression analysis.

The output is shown in Figure 2.24 (several plots are also generated, which we will discuss later). Section 1 provides standard regression statistics. "Multiple R" is the correlation coefficient. Adjusted R Square reflects the sample size and is useful when comparing this model with others that include additional explanatory variables. Section 2 provides an analysis of variance (ANOVA) to test for statistical significance of regression; that is, a hypothesis test as to whether the slope coefficient is zero. This information is explained in the next section.

The values of b_0 and b_1 can be found in section 3 in the column labeled *Coefficients*. Here, $b_0 = 49.49168628$ and $b_1 = 4.694510684$, which are the same as found by the trend-line approach. Section 4 computes the residuals—the difference between the actual observations and those predicted by the regression equation. Standardized residuals are normalized by subtracting the mean and dividing by the standard deviation. Finally, section 5 provides percentiles of the data.

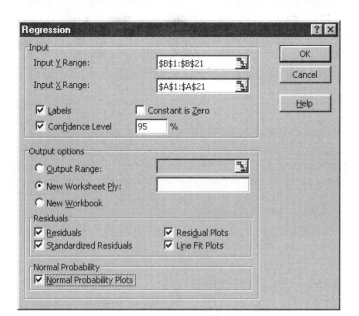

FIGURE 2.23
Regression tool dialog box

FIGURE 2.24 Excel regression output

	A	B	C	D	E	F	G	H	I
3	*Regression Statistics*								
4	Multiple R	0.971710648							
5	R Square	0.944221583							
6	Adjusted R Square	0.941122782							
7	Standard Error	1.366876338							
8	Observations	20							
9									
10	ANOVA								
11		*df*	*SS*	*MS*	*F*	*Significance F*			
12	Regression	1	569.2967384	569.2967384	304.7054659	9.94625E-13			
13	Residual	18	33.63031662	1.868350923					
14	Total	19	602.927055						
15									
16		*Coefficients*	*Standard Error*	*t Stat*	*P-value*	*Lower 95%*	*Upper 95%*	*Lower 95.0%*	*Upper 95.0%*
17	Intercept	49.49168628	0.670726402	73.78818867	8.50567E-24	48.08254131	50.90083125	48.08254131	50.90083125
18	Advertising ($000)	4.694510684	0.268936785	17.45581467	9.94625E-13	4.129495027	5.259526342	4.129495027	5.259526342
19									
20									
21									
22	RESIDUAL OUTPUT					PROBABILITY OUTPUT			
23									
24	*Observation*	*Predicted Sales ($000)*	*Residuals*	*Standard Residuals*		*Percentile*	*Sales ($000)*		
25	1	50.43058842	1.769411583	1.329964927		2.5	51.07		
26	2	51.83894162	-0.768941623	-0.577969196		7.5	52.2		
27	3	53.24729483	1.332705172	1.001717834		12.5	52.86		
28	4	54.18619697	1.703803035	1.280650755		17.5	54.58		
29	5	54.65564803	-1.795648033	-1.349685359		22.5	55.89		
30	6	55.59455017	0.52544983	0.394950418		27.5	56.12		
31	7	57.47235444	0.627645556	0.471765068		32.5	56.4		
32	8	57.47235444	-1.072354444	-0.806027164		37.5	57.49		
33	9	58.88070765	-0.800707649	-0.601845891		42.5	58.08		
34	10	59.81960979	-1.439609786	-1.082071884		47.5	58.1		
35	11	60.28906085	-2.799060855	-2.10389307		52.5	58.38		
36	12	61.22796299	1.172037008	0.880952815		57.5	60.47		
37	13	62.16686513	-1.696865129	-1.27543593		62.5	62.4		
38	14	63.10576727	0.744232735	0.559396946		67.5	63.75		
39	15	64.0446694	-0.294669402	-0.221486043		72.5	63.85		
40	16	64.98357154	-0.263571539	-0.198111568		77.5	64.72		
41	17	65.45302261	0.146977392	0.110474453		82.5	65.6		
42	18	67.33082688	1.769173119	1.329785688		87.5	67.71		
43	19	67.80027795	-0.09027795	-0.067856743		92.5	69.1		
44	20	68.26972902	1.230270982	0.924723944		97.5	69.5		

Statistical Issues in Regression Analysis

In this section we discuss the interpretation of the output from the *Regression* tool.

STRENGTH OF THE REGRESSION RELATIONSHIP The coefficient of determination, or R^2, provides a measure of how well a regression line fits the data. When two variables are not correlated (that is, the slope of the regression line is zero), the best estimate for *any* value of the independent variable is simply the mean, $\bar{y}$. Any variation from the mean is due to random error (see Figure 2.25). However, if the slope is not zero, then a portion of the deviation from the mean is explained by regression, and the remainder is due to error (see Figure 2.25). Squaring these deviations and summing, we define

$$SST = \text{total sum of squares} = \sum_{i=1}^{n}(y_i - \bar{y})^2$$

$$SSE = \text{sum of squared error} = \sum_{i=1}^{n}(y_i - \hat{y}_i)^2$$

$$SSR = \text{sum of squares due to regression} = \sum_{i=1}^{n}(\hat{y}_i - \bar{y})^2$$

It is not difficult to show that $SST = SSE + SSR$. You can see that when the slope of the regression line is zero, $SST = SSE$, whereas a perfect relationship implies that $SST = SSR$. R^2 is a ratio of these sums of squares,

$$R^2 = \frac{SSR}{SST},$$

and indicates *the fraction of the total variation in the dependent variable about its mean that is explained by the regression line.* Clearly, $0 \leq R^2 \leq 1$. An $R^2 = 1$ implies that all of the variation in the data is explained by the regression equation; that is, all of the data are on the regression line. An $R^2 = 0$ means that none of the variation is explained by the regression line. Thus, the higher the value of R^2, the better the line fits the data. How high R^2 has to be before a regression is deemed a "good fit" is subjective, however.

FIGURE 2.25
Illustration of variation about the mean and regression line

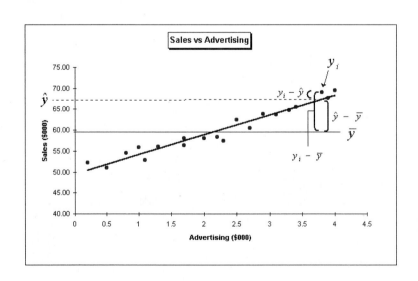

In Example 9, $R^2 = .94$, which indicates that 94 percent of the variation is explained by the regression line. We also note that the sample correlation coefficient calculated in Example 8 for this data is 0.97. The sample correlation coefficient is the square root of the coefficient of determination from the regression analysis ($\sqrt{.94} = 0.97$).

SIGNIFICANCE OF REGRESSION We are often interested in more than simply developing a good-fitting model. The fact that data show a linear relationship might simply be due to chance. A question we would want to answer in the sales-advertising example is: Does advertising really have a significant effect on sales? To test this statistically, we can perform a test of hypothesis to determine if the slope of a regression line is statistically significantly different from zero. This requires us to make certain assumptions concerning the distribution of the errors. Suppose that β_0 and β_1 represent the true parameter values that determine the relationship between y and x (recall that the least squares estimates b_0 and b_1 are only estimates based on sample data), and ε is an error term. The regression model is:

$$y = \beta_0 + \beta_1 x + \varepsilon$$

To test the null hypothesis $\beta_1 = 0$ against the alternative hypothesis that $\beta_1 \neq 0$, we assume the following:

1. The error term ε is normally distributed with mean 0 and variance σ^2.
2. The variance of ε is the same for all values of the independent variable x.
3. The values of ε are independent for different values of the independent variable x.

These assumptions must be checked before tests of significance can be used reliably.

These assumptions can be tested in various ways (some more rigorous than others). A simple visual test is to examine plots of the residuals. Two of these, the residual plot and the normal probability plot, which can be generated by the *Regression* tool, will help us verify these assumptions. These plots for the example are shown in Figures 2.26(a) and 2.26(b). The Normal Probability Plot may be used to test the normality assumption on the error terms (assumption 1). If this plot does not appear linear, then normality is likely to be violated. Here, the plot appears to be linear. The residual plot shows the residual (error term) for each observation. This may be used to test assumptions 2 and 3. For assumptions 2 and 3 to hold, the errors should appear random in uniform band. This appears to be the case for the sales-advertising data, as shown in Figure 2.26(b).

Several residual patterns that would violate these assumptions are shown in Figures 2.27(a) and 2.27(b). You should look for unusual patterns like these before attempting to validate the regression model statistically.

TESTING SIGNIFICANCE OF PARAMETERS The information in sections 2 and 3 of the regression output (Figure 2.24) provides a means of determining the significance of regression. In section 3, a 95 percent confidence interval for β_1 is given based on the regression (we have rounded to 2 digits):

$$4.13 \leq \beta_1 \leq 5.26$$

Notice that this interval does not contain 0. This indicates that at the 95 percent confidence level, we would reject the hypothesis that the slope is zero; that is, advertising *does* affect sales. (The reason that duplicate confidence intervals appear in the figure is that we checked a 95 percent level in the regression dialog box. If we had

FIGURE 2.26
Residual and
normal probabil-
ity plots for
advertising and
sales regression

(a) Residual Plot

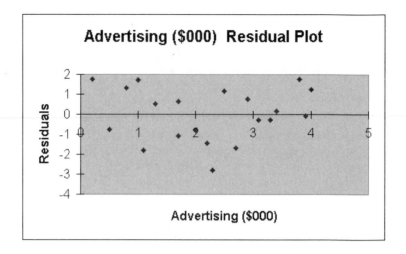

(b) Normal
Probability Plot

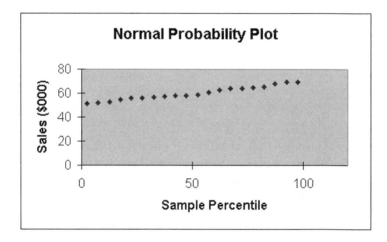

chosen some other confidence level, both the 95 percent interval and the other level chosen would be shown).

A second approach to testing if advertising matters is to use the *t* statistic that appears in section 3. The *t* statistic is calculated as the ratio of the coefficient to its standard error. If $\beta_1 = 0$, then this ratio behaves according to the *t* distribution, and the p-value is the probability of seeing a more extreme value than the observed *t* statistic. Hence, low p-values indicate that the coefficient is different from zero. Here the p-value is extremely small (essentially zero), so we can conclude that $\beta_1 \neq 0$. Thus, we expect an increase in sales of approximately $4.69 for every $1 spent on advertising.

We can interpret the confidence interval and p-value information on β_0 in a similar manner. Hence, it appears that $\beta_0 \neq 0$, so that the intercept is statistically different from 0 at the 95 percent confidence level. Hence, according to the model, we would not expect zero sales if we do not advertise.

Section 2 provides another means of testing the significance of the parameters of the model. ANOVA tests whether or not $\beta_1 \neq 0$ by using an F test. Recall from our discussion of R^2, *SSR* = sum of squares regression and *SSE* = sum squared error (residuals). These are given under the SS column of the ANOVA table. The MS column gives the mean squared error for regression and error (residual), which is simply the sum of squares divided by the appropriate degrees of freedom. Mean squares are

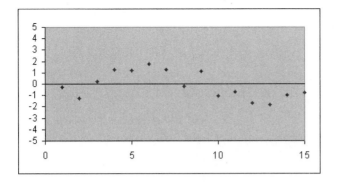

FIGURE 2.27

Residual patterns that violate regression assumptions

(a) Residuals not independent (follow a pattern)

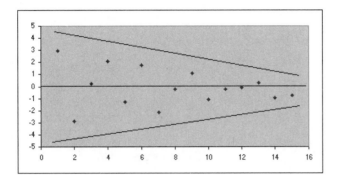

(b) Nonconstant variance

variance estimates, and we may use an F statistic—the ratio of the mean squared errors of regression and error (residual)—to conduct the hypothesis test. Under the assumption that $\beta_1 = 0$, the Significance F gives the same information as a p-value. That is, this is the probability of seeing a more extreme value of F if $\beta_1 = 0$. Here, again, the probability is extremely low (9E-13), so we can conclude that $\beta_1 \neq 0$.

There is an important distinction between the information in sections 2 and 3. The information provided in section 3 is for each individual coefficient. The ANOVA table in section 2 generalizes to test the significance of multiple independent variables simultaneously. Hence, the F test will test if all $\beta_j = 0, j = 1,2,..k$ in a multiple regression with k independent variables.

PREDICTION INTERVALS A confidence interval for the value of y for a given level of x is called a *prediction interval*. A prediction interval gives a measure of the potential variation (from the predicted value on the regression line) that you might see in a particular observed value of y. Exact prediction intervals are a function of the distance the independent variable value is from the mean of the independent variable values used in the regression. This means that the further the value of x is from the mean, the larger the potential error. Hence, using a regression line to predict the dependent variable for an x value outside of the range of the original data can be very risky. Prediction intervals for the sales-advertising data are shown in Figure 2.28. You can see that the intervals expand along the ends of the regression line.

We will not describe the details of calculating exact prediction intervals; however, we may construct an *approximate* prediction interval by calculating a confidence interval using the standard error given in section 1 of the regression output. For example, an approximate 95 percent prediction interval for the observed values is 1.96*(standard error). With this approach, the confidence interval will be the same for all values of x.

FIGURE 2.28

Prediction intervals for advertising and sales regression

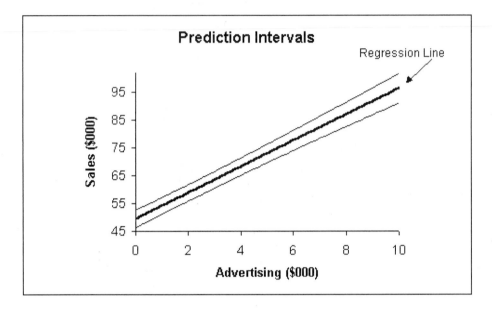

Multiple Linear Regression

Quite often, many independent variables affect a dependent variable. For example, both high-school class rank and SAT score might appear in a model to predict college grade point averages. Square footage, age, and location might be used to predict house prices. Regression models having several independent variables are called multiple linear regression models. A *multiple linear regression model* is of the form

$$\hat{y} = b_0 + b_1 x_1 + b_2 x_2 + \ldots\ldots + b_k x_k$$

where b_0 = the estimated y-intercept

   b_j = the regression coefficient of x_j, j = 1, 2, ...,k

   x_j = the jth independent variable, j = 1, 2, ...,k

   $\hat{y}$ = the estimated value of the dependent variable

Everything discussed previously for simple linear regression is applicable to the multiple regression case. However, the interpretation of the coefficients of the independent variables is slightly different. In the multiple regression case, b_j is the rate of change of the dependent variable with respect to the independent variable x_j *with all other independent variables held at fixed values*. In terms of its use as a predictive tool, the multiple linear regression is a natural extension of the simple linear regression case. For given values of x_1, x_2, ..., x_k, we may estimate the value of y using $\hat{y}$ from the estimated regression equation. For purposes of making conclusions of statistical significance, the assumption on the residuals must be checked, just as in the simple linear regression case.

Example 12: A Multiple Regression Model[13]

An inexpensive motel chain that concentrates its efforts in and around college campuses would like to construct a model to help assess new locations under consideration. Data of interest, collected from its current locations, are shown in Table 2.2. The company would like to develop a predictive model to estimate the profitability of new sites under consideration.

Profit ($000)	Students (000)	Population/Inn (000)
$38	6	9
$40	11	8
$45	11	3.6
$52	15	2
$54	15	3.3
$50	16	6
$55	20	4.3
$53	23	3.5
$52	24	7.2
$60	25	1.1

TABLE 2.2
Motel data

Figure 2.29(a) and 2.29(b) show plots of average profit versus the number of students within five miles and state population per inn in the given motel's state. Both relationships appear relatively linear, indicating that a multiple linear regression of the form

$$\hat{y} = b_0 + b_1 x + b_2 x_2$$

is appropriate.

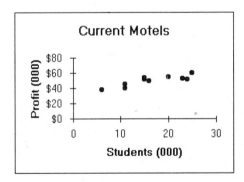

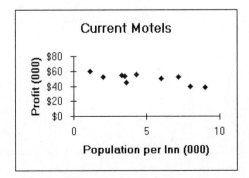

FIGURE 2.29
Profit as a function of independent variables

(a) Average profit vs. students within five miles

(b) Average profit vs. population per inn

In this model:

b_0 = estimated y-intercept

b_1 = change in profits per additional 1,000 students within five miles (population held fixed)

b_2 = change in profits per additional 1,000 in population per inn (students held fixed)

x_1 = number of students (in thousands) within five miles

x_2 = state population (in thousands) per inn

$\hat{y}$ = estimated profitability (in thousands of dollars)

Using the *Regression* tool in Excel, the estimated regression equation for the data in Table 2.2 is

$$\hat{y} = 44.335 + .6897x_1 - 1.226x_2$$

Note that the model has an intuitive appeal, in that a greater number of students leads to higher profits. Higher population per inn leads to less profitability. As a lower-cost motel, it pays to go to lesser-populated states or more populated states with a high number of inns as an alternative for consumers.

The output from the regression model appears in Figures 2.30. This model provides an excellent fit to the data, as evidenced by its R^2, which is .907. In multiple regression models, a more accurate measure of the goodness of fit is the Adjusted R^2, which adjusts for the number of independent variables. Here we see an adjusted R^2 of .881. Note from the ANOVA table that the model is significant, with a Significance F of .0002 (this is the probability of getting these results if all of the coeffi-

FIGURE 2.30 Regression output for the motel example

	A	B	C	D	E	F	G	H	I
1	SUMMARY OUTPUT								
2									
3	*Regression Statistics*								
4	Multiple R	0.952622686							
5	R Square	0.907489982							
6	Adjusted R Square	0.881058548							
7	Standard Error	2.375245157							
8	Observations	10							
9									
10	ANOVA								
11		*df*	*SS*	*MS*	*F*	*Significance F*			
12	Regression	2	387.4074731	193.7037366	34.33374014	0.000240802			
13	Residual	7	39.4925269	5.641789557					
14	Total	9	426.9						
15									
16		*Coefficients*	*Standard Error*	*t Stat*	*P-value*	*Lower 95%*	*Upper 95%*	*Lower 95.0%*	*Upper 95.0%*
17	Intercept	44.33536304	3.585721673	12.3644184	5.20045E-06	35.85648468	52.81424141	35.85648468	52.81424141
18	Students (000)	0.689769536	0.143752629	4.798309025	0.001970131	0.349848827	1.029690245	0.349848827	1.029690245
19	Population/Inn (000)	-1.226153612	0.344957033	-3.554511128	0.009287445	-2.041846795	-0.41046043	-2.041846795	-0.41046043
20									
21									
22									
23	RESIDUAL OUTPUT					PROBABILITY OUTPUT			
24									
25	*Observation*	*Predicted Profit ($000)*	*Residuals*	*Standard Residuals*		*Percentile*	*Profit ($000)*		
26	1	37.43859775	0.561402252	0.268001944		5	38		
27	2	42.11359904	-2.113599039	-1.008988918		15	40		
28	3	47.50867493	-2.508674934	-1.197590063		25	45		
29	4	52.22959886	-0.229598857	-0.109605795		35	50		
30	5	50.63559916	3.364400839	1.606096094		45	52		
31	6	48.01475394	1.985246056	0.947715831		55	52		
32	7	52.85829323	2.141706772	1.022406974		65	53		
33	8	55.908524726	-2.908524726	-1.388470169		75	54		
34	9	52.0615259	-0.061525896	-0.029371203		85	55		
35	10	60.23083247	-0.230832467	-0.110194695		95	60		

cients of the independent variables are really 0). Also, all of the p-values for the individual coefficients are very small, indicating that we may reject the hypotheses that individual coefficients are equal to 0.

In terms of the residual assumptions, the graphs shown in Figure 2.31 indicate that the normal plot of the residuals is somewhat linear, although not as straight as we might like. The residual plots versus each of the independent variables appear fairly random and within a uniform band, although this is difficult to assess with only 10 data points.

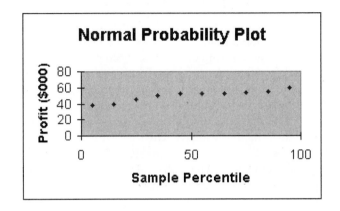

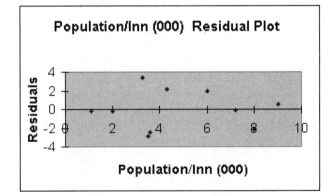

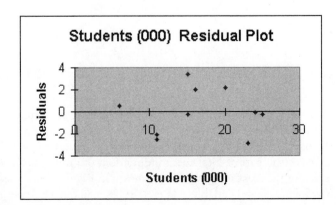

FIGURE 2.31
Verifying assumptions for the motel regression example

(a) *Normal Probability Plot*

(b) Residual plot (population/inn)

(c) Residual plot (students)

Excel Exercise 2.11

Open the spreadsheet *Motels.xls*. Use the *Correlation* option in the Excel *Data Analysis* tool to calculate the correlation between profit and number of students, and between profit and population per inn. Do these two factors appear to have a strong linear relationship with profit?

Building Regression Models

The goal of most regression applications is to determine what variables or factors are the best predictors of the dependent variable, or what factors have the most significant influence on the dependent variable. Hence, from perhaps a long list of possible independent variables, we seek the set of variables that gives us the "best model." One approach would be to run all possible models, that is, perform a separate regression for each possible subset of variables. The best model would be the one with the highest adjusted R^2. The simple R^2 measure will always go up for an increase in the number of independent variables, which is why the adjusted R^2 is a better measure.

If the number of possible independent variables is large, it might be too time consuming to run regressions for all possible subsets (for n variables, there are $2^n - 1$ subsets!). In these cases, good judgment and intuition must be used to decide which models to test. Also, the correlation of variables can be used to guide variable selection. If two possible variables are known to be highly correlated, there is likely to be little benefit from including them both in the model, since they will correlate with the dependent variable in the same manner. For example, in our advertising example, we would not want to include dollars spent on advertising and number of television commercial spots both as independent variables. In fact, using strongly correlated independent variables, a condition referred to as *multi-collinearity*, can cause unreliable coefficients in a regression.

Using Regression in Optimization Models

Many optimization models require costs as input data (either directly or as part of a contribution calculation). In Chapters 4 and 5, for example, we will discuss optimization models to help Hanover, Inc., decide on a location for a new plant. A model might require specific values of the fixed and variable costs at the plant. Regression analysis can be useful in identifying these parameters and determining an appropriate objective function.

Example 13: Separating Fixed and Variable Costs

The worksheet *Austin.xls* contains historical data on production levels and total cost incurred by Hanover's Austin plant. Define y = total cost and x = production volume. If we estimate the regression $y = b_0 + b_1x$, b_1 will be the estimate of variable cost per unit (that is the part of total cost that varies by production level) and b_0 will be the estimate of fixed cost (the part of total cost independent of production volume). The regression results for the Austin plant data are shown in Figure 2.32. The estimate of the fixed cost is $605,005, and variable cost is $1.1989, or about $1.20 per unit.

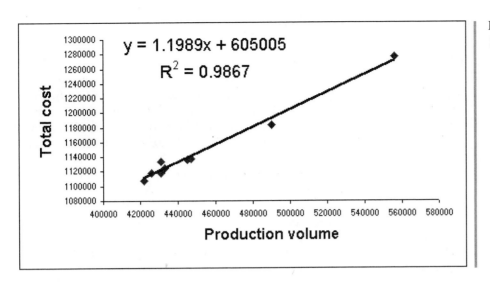

FIGURE 2.32
Regression line
for separating
fixed and vari-
able costs

TECHNOLOGY FOR STATISTICAL ANALYSIS

Because statistics is such a well-established discipline and is used in many different fields of application, it is no wonder that many different statistical software packages with a wide range of prices and capabilities are available. Excel provides many of the basic capabilities that are needed to support the development and analysis of management science models, but it is not a high-performance statistical package. Dedicated statistical software provides more advanced statistical support. Table 2.3 summarizes some of the more popular software.

Product	Company	Description
BMDP Classical Statistical Software	BMDP Statistical Software, Inc. Los Angeles, Calif.	Features one of the most comprehensive statistical libraries for the professional researcher and master statistician
BMDP New System Personal Edition	BMDP Statistical Software, Inc. Los Angeles, Calif.	Features the most common techniques for non-statisticians
MINITAB	Minitab, Inc. State College, Penn.	General-purpose statistical package
The SAS System	SAS Institute, Inc. Cary, N.C.	Integrated suite of information delivery software for data warehousing, analysis, database access, and more
SPSS	SPSS Inc. Chicago, Ill.	Statistical data analysis product for marketing, survey, and other types of research

TABLE 2.3
Some popular
statistical soft-
ware packages

PROBLEMS

1. Bob's Bookstore is considering a magazine advertising campaign. In an effort to ascertain what magazines its customers (and presumably potential customers) read, a survey was sent to 100 customers. Bob would like to place ads in two of ten magazines. His goal is to reach as many customers as possible with at least one ad. Hence, overlap in readership is an important consideration. The survey listed ten magazines and asked customers to indicate which (if any) they subscribed to. Fifty-three customers responded. The data are stored in the worksheet *Bookstore.xls*. Open this worksheet and use the *AutoFilter* to answer the following questions:

    a. Which customers subscribe to magazine 5?
    b. Which customers subscribe to magazine 6?
    c. Which customers subscribe to no magazines in the list of 10?

2. Use the *Custom AutoFilter* option with the Bob's Bookstore data to obtain a list of all customers who subscribe to either magazine 5 or 6 (or both). How many customers subscribe to both magazines 5 and 6?

3. Use the *Pivot Table* tool to construct a one-way pivot table from the Bob's Bookstore data to give you the number of customers subscribing to each magazine.

4. If we can choose only 2 magazines for ads so as to maximize reach, we want magazines that have many readers and are negatively correlated with one another. We can use the *Pivot Table* option to convert these data to a form that will allow us to calculate correlations. Use the *Pivot Table* tool to construct a table of readership from the Bob's Bookstore data (construct the table using customer as the row, magazine as the column and count of customer in the body of the table). Once the table is constructed (it should have 53 rows and 10 columns), copy the body of the table to a new worksheet. Convert all blank cells in this matrix to 0 (you can do this by using *Edit, Replace*, leaving *Find what* blank, and entering 0 in the *Replace* box). From this matrix of zeros and ones, use the data analysis tool correlation option to calculate the correlation of readership between magazines. Which magazines are most strongly correlated? Negatively correlated? Which two magazines would you recommend to Bob?

5. Open the workbook *Hanover.xls* and select the Freight & Warehousing Cost worksheet. Use the *AutoFilter* to get a listing of these costs for the proposed plant in Spain. Copy this list to another worksheet. Use the descriptive statistics option of the *Data Analysis* tool to get statistics on these costs. How do these costs compare to the statistics over all plants, as shown in Figure 2.14?

6. RestEZ Inc. produces hospital beds in its factory just west of Atlanta, Georgia. RestEZ has contracted demand for the next six months. These data are in the worksheet *Demand* in the file *RestEZ.xls*. The data are in truckloads by customer and city.

    a. Construct a *Pivot Table* to give total demand by state.
    b. Use the Excel mapping tool to display the demand on a map of the United States. Which states are the high volume states?

7. Refer again to *RestEZ.xls* (Problem 6). RestEZ recently received bids from five common carriers for the transport of its product from plant to customer. These bids are in file *RestEZ.xls* in worksheet *Bids*. The bid matrix contains rows corresponding to 99 customers, a column for each carrier, and the matrix element is the bid per truckload from that carrier for transporting from the plant to that customer (not all carriers bid on every lane). RestEZ would like to select a set of 2 or 3 carriers (called core carriers) to which to give its business. Use the descriptive statis-

tics option in the *Data Analysis* tool to get statistics on each carrier's bids. Which carrier bid on the most lanes? Which carriers appear to be the most cost effective, based on average bid? How might you use these statistics and the map generated in problem 6 to pick a cost-effective set of 3 carriers and assign customer cities to these carriers?

8. Suppose you are trying to decide which of four candidates we should hire for an open faculty position at State U. Each candidate received a score from 1 (great) to 5 (weak) on seven different categories, as shown in Table 2.4 (averaged over 10 scores from the current faculty). These data can be found in file *Hire.xls*. Construct a radar chart of these data using the Excel chart option. Which candidate would you choose?

	Teaching	Research	Professional	Funding	Leadership	Facilitator	Fit
Crosby	3.7	2.1	2.8	2.8	3.6	2.5	2.6
Stills	3.5	2.6	2.7	1.1	2.6	3	3.5
Nash	2	2	1.2	2.1	1.3	1.2	2
Young	1.2	1	2.9	2.9	3	3.1	2

TABLE 2.4
Candidate scores on hiring criteria

9. Suppose you are interested in modeling the shop floor of a machine shop. For simplicity, assume we have a single machine and that we have collected data on time between failures (measured in days). These data can be found in file *machine.xls*. Use the data-fitting option in Crystal Ball to fit a distribution to the data. What distribution gives the best fit? Is it a good fit according to the Chi-square test?

10. Open the worksheet *Portfolio.xls*. This worksheet contains ten years' worth of data on the return of the S&P 500 and three individual stocks. The returns are total year-end return. For example, in year 1, Stock A returned 30 percent. For this problem, you can ignore the data on the S&P 500.
   a. Calculate the average return for each of the three stocks.
   b. Use the covariance option of the Analysis Toolpak to obtain the covariance matrix for these three stocks.
   c. Use the expected returns and the covariance matrix to construct a Markowitz model for these three stocks. Assume we would like an expected return of 20 percent from our portfolio.
   d. Use trial and error to find a good solution (that is, a decision on the percentage of the portfolio to invest in each stock with expected return of at least 20 percent and small variance).

11. Open the worksheet *Portfolio.xls*.
   a. Graph an *x-y* chart of the data (including the S&P 500 data). Which stock looks to be the most correlated with the market?
   b. Use the correlation option of the Analysis Toolpak to calculate the correlation for each of the three stocks with the S&P 500.

12. The *Capital Asset Pricing Model* (CAPM) relates the return of a given stock to the return of the market via a simple linear regression. The CAPM model is: $y = \alpha + \beta x + \varepsilon$, where y = return of the stock, x = return of the market, and α and β are unknown parameters. The true value of β is known as the "beta" of that stock. It measures how the return on the stock moves with the return in the market. We estimate the value of the stock's beta by running a simple linear regression on historical data. Stocks with larger beta values are presumably riskier than those having smaller betas. Open the worksheet *Portfolio.xls*. Use the regression option in

the *Data Analysis* tool to obtain betas for each of the three stocks in *Portfolio.xls*. Which stock has the largest beta value? Do these results make sense based on what you found in Problem 11? Discuss.

13. For the three regressions of Problem 12, answer the following:
    a. Are the parameters statistically significant?
    b. How much of the variation in the data is explained by the regression?
    c. Are the residual assumptions satisfied?

14. The basis of learning curve theory in production is that in highly labor-intensive production, the time to produce a unit decreases in a systematic fashion as production volume increases. Simply stated, the more of a product you make, the faster you become in making that product. This has important implications for production planning and resource allocation decisions. The classical learning curve model is of the form: $y_x = ax^b$ where y_x is the number of labor hours required to produce the xth unit, a is called the first unit cost, and b the learning slope parameter. The learning phenomenon can be modeled by a data transformation, using simple linear regression. Taking the natural log (denoted *ln*) of each side of the exponential model gives: $ln(y_x) = ln(a) + bln(x)$.
    a. The file *learning.xls* contains 10 observations of production. Graph labor hours versus unit production number.
    b. Using the Excel function LN( ), transform the data by taking the natural log of unit number and labor hours. Graph the transformed data. Does it appear more linear?
    c. Run a simple linear regression on the transformed data. What are the estimated values of $ln(a)$ and b? Are they statistically significant?
    d. Do the residual assumptions appear to be satisfied?

15. Lovely Lawns is considering expanding into the landscaping business. LL will focus on upscale homes and has obtained the data in file *LL.xls* from a landscaping company in a nearby state. LL would like to develop a model that will help estimate landscape expenditures as a function of household income and mortgage amount. Develop this model, using the data in the file *LL.xls* and multiple linear regression. Estimate the landscape expenditures for a household income of $420,000 and a mortgage amount of $510,000.

NOTES

1. R. Sharma, "Decision Enabling in the Retail Information Revolution," *Canadian Manager*, Fall 1997, 17–18.
2. C. Stedman, "Wal-Mart Triples Data Warehouse," *Computerworld*, Vol. 31, No. 7, 1997, 8.
3. J. Zicker, "Data Marts vs. Data Warehouses: Have Your Cake and Eat It Too, *Computer Technology Review*, Vol. 18, No. 2, 1998, 48.
4. Coaxing Meaning Out of Raw Data," *Business Week*, February 3, 1997, 134–138.
5. NeuralWare, Inc., advertisement in *OR/MS Today*, February 1997, 13.
6. Mark Dodge, Chris Kinata, and Craig Stinson, *Microsoft Excel 97* (Redmond, Wash.: Microsoft Press, 1997), 854.
7. E. Kay, "Dirty Data Challenges Warehouses," *Software Magazine*, October 1, 1997, S5–S8.
8. J. Camm et al., "Blending OR/MS, Judgment and GIS: Restructuring P&G's Supply Chain," *Interfaces*, Vol. 27, No. 1, 1997, 128–142.
9. Notice that the misspelling of Mobile as Mobil appears in the output as a separate entry. This is the error you should have corrected in Excel Exercise 2.4. If you have not made the correction, open file *Hanover.xls* and correct this spelling error now.
10. These descriptions are adapted from the Crystal Ball user's manual. A student version of Crystal Ball Pro is included with this text and will be used in this chapter and others later in this book.

11. Adam Bryant, "CEOs' Golf Games Linked to Companies' Performance," *Cincinnati Enquirer,* June 7, 1998, E1.
12. H. M. Markowitz, *Portfolio Selection: Efficient Diversification of Investments* (New York: John Wiley & Sons, 1959).
13. Based on Sheryl E. Kimes, and James A. Fitzsimmons, "Selecting Profitable Hotel Sites at La Quinta Motor Inns," *Interfaces*, Vol. 20, No. 2, March–April 1990, 12–20.

BIBLIOGRAPHY

Atre, Shaku. 1998. "Problem-solving." *ComputerWorld*, Vol. 32, No. 3, pp. 71–72.

Cafasso, R. 1996. "Sears, Roebuck and Company: Come See the Smarter Side of Sears as It Plays for Keeps in the Competitive Retail Industry." *ComputerWorld,* Vol. 4, No. 5, pp. 12–13.

Dodge, M., C. Kinata, and C. Stinson. 1997. *Running Microsoft Excel 97.* Redmond, Wash.: Microsoft Press.

Fortune. 1997. "Bringing Sears into the New World," October 13, pp. 183–184.

Helmer, F., E. Oppermann, and J. Suver. 1980. "Forecasting Nursing Staffing Requirements by Intensity-of-Care Level." *Interfaces*, Vol. 10, No. 3, June, pp. 50–56.

Kay, E. 1997. "Dirty Data Challenges Warehouses." *Software Magazine*, October 1, pp. S5–S8.

Kimes, Sheryl E., and James A. Fitzsimmons. 1990. "Selecting Profitable Hotel Sites at La Quinta Motor Inns." *Interfaces*, Vol. 20, No. 2, March–April, pp. 12–20.

Sharma, R. 1997. "Decision Enabling in the Retail Information Revolution." *Canadian Manager,* Fall, pp. 17–18.

Software Magazine. 1996. "Land O'Lakes Butters Up Brokers," Vol. 16, No. 12, p. S15.

Stedman, C. 1997. "Wal-Mart Triples Data Warehouse." *ComputerWorld*, Vol. 31, No. 7, p. 8.

Swain, James J. 1996. "Number Crunching: 1996 Statistics Survey." *OR/MS Today*, February, pp. 42–55.

Tufte, E. R. 1983. *The Visual Display of Quantitative Information.* Cheshire, Conn.: Graphics Press.

———— 1997. *Visual Explanations.* Cheshire, Conn.: Graphics Press.

Vijayan, J. 1997. "Data Warehouse Gives Delaware Police New Weapon to Fight Crime." *ComputerWorld*, Vol. 31, No. 17, pp. 65–67.

Zicker, J. 1998. "Data Marts vs. Data Warehouses: Have Your Cake and Eat It Too." *Computer Technology Review*, Vol. 18, No. 2, p. 48.

Forecasting

O U T L I N E

Applications of Forecasting
Plant Closure at Allied-Signal
Call-Center Demand at L. L. Bean
Inventory Management at IBM

Time Series Components

Models for Time Series with Only an Irregular Component
Single Moving Average
Measuring Forecast Accuracy
Single Exponential Smoothing

Models for Time Series with a Trend Component
Double Moving Average
Double Exponential Smoothing

Models for Time Series with a Seasonal Component
Additive Seasonality
Multiplicative Seasonality

Models for Time Series with Trend and Seasonal Components
Holt-Winters Additive Model
Holt-Winters Multiplicative Model

Forecasting with Regression Models
Management Science Practice: A Forecasting Model for Supermarket Checkout Services
Linear Regression for Time Series Forecasting
Using Indicator Variables in Regression Models

Forecasting Technology

Choosing the Best Forecasting Method

Forecasting with CB Predictor
Interpreting CB Predictor Outputs

Applications of Forecasting to Model Building

Problems

Forecasting is the general term used to denote the estimation of some unknown variable in the future. All organizations need forecasts for planning purposes. While most applications of forecasting relate to the future demand for goods and services, organizations must also forecast various economic indicators, such as interest rates and production costs, demand for energy, and changes in consumer demographics. Forecasts are necessary inputs to decision models; in particular, they are key building blocks for optimization and simulation models that we will discuss later in this text.

Forecasting is typically approached in two ways: judgmentally and quantitatively. Many forecasts are obtained by querying "experts," such as field sales managers who may have many years of experience and close relationships with customers that allow them to develop very good estimates of future demand. *Fortune* magazine publishes a "business confidence index," which is an opinion-based forecast of the future state of the U.S. economy. Judgmental forecasting is useful and often the only alternative for forecasting such things as changes in technology or the demand for unique and innovative products with no prior history.

For many forecasts, such as product demand or energy requirements, some type of historical data are usually available. A stream of historical data collected at different points in time is called a **time series.** You see time series data every day in the daily newspaper; examples are the closing prices of the Dow Jones Industrial Average and daily temperature highs and lows. Most businesses maintain databases of time series data, such as weekly or monthly sales. Working from the assumption that the future will be an extrapolation of the past (which all mutual fund prospectuses are careful *not* to suggest!), quantitative forecasting methods generate forecasts based on historical time series.

In this chapter, we will describe various quantitative methods for forecasting time series data. In practice, businesses often begin with such quantitative forecasts and adjust them judgmentally to reflect their experience and intuition. We will also discuss the use of regression analysis as a forecasting tool and approaches for selecting the best forecasting method for a specific application.

APPLICATIONS OF FORECASTING

The following examples illustrate the impact of reliable forecasting.

Plant Closure at Allied-Signal

In 1989, a decision was made to close a plant run by Allied-Signal for the U.S. Department of Energy (DOE). The Albuquerque Microelectronics Operation (AMO) produced radiation-hardened microchips. Originally, private sector companies were unable to produce chips that met the radiation tolerance levels required by the DOE. Gradually, the private sector production capabilities improved so that DOE standards were met. As a result, operations once solely conducted at AMO could be outsourced. However, operations at AMO had to be phased out over several years because of long-term obligations.

AMO experienced fairly erratic yields in the production of some of its complex microchips. Because contractual obligations involved these chips, accurate forecasts of yields were critical. Overestimating yields could lead to an inability to meet contractual obligations in a timely manner, requiring the plant to remain open longer. Underestimates would cause AMO to produce more chips than actually needed. AMO's yield forecasts had previously been made by simply averaging all historical data. More sophisticated forecasting techniques were implemented, resulting in improved forecasts of wafer fabrication. Using more accurate yield forecasts and linear programming planning models, AMO was able to close the plant sooner, resulting in significant cost savings (Clements and Reid, 1994).

Call-Center Demand at L. L. Bean

L. L. Bean is a retailer of high-quality outdoor gear. A large percentage of the company's sales are generated through orders to its call center (the call center can account for over 70 percent of the total sales volume). Calls to the L. L. Bean call center are classified into two types: telemarketing (TM) and telephone-inquiry (TI). TM calls are calls for placing an order, whereas TI calls involve customer inquiries, such as order status or order problems. TM calls and TI calls differ in duration and volume. The annual call volume of TM calls is much higher than that of TI calls, but the duration of a TI call is generally much longer than the duration of a TM call.

Accurately forecasting the demand of TM and TI calls is very important to L. L. Bean to reduce costs. Accurate forecasts allow for properly planning the number of agents to have on hand at any point in time. Too few agents result in lost sales, loss of customer loyalty, excessive queue times, and increased phone charges. Too many agents obviously result in unnecessary labor costs.

L. L. Bean developed analytical forecasting models for both TM and TI calls. These models take into account historical trends, seasonal factors and external explanatory variables, such as holidays and catalog mailings. The estimated benefit from better precision from the two forecasting models is approximately $300,000 per year (Andrews and Cunningham, 1995).

Inventory Management at IBM

IBM developed a system called Optimizer to control service levels and spare parts inventory in its U.S. network for service support. IBM's National Service Division (NSD) is responsible for after-market support. The geographic dispersion of customers and the need for quick response necessitate an accurate system for the control and distribution of spare parts inventory.

A key component of the Optimizer system is the forecasting module. This module estimates the failure rates of individual parts included in each product, which serves as input to the multi-echelon stock-control planning models. The use of Optimizer has resulted in numerous benefits, including a reduction in required inventory investment, improved service, and increased efficiency of human resources (Cohen et al., 1990).

TIME SERIES COMPONENTS

A basic approach to analyzing a time series is to assume that it consists of four basic components: trend, seasonal, cyclical, and irregular.

- *Trend* is the gradual increase or decrease of the time series over a long period of time. For example, the stock market as a whole has exhibited an upward trend for many decades. The populations of some cities show an upward trend, while others show a downward trend over many years. Trend is often estimated using regression analysis (introduced in Chapter 2) with time as the independent variable.
- A repeating pattern from one year to the next is known as the *seasonal component* of a time series. For example, products such as hot dogs, cold medicine, air conditioners, and beer typically have demand patterns that vary over the course of a year, but repeat year after year. A seasonal component might also reflect other appropriate time intervals. For example, shopping patterns often exhibit "seasonality" over the course of a week.
- A *cyclical component* of a time series is a longer-term up or down pattern that may vary in length from as few as 2 to 10 years or more. Some examples would be interest rates and housing prices. Cyclical components are usually due to business cycles or the state of the economy.
- Finally, the *irregular component* of a time series is the remaining variation in the series that cannot be described as a trend, cyclical, or seasonal component. These fluctuations are due to random variability, occur over the short term, and are non-repetitive.

The classical **multiplicative forecasting model** assumes that time series data are represented as a product of these components:

$$Y_t = T_t \times S_t \times C_t \times I_t$$

where Y_t is the value of the tth observation, and $= T_t, S_t, C_t,$ and I_t are values for the trend, seasonal, cyclical, and irregular components in time period t. The following example provides some insight into this model.

Example 1: Components of a Multiplicative Time Series Model

Figure 3.1 shows an example of a time series over 48 months. While the time series seems to have an upward trend and has some irregular fluctuation, it is difficult to clearly recognize any trend and cyclical components because of the interactions among them. Figure 3.2 shows the four components on which the time series in Figure 3.1 was built. Superimposed on the trend are an annual seasonal component that reaches its peak in month 3 and lowest point in month 10 of each year, a 2-year cyclical component, and some irregular fluctuation in each month.

Figure 3.3 shows how the time series was constructed using the multiplicative model. First examine the trend-seasonal line. Note that because the values of the time series are increasing due to the trend, the amplitude of the seasonal effect becomes larger as time progresses. When the cyclical component is added, a similar amplification exists over time, but the interaction with the seasonal component makes it difficult to see the cyclical effect clearly. Finally, the irregular component effect simply causes some random fluctuation around the trend-seasonal-cyclical line.

FIGURE 3.1
Example of a
time series

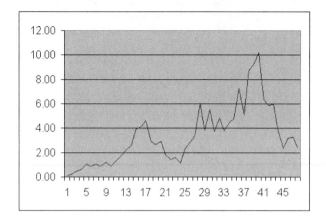

FIGURE 3.2
Components of
time series in
Figure 3.1

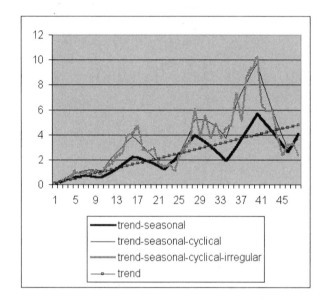

FIGURE 3.3
Partial compo-
nents of time
series in
Figure 3.1

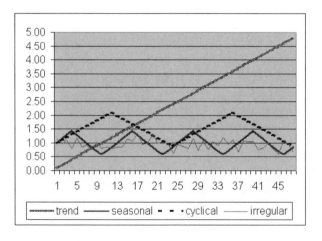

Classical methods, often used by economists to understand and explain historical time series, rely on "decomposing" the time series into these components. A simple example illustrates this idea.

Example 2: Decomposing Time Series Components

Figure 3.4 shows a time series for quarterly sales data over 5 years that appears to have both trend and cyclical components, suggesting a model $Y_t = T_t \times C_t$. We may estimate the trend line by fitting a regression line with sales as the dependent variable and quarter as the independent variable, using the techniques described in Chapter 2. The estimated trend line *is* $T_t = 81.758 + 3.799t$, where t is the quarter. Figure 3.5 shows the output of a spreadsheet that computes the value of the trend line for each quarter, and then divides it into the actual observation, leaving a residual without the trend influence. Note that the actual observations are products of the trend component and the residual. A graph of the residual shows cyclical, not irregular, behavior. The average of the residuals is 1.0; if neither the cyclical component nor irregular components existed, the residuals would all have been one, and the model would simply be $Y_t = T_t$. On the other hand, if the residuals did not have a cyclical pattern, then we would probably attribute them to an irregular component.

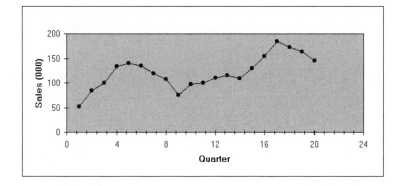

FIGURE 3.4
Sales data

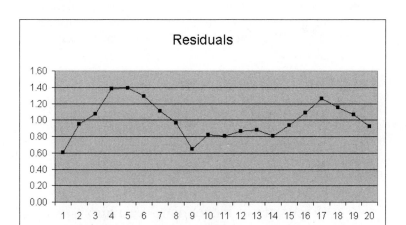

FIGURE 3.5
Residuals after removing trend

While the multiplicative model is useful in understanding some of the essential characteristics of time series, particularly in retrospect, it is not used often today as a forecasting tool, primarily because many forecasts used in business operations are short term in nature and do not include cyclical components. For further discussion of the cyclical component of time series data, we recommend the more advanced text by Makridakis, Wheelwright, and Hyndman, 1998.

Many different techniques for time series forecasting have been developed. The key difference among methods is their appropriateness for handling trends and seasonal factors. Table 3.1 summarizes the models that we discuss.

TABLE 3.1
Classification of time series forecasting models

	No Seasonal Component	Seasonal Component
No Trend Component	Moving average model Exponential smoothing model	Additive seasonal model Multiplicative seasonal model
Trend Component	Double moving average Double exponential smoothing	Holt-Winters additive model Holt-Winters multiplicative model

MODELS FOR TIME SERIES WITH ONLY AN IRREGULAR COMPONENT

Single moving average and single exponential smoothing models are most appropriate for relatively stable time series—those not exhibiting significant trend or seasonal components. This is not to say that such time series may not exhibit short-term movements in either direction, just that they do not have long-term trends or repeatable seasonal components (otherwise there would be little incentive to forecast—just use the mean!).

Single Moving Average

Single (or *simple*) *moving average* is a forecasting technique that uses the average of the last k (some prespecified number) data points as the forecast for the next period. The idea is simply to smooth out the irregular fluctuations in the data (in the multiplicative model without trend, seasonal, or cyclical components, we simply have $Y_i = I_i$). Let y_t be the known value of the time series in time period t. Suppose we have historical data for periods $t = 1, 2, 3, ..., T$. Then the moving average forecast for period $T + 1$ is:

$$\hat{y}_{T+1} = \frac{\sum_{t=T-k+1}^{T} y_t}{k}$$

The single moving average forecast simply averages the last k data points to forecast for the next period. The larger the value of k selected, the more historical data are reflected in the forecast. Before we illustrate this method, we need to discuss measures of forecast accuracy.

Measuring Forecast Accuracy

The most important thing to remember in forecasting is that *no forecast is accurate*. However, some forecasts are better than others. For example, in the moving average technique, different values of k will yield different forecasts. There is no way of determining, *a priori*, the best value of k that provides the best forecasts. A common approach is to simply select different values of k, and determine how well the models would have predicted the historical data by computing a measure of how well the forecasted values compare to the actual time series. This is referred to as *goodness of fit*. The forecaster would then use the value of k that provides the best fit.

Several goodness-of-fit measures are commonly used. These are based on the forecast error, which, for period t, is the difference between the observed and forecasted value:

$$e_t = y_t - \hat{y}_t$$

One popular goodness-of-fit measure is the **mean absolute deviation (MAD)**. This is the average of the absolute value of the errors for each data point for which we have observed and forecasted data. If we have n such data points:

$$\text{MAD} = \frac{\sum_{t=1}^{n} |e_t|}{n}$$

A second measure of goodness of fit is **root mean square error (RMSE)**, which is defined as follows:

$$\text{RMSE} = \sqrt{\frac{\left(\sum_{t=1}^{n} e_t^2\right)}{n}}$$

You might note that RMSE is similar to a standard deviation. Finally, a third measure of fit is the **mean absolute percentage error (MAPE)**. The percentage error for period t is the difference between the actual and forecasted values divided by the actual value:

$$p_t = 100 \times \frac{(y_t - \hat{y}_t)}{y_t}$$

MAPE is the average of these percentage errors:

$$\text{MAPE} = \frac{\sum_{t=1}^{n} |p_t|}{n}$$

Which measure of fit one uses is largely a matter of user preference. Note however, that since RMSE uses the square of error, it places more emphasis on large errors than MAD. Also, since MAPE uses percentage errors, it is less appropriate if there are zeros or values close to zero in the data (since dividing by the actual data to get p_t will result in very large numbers).

Example 3: Forecasting with Single Moving Averages

Atlas Dry Cleaning is a full-service dry-cleaning business. Atlas has been in business for over 40 years and has a relatively stable base of satisfied customers. A major portion of Atlas's business is cleaning and starching men's shirts. Data on the number of shirts cleaned in each of the last 10 weeks are shown in Figure 3.6, along with a plot of these data. The data do not appear to have any trend, cyclical, or seasonal components.

FIGURE 3.6
Atlas Cleaners
shirt data

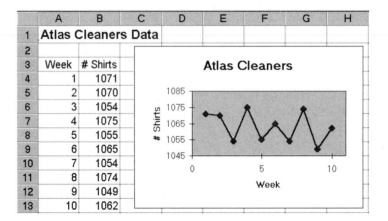

	A	B	C
1	**Atlas Cleaners Data**		
2			
3	Week	# Shirts	
4	1	1071	
5	2	1070	
6	3	1054	
7	4	1075	
8	5	1055	
9	6	1065	
10	7	1054	
11	8	1074	
12	9	1049	
13	10	1062	

We will use a single moving average to forecast the number of shirts needing cleaning next week. Suppose we choose the number of periods to be used in the moving average to be three ($k = 3$). Then the forecast for week 11 will be the average of the demand for weeks 8, 9, and 10: $(1{,}074 + 1{,}049 + 1{,}062)/3 = 1{,}061.67$ or 1,062 shirts.

Figure 3.7 shows a spreadsheet designed to compute the moving average forecasts beginning with week 4 (since this is the first week we may forecast the historical data with $k = 3$), and a graph of how the data and forecasts compare. You can see that the moving average forecasts show much less fluctuation; in essence, they *smooth out* the irregular component of the time series. Measures of forecast error—MAD, RMSE, and MAPE—are computed near the bottom of the spreadsheet.

FIGURE 3.7 Spreadsheet for computing simple moving average forecasts

	A	B	C	D	E	F	G
1	**Atlas Cleaners**				Key cell formulas		
2	3-Period Simple Moving Average Model						
3							
4				Absolute	Squared	Percentage	
5	Week	# Shirts	Forecast	Error	Error	Error	
6	1	1071					
7	2	1070					
8	3	1054					
9	4	1075	1065.00	10.00	100.00	0.93%	
10	5	1055	1066.33	11.33	128.44	1.07%	
11	6	1065	1061.33	3.67	13.44	0.34%	
12	7	1054	1065.00	11.00	121.00	1.04%	
13	8	1074	1058.00	16.00	256.00	1.49%	
14	9	1049	1064.33	15.33	235.11	1.46%	
15	10	1062	1059.00	3.00	9.00	0.28%	
16	11		1061.67				
17							
18			MAD	10.05			
19			RMSE		11.10		
20			MAPE			0.95%	

Excel Exercise 3.1

Open the worksheet *Atlas SMA.xls*, shown in Figure 3.7. Modify the spreadsheet so that it computes both 2- and 4-period simple moving averages in addition to the 3-period moving average. Note that the calculations of MAD, RMSE, and MAPE must also be updated, since there will be fewer periods with forecasted values. How do the error measures for the 2- and 4-period models compare to that of the 3-period model? Is one of these the best for all error measures?

Single Exponential Smoothing

Single (or *simple*) *exponential smoothing* is another very common short-term forecasting technique. Let y_t be the actual value of the time series and $\hat{y}_t$ be the forecasted value for time period t. Exponential smoothing uses the following model:

$$\hat{y}_2 = y_1$$
$$\hat{y}_{t+1} = \hat{y}_t + \alpha(y_t - \hat{y}_t)$$
$$= \alpha y_t + (1 - \alpha)\,\hat{y}_t \ \text{ for } t = 2, 3, 4,.....$$

where $\alpha(0 \leq \alpha \leq 1)$ is called the *smoothing constant*. To begin, the forecast for period 2 is equal to the actual time series value for period 1. After that, the exponential smoothing model takes the forecast from the previous period and adjusts it by adding a fraction, α, of the forecast error as shown in the second equation. If we overestimated in period t, then $(y_t - \hat{y}_t)$ is negative and the forecast for period $t + 1$ is lower than for period t. If we underestimated in period t, $(y_t - \hat{y}_t)$ is positive and the forecast for period $t + 1$ is higher than for period t.

By combining the terms in a different fashion, as shown in the last part of the equation above, we see that the forecast for the next period is a weighted average of the actual time series value and its forecasted value from the previous period, where α is the weight placed on the actual value. Thus, higher values of α place more weight on the actual value and less weight on the previous forecast. Like the moving average technique, we cannot tell ahead of time what the best value of α might be; it must be chosen by experimentation.

With a little algebraic substitution, it is easy to show that exponential smoothing forecasts are simply weighted averages of *all* the prior time series values, with smaller weights associated with older data. To see this, consider that $\hat{y}_t = \alpha y_{t-1} + (1 - \alpha)\hat{y}_{t-1}$. By substituting this into the forecast model for period $t + 1$ we have

$$\hat{y}_{t+1} = \alpha y_t + (1 - \alpha)\hat{y}_t$$
$$= \alpha y_t + (1 - \alpha)[\alpha y_{t-1} + (1 - \alpha)\hat{y}_{t-1}]$$
$$= \alpha y_t + \alpha(1 - \alpha)y_{t-1} + (1 - \alpha)^2\hat{y}_{t-1}$$

This shows that the forecast for period $t + 1$ is a weighted average of the prior two time series values and the forecast for period $t - 1$. If we continue in this fashion to substitute the formulas for the forecasts for the other earlier periods we obtain

$$\hat{y}_{t+1} = \alpha y_t + \alpha(1-\alpha)y_{t-1} + \alpha(1-\alpha)^2 y_{t-2}$$
$$+ \alpha(1-\alpha)^3 y_{t-3} + \alpha(1-\alpha)^4 y_{t-4}$$
$$+ \ldots\ldots + \alpha(1-\alpha)^n y_{t-n}$$

This verifies that the forecast for period $t + 1$ is just a weighted average of all the previous time series values. Since $1 - \alpha$ is less than 1, then higher powers of this term are smaller; hence, less weight is placed on older data.

Example 4: Forecasting with Exponential Smoothing

We will apply exponential smoothing to the Atlas Cleaners data given in Example 3. Suppose we choose $\alpha = 0.7$. Then the forecast for period 2 is equal to the actual value for period 1: 1,071. Once the actual time series value for period 2 is known, we can compute a forecast for period 3: $1,071 + 0.7(1,070 - 1,071) = 1,070.3$.

Figure 3.8 shows a spreadsheet implementation of this technique and a plot of the time series and forecasts along with the associated MAD, RMSE, and MAPE.

FIGURE 3.8 Spreadsheet for computing single exponential smoothing forecasts

	A	B	C	D	E	F	G
1	**Atlas Cleaners**			Key cell formulas			
2	**Single Exponential Smoothing**						
3							
4	*Alpha*	0.7					
5					Absolute	Squared	Absolute
6	**Week**	**# Shirts**	**Forecast**	**Error**	**Error**	**Error**	**% Error**
7	1	1071					
8	2	1070	1071.00	-1.00	1	1.00	0.09%
9	3	1054	1070.30	-16.30	16.3	265.69	1.55%
10	4	1075	1058.89	16.11	16.11	259.53	1.50%
11	5	1055	1070.17	-15.17	15.167	230.04	1.44%
12	6	1065	1059.55	5.45	5.4499	29.70	0.51%
13	7	1054	1063.37	-9.37	9.36503	87.70	0.89%
14	8	1074	1056.81	17.19	17.19049	295.51	1.60%
15	9	1049	1068.84	-19.84	19.84285	393.74	1.89%
16	10	1062	1054.95	7.05	7.047144	49.66	0.66%
17	11		1059.89				
18				MAD	11.94138		
19				RMSE		13.39	
20				MAPE			1.13%

Exponential Smoothing chart (# Shirts and Forecast plotted against Week); y-axis # Shirts from 1045 to 1080, x-axis Week from 0 to 10.

Excel Exercise 3.2

Open the worksheet *Atlas SES.xls*. Change the value of α (found in cell B4), varying it from 0 to 1 by .1. What value of α provides the best MAD? Relate the values of α to what you see happening to the graphed forecasts (fitted values). Perform similar analyses using RMSE and MAPE.

MODELS FOR TIME SERIES WITH A TREND COMPONENT

Single moving average and single exponential smoothing can both be modified to better handle trends in time series data. These adjusted methods are called *double moving average* and *double exponential smoothing* models. These techniques are most appropriate for relatively stable time series with some trend, but no significant cyclical or seasonal components.

Both double moving average and double exponential smoothing forecasts are based on the linear trend equation:

$$\hat{y}_{t+k} = a_t + b_t k$$

This may look familiar from simple linear regression. That is, the forecast for k periods into the future from period t is a function of some base a_t, also known as the *level*, and a *trend* or slope b_t. Double moving average and double exponential smoothing differ in how the data are used to arrive at appropriate values for a_t and b_t.

Double Moving Average

As the name implies, double moving average involves taking averages of averages. Let M_t be the simple moving average for the last k periods (including period t):

$$M_t = \frac{[y_{t-k+1} + y_{t-k+2} + \dots y_t]}{k}$$

The double moving average, D_t, for the last k periods (including period t) is the average of the simple moving averages:

$$D_t = \frac{[M_{t-k+1} + M_{t-k+2} + \dots M_t]}{k}$$

Using these values, the double moving average method estimates the values of a_t and b_t in the linear trend model as

$$a_t = 2M_t - D_t$$

$$b_t = \left(\frac{2}{k-1}\right)[M_t - D_t]$$

These equations are derived essentially by minimizing the sum of squared errors using the last k periods of data. (Recall that the sum of squared errors was also used in our development of regression in Chapter 2). Once these parameters are determined, forecasts beyond the end of the observed data (time period T) are calculated using the linear trend model with values of a_T and b_T. That is, for k periods beyond period T, the forecast is $\hat{y}_{T+k} = a_T + b_T k$. The forecast for the next period would be $\hat{y}_{T+1} = a_T + b_T$.

Example 5: Forecasting with Double Moving Averages

In addition to the shirt data previously discussed, Atlas Dry Cleaning has also kept data on the number of dresses cleaned over the last 10 weeks. Atlas has noticed that the number of dresses cleaned has been increasing. Data on the number of dresses cleaned in each of the last 10 weeks are shown in Figure 3.9 along with a plot of these data and a forecast based on 3-period, double moving average (these are in file *Atlas DMA.xls*).

The intermediate calculations for the first forecast (period 6) are as follows:

$$M_3 = \frac{(762 + 770 + 761)}{3} = 764.333$$

$$M_4 = \frac{(770 + 761 + 772)}{3} = 767.667$$

$$M_5 = \frac{(761 + 772 + 767)}{3} = 766.667$$

$$D_5 = \frac{(M_3 + M_4 + M_5)}{3} = \frac{(764.333 + 767.667 + 766.667)}{3} = 766.222$$

$$a_5 = 2M_5 - D_5 = 2(766.667) - 766.222 = 767.112$$

$$b_5 = \left(\frac{2}{(3-1)}\right)[M_5 - D_5] = 766.667 - 766.222 = .445$$

The forecast for period 6 is

$$\hat{y}_6 = a_5 + b_5 = 767.112 + .445 = 767.557$$

Other forecasts are calculated in a similar manner. The model for predicting future dress demand for period 11 and beyond (using the computed values for the level and trend in week 10) is

$$\hat{y}_{10+k} = 777.78 + 2.78k$$

Note that the forecast for period 11 is 780.56, or 781 dresses.

For comparison purposes, we have also calculated and plotted the 3-period single moving average. Notice from the graph that the single moving average appears to lag the data and underestimate the trend, and that the double moving average tracks the trend better.

FIGURE 3.9 Spreadsheet for computing double moving average forecasts

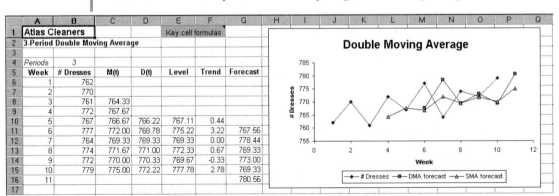

Double Exponential Smoothing

Like double moving average, double exponential smoothing is also based on the linear trend equation, $\hat{y}_{t+k} = a_t + b_t k$, but the estimates of a_t and b_t are obtained from the following equations:

$$a_t = \alpha y_t + (1 - \alpha)(a_{t-1} + b_{t-1})$$
$$b_t = \beta(a_t - a_{t-1}) + (1 - \beta)b_{t-1}$$

Here we are smoothing both parameters of the linear trend model that defines the forecast. From the first equation, the estimate of the level in period t is a weighted average of the observed value at time t and the predicted value at time t, $a_{t-1} + b_{t-1}(1)$ based on single exponential smoothing. For large values of α, more weight is placed on the observed value. Lower values of α put more weight on the smoothed predicted value. Similarly, from the second equation, the estimate of the trend in period t is a weighted average of the differences in the estimated levels in periods t and $t - 1$ and the estimate of the level in period $t - 1$. Larger values of β place more weight on the differences in the levels, whereas lower values of β put more emphasis on the previous estimate of trend.

To initialize the double exponential smoothing process, we need values for a_1 and b_1. There are a variety of approaches for initialization. We will use the following: let $a_1 = y_1$ and $b_1 = y_2 - y_1$, that is, estimate the initial level with the first observation and the initial trend with the difference in the first two observations. As with single exponential smoothing, we are free to choose the values of α and β. MAD or other measures of fit may be used to find good values for these smoothing parameters.

Example 6: Forecasting with Double Exponential Smoothing

Let us apply double exponential smoothing to the dress forecast for Atlas Cleaners (Example 5). Figure 3.10 shows a spreadsheet implementation of this, using $\alpha = .7$ and $\beta = .7$. The calculations for the first forecast (period 3 is the first period that can be based on smoothing) are as follows:

$$a_1 = y_1 = 762.00$$
$$b_1 = y_2 - y_1 = 770 - 762 = 8.00$$
$$a_2 = \alpha y_2 + (1 - \alpha)(a_1 + b_1) = .7(770) + .3(762.00 + 8.00) = 770.00$$
$$b_2 = \beta(a_2 - a_1) + (1 - \beta)b_1 = .7(770.00 - 762.00) + .3(8.00) = 8.00$$

Using these parameters, the forecast for period 3 is

$$\hat{y}_3 = 770.00 + 8.00 = 778.00$$

The other forecasts are calculated in a similar manner. The forecasting model for k periods beyond period 10 is

$$\hat{y}_{10+k} = 777.22 + 3.82k$$

For example, the forecast for period 11 is 777.22 + 3.82(1) = 781.04.

The value of MAD for these forecasts is 7.46. Are there values of α and β that provide a better fit? We can answer this using a two-way data table, as shown in Figure 3.11. Based on increments of .1 in α and β, the best MAD is 6.25, which is achieved at α = .5 and β = .5.

FIGURE 3.10 Spreadsheet for computing double exponential smoothing forecasts

	A	B	C	D	E	F
1	Atlas Cleaners				Key cell formulas	
2	Double Exponential Smoothing					
3						
4	*Alpha*	0.7				
5	*Beta*	0.7				
6						Absolute
7	Week	# Dresses	Level	Trend	Forecast	Error
8	1	762	762.00	8.00		
9	2	770	770.00	8.00	770.00	0.00
10	3	761	766.10	-0.33	778.00	17.00
11	4	772	770.13	2.72	765.77	6.23
12	5	767	768.76	-0.15	772.85	5.85
13	6	777	774.48	3.97	768.61	8.39
14	7	764	768.33	-3.11	778.45	14.45
15	8	774	771.37	1.19	765.22	8.78
16	9	772	772.17	0.92	772.55	0.55
17	10	779	777.22	3.82	773.08	5.92
18	11				781.04	
19						
20					*MAD*	7.46

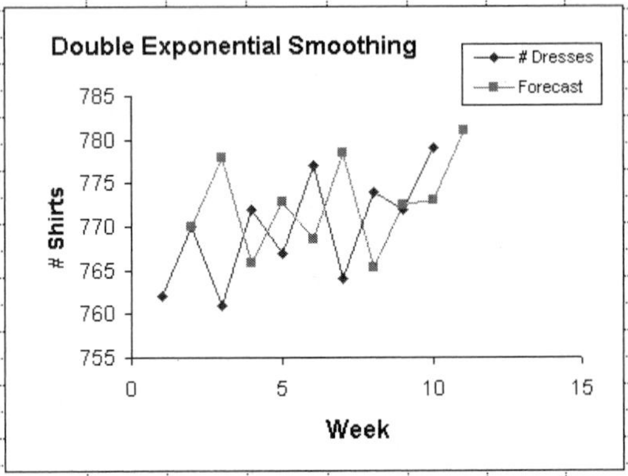

FIGURE 3.11 Data table for optimizing double exponential smoothing forecasts

7.46	0.00	0.10	0.20	0.30	0.40	0.50	0.60	0.70	0.80	0.90	1.00	Beta
0.00	31.33	24.52	19.58	15.97	13.29	11.28	9.73	8.52	7.55	7.21	7.67	
0.10	31.33	22.94	17.19	13.23	10.47	8.52	7.10	6.33	6.63	7.28	8.05	
0.20	31.33	21.44	15.04	10.91	8.27	6.91	6.29	6.40	6.99	7.70	8.58	
0.30	31.33	20.02	13.12	9.10	7.22	6.32	6.28	6.65	7.29	8.08	9.09	
0.40	31.33	18.67	11.40	8.18	6.42	6.29	6.29	6.86	7.57	8.47	9.64	
0.50	31.33	17.40	10.13	7.32	6.39	**6.25**	6.43	7.05	7.85	8.87	10.24	
0.60	31.33	16.19	9.41	6.71	6.52	6.28	6.60	7.25	8.14	9.32	10.93	
0.70	31.33	15.04	8.71	6.78	6.57	6.33	6.75	7.46	8.46	9.82	11.72	
0.90	31.33	12.94	7.48	7.00	6.43	6.48	7.03	7.93	9.20	11.02	13.72	
1.00	31.33	11.98	7.48	6.98	6.29	6.54	7.19	8.19	9.63	11.75	15.00	
Alpha												

Excel Exercise 3.4

Open the worksheet *Atlas DES.xls*. Use a data table to find values of α and β that give the lowest RMSE. Also find the values of α and β that give the lowest MAPE.

MODELS FOR TIME SERIES WITH A SEASONAL COMPONENT

In this section we consider techniques that are appropriate for time series with seasonal components that exhibit no trend. Seasonal factors can be incorporated in one of two ways, as an additive factor using the model

$$\hat{y}_{t+k} = a_t + S_{t-s+k}$$

or as a multiplicative factor using the model

$$\hat{y}_{t+k} = a_t S_{t-s+k}$$

In both models, S_j is the seasonal factor for period j and s is the number of periods in a season. A "season" can be a year, quarter, month, or even a week, depending on the application. In any case, the forecast for period $t + k$ is adjusted up or down from a level (a_t) by the seasonal factor. The multiplicative model is perhaps more appropriate when the seasonal factors are increasing or decreasing over time.

Additive Seasonality

The level and seasonal factors are estimated in the additive model, using the following equations:

$$a_t = \alpha(y_t - S_{t-s}) + (1 - \alpha)a_{t-1}$$
$$S_t = \gamma(y_t - a_t) + (1 - \gamma)S_{t-s}$$

where α and γ are smoothing constants. The first equation estimates the level for period t, as a weighted average of the deseasonalized data for period t, $(y_t - S_{t-s})$, and the previous period's level. The seasonal factors are updated as well, using the second equation. The seasonal factor is a weighted average of the estimated seasonal component for period t, $(y_t - a_t)$, and the seasonal factor for the last period of that season type. Then the forecast for the next period is $\hat{y}_{t+1} = a_t + S_{t-s+1}$. For k periods out from the final observed period T, the forecast is

$$\hat{y}_{T+k} = a_T + S_{T-s+k}$$

As in the case for double exponential smoothing, initialization issues need to be addressed. We need to estimate the level and seasonal factors for the first s periods (e.g., for an annual season with quarterly data this would be the first 4 periods, for monthly data it would be the first 12 periods, etc.) before we can use the smoothing equations. Again, there are numerous ways to initialize the process. We will use the following:

$$a_s = \frac{\sum\limits_{t=1}^{s} y_t}{s}$$

$$a_t = a_s \qquad t = 1, 2,..., s - 1$$

and

$$S_t = y_t - a_t \qquad t = 1, 2,..., s$$

That is, we initialize the level for the first s periods to the average of the observed values over these periods and the seasonal factors to the difference between the observed data and the estimated levels. Once these have been initialized, the smoothing equations can be implemented for updating.

Example 7: Forecasting a Seasonal Time Series Using the Additive Model

Atlas Cleaners has collected data on the demand for sweater cleaning. Five years of data (20 quarters, 1-Fall, 2-Winter, 3-Spring, 4-Summer, etc.) are shown in Figure 3.12 (*Atlas Sweaters Additive Seasonal.xls*). The data are clearly seasonal, with medium demand in fall and spring, high demand in winter, and low demand in summer. The startup calculations needed for the forecasts for periods 5 and 6 are as follows:

$$a_1 = a_2 = a_3 = a_4 = \frac{\sum_{t=1}^{4} y_t}{4} = \frac{(120 + 284 + 123 + 22)}{4} = 137.25$$

$$S_1 = y_1 - a_1 = 120 - 137.25 = -17.25$$

$$S_2 = y_2 - a_2 = 284 - 137.25 = 146.75$$

$$S_3 = y_3 - a_3 = 123 - 137.25 = -14.25$$

$$S_4 = y_4 - a_4 = 22 - 137.25 = -115.25$$

Using the seasonal factor for the first period, the forecast for period 5 is

$$\hat{y}_5 = a_4 + S_1 = 137.5 - 17.25 = 120$$

We may now update the parameters based on the new data:

$$a_5 = \alpha(y_5 - S_1) + (1 - \alpha)a_4 = .7(136.00 - (-17.25)) + .3(137.25) = 148.45$$

$$S_5 = \gamma(y_5 - a_5) + (1 - \gamma)S_1 = .7(136 - 148.45) + .3(-17.25) = -13.89 \text{ (which will be used in later forecasts)}$$

The forecasted value for period 6 is then

$$\hat{y}_6 = a_5 + S_2 = 148.45 + 146.75 = 295.2$$

Other forecasted values are calculated in a similar manner. The forecast for the four quarters beyond the data are obtained from the last computed value of the level and the past four seasonal factors:

Quarter	Forecasted Demand	
21	268.26 + 3.32	= 271.58
22	268.26 + 163.73	= 431.99
23	268.26 − 9.69	= 258.57
24	268.26 − 118.06	= 150.2

FIGURE 3.12 Additive seasonal forecasting model

	A	B	C	D	E	F
1	Atlas Cleaners			Key cell formulas		
2	Additive Seasonal Model					
3	Alpha	0.7				
4	Gamma	0.7				
5						
6	Quarter	#Sweaters	Level	Seasonality	Forecast	
7	1	120	137.25	-17.25		
8	2	284	137.25	146.75		
9	3	123	137.25	-14.25		
10	4	22	137.25	-115.25		
11	5	136	148.45	-13.89	120.00	
12	6	306	156.01	149.02	295.20	
13	7	138	153.38	-15.04	141.76	
14	8	35	151.19	-115.91	38.13	
15	9	147	157.98	-11.85	137.30	
16	10	327	171.98	153.22	307.00	
17	11	155	170.62	-15.45	156.94	
18	12	59	173.62	-115.01	54.72	
19	13	183	188.48	-7.39	161.77	
20	14	370	208.29	159.16	341.70	
21	15	197	211.20	-14.57	192.84	
22	16	95	210.37	-115.26	96.19	
23	17	254	246.09	3.32	202.97	
24	18	427	261.31	163.73	405.25	
25	19	270	277.60	-9.69	246.74	
26	20	149	268.26	-118.06	162.34	
27	21				271.58	

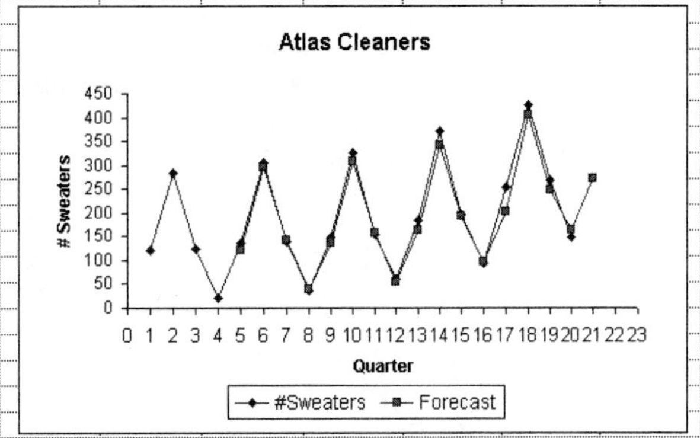

Atlas Cleaners chart (#Sweaters and Forecast by Quarter)

Multiplicative Seasonality

The multiplicative seasonal model has the same basic smoothing structure as the additive seasonal model:

$$a_t = \alpha\left(\frac{y_t}{S_{t-s}}\right) + (1 - \alpha)a_{t-1}$$

$$S_t = \gamma\left(\frac{y_t}{a_t}\right) + (1 - \gamma)S_{t-s}$$

where α and γ are again the smoothing constants. Here, y_t / S_{t-s} is the deseasonalized estimate for period t. Large values of α put more emphasis on this term in estimating the level for period t. The term y_t / a_t is an estimate of the seasonal factor for period t. Large values of γ put more emphasis on this in the estimate of the seasonal factor.

The forecast for the next period is $\hat{y}_{t+1} = a_t S_{t-s+1}$. For k periods out from the final observed period T, the forecast is

$$\hat{y}_{T+k} = a_T S_{T-s+k}$$

As in the additive model, we need initial values for the level and seasonal factors. We do this as follows:

$$a_s = \frac{\displaystyle\sum_{t=1}^{s} y_t}{s}$$

$$a_t = a_s \qquad t = 1, 2, ..., s - 1$$

and

$$S_t = \frac{y_t}{a_t} \qquad t = 1, 2, ..., s$$

That is, we initialize the level for the first s periods to the average of the observed values over these periods and the seasonal factors to the ratio between the observed data and the estimated levels. Once these have been initialized, the smoothing equations can be implemented for updating.

Example 8: Forecasting a Seasonal Time Series Using the Multiplicative Model

Figure 3.13 shows the results of applying the multiplicative seasonal model to the sweater data for Atlas Cleaners with $\alpha = .7$ and $\gamma = .7$ (worksheet *Atlas Sweaters Multiplicative Seasonal.xls*).

$$a_1 = a_2 = a_3 = a_4 = \frac{\sum_{t=1}^{4} y_t}{4} = \frac{(120 + 284 + 123 + 22)}{4} = 137.25$$

$$S_1 = \frac{y_1}{a_1} = \frac{120}{137.25} = .8743$$

$$S_2 = \frac{y_2}{a_2} = \frac{284}{137.25} = 2.0692$$

$$S_3 = \frac{y_3}{a_3} = \frac{123}{137.25} = .896$$

$$S_4 = \frac{y_4}{a_4} = \frac{22}{137.25} = .160$$

Then the forecast for period 5 is

$$\hat{y}_5 = a_4 S_1 = 137.25(.8743) = 119.997 \text{ or } 120$$

Updating the calculations, we obtain

$$a_5 = \alpha\left(\frac{y_5}{S_1}\right) + (1 - \alpha)a_4 = .7\left(\frac{136.00}{.8743}\right) + .3(137.25) = 150.062$$

$$S_5 = \gamma\left(\frac{y_5}{a_5}\right) + (1 - \gamma)S_1 = .7\left(\frac{136.00}{150.062}\right) + .3(.8743) = .89669 \text{ or } .90 \text{ (to be used in}$$

later forecasts)

The forecasted value for period 6 is

$$\hat{y}_6 = a_5 S_2 = 150.062(2.0692) = 310.508 \text{ or } 310.51.$$

Other forecasted values are calculated in a similar manner.

Note that the forecast for period 21 is much higher than that produced by the additive model (420.38 versus 271.58). Also, comparing the two graphs from additive and multiplicative models shows how the multiplicative model intensifies the seasonal factors, suggesting that the additive model might be better for these data.

FIGURE 3.13 Multiplicative seasonal forecasting model

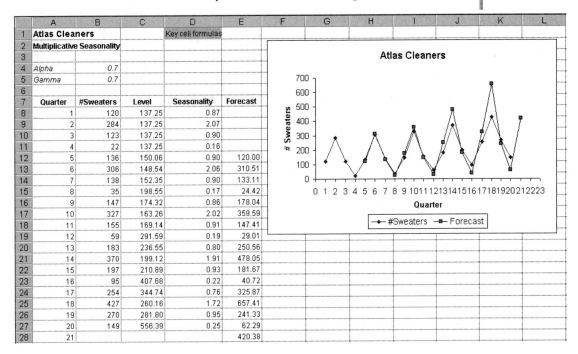

MODELS FOR TIME SERIES WITH TREND AND SEASONAL COMPONENTS

The additive and multiplicative models discussed in the last section for seasonal components can both be extended to incorporate a linear trend. These models are often referred to as the Holt-Winters Additive and Multiplicative models (Charles Holt developed double exponential smoothing, which P. R. Winters extended to incorporate seasonality).

Holt-Winters Additive Model

The Holt-Winters additive model is the same model as the simple additive model incorporating seasonality, but with the addition of a trend component:

$$a_t = \alpha(y_t - S_{t-s}) + (1 - \alpha)(a_{t-1} + b_{t-1})$$

$$b_t = \beta(a_t - a_{t-1}) + (1 - \beta)b_{t-1}$$

$$S_t = \gamma(y_t - a_t) + (1 - \gamma)S_{t-s}$$

Here, α, β, and γ are the smoothing parameters for level, trend, and seasonal components, respectively. The forecast for period $t + 1$ is

$$\hat{y}_{t+1} = a_t + b_t + S_{t-s+1}$$

The forecast for k periods beyond the last period of observed data (period T) is

$$\hat{y}_{T+k} = a_T + b_T k + S_{T-s+k}$$

The initial values of level and trend are estimated as in the simple additive model for seasonality. The initial values for the trend are $b_t = b_s$, for $t = 1,2,...s - 1$, where

$$b_s = \frac{\left[\dfrac{(y_{s+1} - y_1)}{s} + \dfrac{(y_{s+2} - y_s)}{s} + ... + \dfrac{(y_{s+s} - y_s)}{s}\right]}{s}$$

Hence, we initialize the trend by averaging estimates of trend over each season for the initial $2s$ periods (each term inside the brackets is an estimate of trend using a given season and we average these estimates).

Holt-Winters Multiplicative Model

The Holt-Winters multiplicative model parallels the additive model:

$$a_t = \alpha\left(\frac{y_t}{S_{t-s}}\right) + (1 - \alpha)(a_{t-1} + b_{t-1})$$

$$b_t = \beta(a_t - a_{t-1}) + (1 - \beta)b_{t-1}$$

$$S_t = \gamma\left(\frac{y_t}{a_t}\right) + (1 - \gamma)S_{t-s}$$

The forecast for period $t + 1$ is

$$\hat{y}_{t+1} = (a_t + b_t)S_{t-s+1}$$

The forecast for k periods beyond the last period of observed data (period T) is

$$\hat{y}_{T+k} = (a_T + b_T k)\, S_{T-s+k}$$

The initial values of level and trend are estimated as in the simple multiplicative model for seasonality and the trend component, as just described for the Holt-Winters additive model.

Example 9: Forecasting with Holt-Winters Models

We will apply the Holt-Winters additive model to the Atlas Cleaners sweater data used in previous examples. Figure 3.14 shows the results for the Holt-Winters additive model (*Atlas Sweaters Holt Winters Additive.xls*) with $\alpha = .7$, $\beta = .7$, and $\gamma = .7$. The level and seasonal factors are initialized as in Example 7. The trend factors are initialized as follows:

$$b_1 = b_2 = b_3 = b_4 = \frac{\left[\dfrac{(136 - 120)}{4} + \dfrac{(306 - 284)}{4} + \dfrac{(138 - 123)}{4} + \dfrac{(35 - 22)}{4}\right]}{4}$$

$$= 4.125$$

The forecast for period 5 is

$$\hat{y}_5 = a_4 + b_4 + S_1 = 137.25 + 4.125 + (-17.25) = 124.125$$

The following calculations update the parameters for period 5:

$$a_5 = \alpha(y_5 - S_1) + (1 - \alpha)(a_4 + b_4) = .7(136 - (-17.25)) + .3(137.25 + 4.125)$$
$$= 149.6875$$

$$b_5 = \beta(a_5 - a_4) + (1 - \beta)b_4 = .7(149.6875 - 137.25) + .3(4.125) = 9.94375$$

$$S_5 = \gamma(y_5 - a_5) + (1 - \gamma)S_1 = .7(136 - 149.6875) + .3(-17.25) = -14.75625$$

Then the forecast for period 6 is

$$\hat{y}_6 = a_5 + b_5 + S_2 = 149.6875 + 9.94375 + 146.75 = 306.38125$$

Other calculations are similar. Notice from the graph that the additive model appears to fit the data extremely well.

FIGURE 3.14 Holt-Winters additive seasonal forecasting model

	A	B	C	D	E	F
1	**Atlas Cleaners**				Key cell formulas	
2	Holt-Winters Additive Seasonality Model					
3						
4	Alpha	0.7				
5	Beta	0.7				
6	Gamma	0.7				
7						
8	Quarter	# Sweaters	Level	Trend	Seasonality	Forecast
9	1	120	137.2500	4.1250	-17.2500	
10	2	284	137.2500	4.1250	146.7500	
11	3	123	137.2500	4.1250	-14.2500	
12	4	22	137.2500	4.1250	-115.2500	
13	5	136	149.6875	9.9438	-14.7563	124.1250
14	6	306	159.3644	9.7569	146.6699	306.3813
15	7	138	157.3114	1.4900	-17.7930	154.8713
16	8	35	152.8154	-2.7002	-117.0458	43.5514
17	9	147	158.2639	3.0039	-12.3116	135.3590
18	10	327	174.6114	12.3444	150.6730	307.9378
19	11	155	177.0418	5.4046	-20.7672	169.1628
20	12	59	177.9660	2.2683	-118.3899	65.4006
21	13	183	190.7884	9.6562	-9.1454	167.9226
22	14	370	213.6623	18.9086	154.6383	351.1176
23	15	197	222.2083	11.6548	-23.8759	211.8037
24	16	95	219.5319	1.6229	-122.6893	115.4731
25	17	254	250.5482	22.1983	-0.3274	212.0094
26	18	427	272.4772	22.0098	154.5575	427.3848
27	19	270	294.0592	21.7104	-24.0042	270.6110
28	20	149	284.9134	0.1110	-131.9461	193.0803
29	21					284.6970

Excel Exercise 3.5

The worksheet *Atlas Sweaters Holt Winters Additive.xls* (discussed in Example 9) contains the Holt-Winters additive model with seasonality and trend. Modify this spreadsheet to apply the multiplicative form of this model, using $\alpha = .7$, $\beta = .7$, and $\gamma = .7$. Compare the additive and multiplicative approaches. Which appears to provide a better fit?

FORECASTING WITH REGRESSION MODELS

In addition to the eight different forecasting models we have discussed so far in this chapter, linear regression models may also be used for forecasting. For pure time series, we can use various functions of time as independent variables (see the

Management Science Practice box). However, regression also allows us to use other independent variables to account for other factors besides time that may influence forecasts of the dependent variable. For example, we may see a spike in sales depending on whether or not a sale is on, demand for emergency services may drastically increase during a full moon, and beer consumption increases on St. Patrick's Day. Such factors can be modeled using indicator variables (variables that take on a value of zero or one).

Management Science Practice: A Forecasting Model for Supermarket Checkout Services

Customer service is a critical aspect in the management of grocery stores. Because of highly variable demand, store managers must make frequent, short-term decisions about staffing checkout counters. The number of checkers at any time can be adjusted by the store manager to control the length of the waiting lines and hence customer waiting time. It is usually possible to assign an employee who is working in another part of the store—the produce department, for example—to work temporarily at a checkout counter during periods of high demand.

One large grocery chain was investigating the use of a laser-based scanner at the doors of the store to count arriving and departing customers. Company managers wanted to use this information to forecast demand at the checkout counters about 30 minutes in the future to make staffing adjustments before long lines developed.

Figure 3.15 is a schematic diagram of the system operation. Customers enter the store, shop, wait in line at one of several checkout counters, and then leave the store. The uncertainty in arrival rates at the checkout counters is due to not knowing how long customers will shop. Clearly, if the store arrival time and shopping time were known, the time at which a customer will arrive at the checkout counter could be computed.

A forecasting model was developed on the assumption that the demand for checkout services is closely related to the number of shoppers in the store; the larger the population of shoppers, the larger will be the demand for checkout services. If the probability distribution of shopping times is relatively stable, the number of customers demanding checkout service will be proportional to the number of customers in the store. Over a one-week period, data were collected on store arrivals, departures, lengths of checkout lines, and the number of cashiers working at the end of fixed time intervals. In effect, these provided "snapshots" of the state of the store over time. By keeping a running total of arrivals less departures, the store could calculate the number of customers in the store at any time.

continued

FIGURE 3.15 Grocery-store customer flow

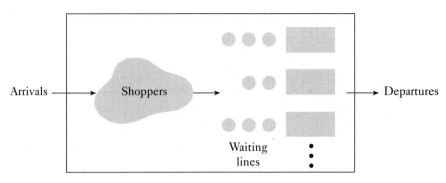

Grocery store

From the arrival and departure data, a variable representing checkout service demand in any given period was created as follows. Let

y = number of customers demanding checkout service during a time period,

Q = number of customers in line at the end of the period,

C = number of checkers working at the end of the period, and

d = number of departures during a time period.

An estimate of the demand during a given period is then given by

$$y = d + Q + C.$$

The rationale for this equation is that the demand is equal to the number of departures (people who actually obtained service), plus the number waiting to be served (current demand), plus the number of checkers (assuming that all are busy).

The first analysis conducted was an investigation of demand fluctuation over time. Figure 3.16 is an example of data collected during one day. In general, demand follows a bimodal distribution, with peaks at roughly noon and 6 p.m. each day. The next step was to identify quantities that would effectively predict demand for future time periods. Intuition would dictate that the number of customers in the store in earlier periods would be of importance. Likewise, the arrival rate of customers would be a significant variable. Figure 3.17 shows demand as a function of the number of arrivals to the store three time periods earlier. (Each period was 15 minutes, so this provides demand data for 30 to 45 minutes in the future.) From this figure, we see that as the number of arriving customers increases, so does demand for checkout services. This relationship was not surprising. What was surprising was the highly linear relationship found for many of the days. This indicated that regression models might work extremely well.

The choice of prediction variables was determined by intuition, the strong relationships suggested by the data, and by extensive analysis of certain statistical measures of model adequacy. The model that was ultimately developed

continued

FIGURE 3.16 Customer demand distribution during a day

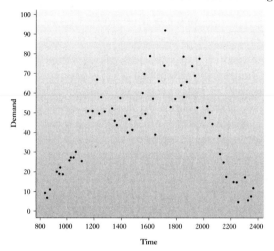

FIGURE 3.17 Projected customer demand three time periods in the future

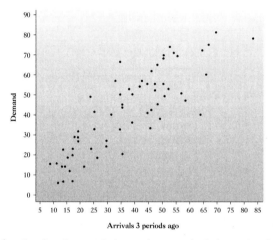

used y_{t+3}—that is, the demand three time periods into the future—as the dependent variable, with the following independent variables:

N_t = number of customers in the store in period t

N_{t-1} = number of customers in the store in period $t - 1$

a_t = number of customers arriving in period t

Thus, if t is the time period 2:15 to 2:30, the model includes the number of customers in the store at 1:30 to 1:45 and at 1:15 to 1:30 and the number of arrivals during the period 1:30 to 1:45. Therefore, a forecast for the period 2:15 to 2:30 can be made at 1:45. An example of the actual model for one day is

$$Y_{t+3} = 0.34431 - 0.12760N_t + 0.31627N_{t-1} + 0.90634a_t$$

The coefficient of determination, R^2, was 0.80, which indicated that a high percentage of the variation in demand was explained by the variables chosen.

Source: Based on a project in which the authors were involved.

Linear Regression for Time Series Forecasting

Linear regression can be used as a forecasting technique by using time (or some function of time) as the independent variable(s). For example, we might use one of the following models:

Model 1: $\hat{y}_{t+1} = bt + a$

Model 2: $\hat{y}_{t+1} = at^2 + bt + c$

Model 1 is a simple linear regression model with time as the independent variable, and model 2 is a multiple linear regression model using functions of time as the independent variables. You might think that model 2 is nonlinear because of the t^2 term. However, t^2 is just data in this model—the model is still linear in its parameters a, b, and c—and therefore, linear regression can still be used.

As we saw in Chapter 2, Excel has a built-in capability for fitting various regression models to data for forecasting. These can be found by creating a chart for the data, right-clicking the mouse on the data series, and choosing *Add Trendline*. This displays the *Add Trendline* dialog box shown in Figure 3.18. Six options for fitting data are available: five linear regression models and simple moving average. The regression models are based on minimizing the sum of squared errors, as discussed in Chapter 2, and the simple moving average is as discussed earlier in this chapter. The models used in each case are as follows (where x is the independent variable, e.g., time):

Linear	$y = ax + b$	
Logarithmic	$y = aLN(x) + b$	where $LN(x)$ is the natural log of x
Polynomial	$y = ax^2 + bx + c$	second order
	$y = ax^3 + bx^2 + cx + d$	third order
		etc....
Power	$y = ax^b$	
Exponential	$y = ae^{bx}$	where e is the base of natural logarithms, $e \approx 2.71828...$

While the power and exponential models are nonlinear in the parameters, they can be estimated using linear regression by transforming the data by taking logarithms; however, we will not describe the details of that here. You may consult an appropriate statistics textbook for the details.

The pictures in Figure 3.18 give an indication of the types of time series for which each method is appropriate. We have already discussed the moving average approach (note that this approach, unlike the others listed, does not result in a fitted equation) and the linear case is fairly straightforward. From Figure 3.18 you can see that the Logarithmic option might be most appropriate for situations exhibiting decreasing returns, that is, the relationship is one that is increasing, but at a decreasing rate. For example, the impact of investment in advertising on revenue typically follows this type of pattern. The Power and Exponential options are increasing, but at an increasing rate. For example, certain technology-based stock prices over time fall into this category. Finally, a Polynomial fit is more appropriate perhaps for less well-behaved data.

The options tab allows you to display the fitted equation and its R^2 directly on the graph (recall from Chapter 2 that R^2 is the percentage of the variation in the data explained by the regression equation).

FIGURE 3.18
Add Trendline
dialog box

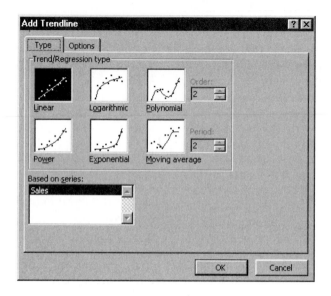

Example 10: Forecasting with the *Add Trendline* Tool

Rick's Barbeque has been in business for 15 years. Sales revenue has increased every year, as shown in the spreadsheet in Figure 3.19 (*Ricks.xls*). Rick notes that the fast growth of the early years is over. Nonetheless, he would like to develop a forecasting model to help predict sales revenue in the future.

By right-clicking on the data in the graph, we add a trend line to the data. From Figure 3.18, it is clear that the Logarithmic option is a natural choice for these data. We choose to display the equation and R^2 value from the *Options* tab in the *Add Trendline* dialog box. The results are shown in Figure 3.20. The fitted equation is $y = 24,846\ LN(x) + 14,954$, with an R^2 of .97. The forecasted revenue for year 16 is $24,846(LN[16]) + 14,954 = 24,846(2.772589) + 14,954 = \$83,841.739$, or $83,841.74.

FIGURE 3.19
Sales data for
Rick's
Barbeque

	A	B
1	Rick's Barbeque	
2	Year	Sales
3	1	$21,400
4	2	$33,400
5	3	$36,700
6	4	$44,500
7	5	$53,500
8	6	$58,800
9	7	$59,500
10	8	$64,000
11	9	$70,100
12	10	$74,000
13	11	$75,800
14	12	$77,200
15	13	$82,500
16	14	$82,700
17	15	$83,400

Rick's Barbeque chart: Sales ($) vs. Year, plotting points from $20,000 to $90,000 across years 0 to 20.

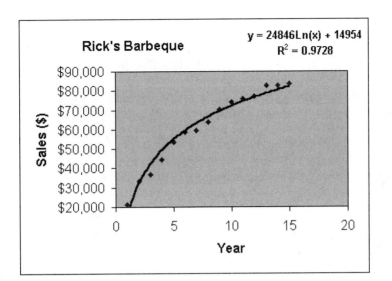

FIGURE 3.20
Logarithmic
trend line

Excel Exercise 3.6

Open the worksheet *Ricks.xls*. Right-click on the data in the graph and invoke the *Add Trendline* option. Fit separately the linear, power, exponential, and polynomial regressions, displaying the equation and R^2 for each. Which method's equation has the highest R^2 value?

Using Indicator Variables in Regression Models

The presence (or absence) of external factors can have an impact on the quantity we would like to forecast. For example, a sale can drastically increase the demand for goods. This type of phenomenon can be captured in a regression model using an indicator variable. Consider the following regression model:

$$DEMAND = c(SALE) + b(PERIOD) + a$$

where DEMAND is quantity of the good sold in time period PERIOD and SALE = 1 if a sale was held during the period, 0 if not. The quantity $b(PERIOD) + a$ is an estimate of the demand in the time period in the absence of a sale on the good. The parameter c is an estimate of the "lift" in demand due to a sale. When a sale is planned for PERIOD, the estimate for demand is $c(1) + b(PERIOD) + a = c + b(PERIOD) + a$.

Example 11: Forecasting with Indicator Variables

Nate's Components, Inc., a retail chain that sells computer components, has been incorporated for two years. Nate has kept data on the number of pairs of high-grade speakers sold per month over the last 24 months. He has noticed that demand for speakers seems to increase rather dramatically in months in which a faster chip is released. Figure 3.21 shows the demand data for pairs of speakers.

FIGURE 3.21
Data for Nate's Components regression model

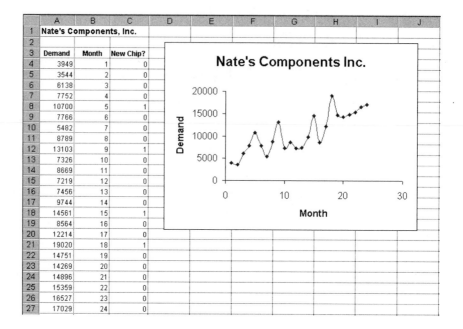

We would like to fit a model of the form $y = cz_t + bt + a$, where $z_t = 1$ if a new chip was introduced in month t, 0 if not. We invoke the Excel regression option by selecting *Tools*, *Data Analysis*, *Regression*, as discussed in Chapter 2. The independent variable is in column 1 (demand) with two independent variables in columns 2 and 3 (month and whether or not a new chip was released during that month). The regression results appear in Figure 3.22.

FIGURE 3.22 Regression output for Nate's Components

	A	B	C	D	E	F	G	H	I
1	SUMMARY OUTPUT								
2									
3	*Regression Statistics*								
4	Multiple R	0.944266829							
5	R Square	0.891639844							
6	Adjusted R Square	0.881319829							
7	Standard Error	1513.459395							
8	Observations	24							
9									
10	ANOVA								
11		*df*	*SS*	*MS*	*F*	*Significance F*			
12	Regression	2	395804463.8	197902231.9	86.39909	7.34735E-11			
13	Residual	21	48101746.12	2290559.339					
14	Total	23	443906210						
15									
16		*Coefficients*	*Standard Error*	*t Stat*	*P-value*	*Lower 95%*	*Upper 95%*	*Lower 95.0%*	*Upper 95.0%*
17	Intercept	3097.183006	658.7938872	4.701292872	0.000122	1727.14588	4467.220132	1727.14588	4467.220132
18	Month	535.5705134	44.68196631	11.98627898	7.44E-11	442.6492615	628.4917652	442.6492615	628.4917652
19	New Chip?	4955.863462	829.9306893	5.971418488	6.31E-06	3229.927811	6681.799113	3229.927811	6681.799113

The estimated model of $y = 4{,}955.86z_t + 535.57t + 3{,}097.18$ has an adjusted R^2 of 0.88. Hence, based on past data, Nate can expect demand to increase by about 536, with an increase in chip release months of about 4,956.

We have seen eight time series methods for forecasting, along with how regression may be used for external factors. Shortly, we will discuss how to choose among these methods for a given data series.

FORECASTING TECHNOLOGY[1]

Forecasting software may be classified into three groups:

1. "Automatic" software, in which the user enters or imports data and the program automatically performs various diagnostic tests and recommends the best approach. The user may accept the recommendation, or override it and select an approach. The software then finds the optimal parameters, makes forecasts, and provides statistics.
2. "Semiautomatic" software, in which the program does not recommend a procedure, but will find the optimal parameters and provide forecasts and statistics.
3. "Manual" software, in which the user must specify both the method and the parameters, usually requiring many runs to find the best set. Few products today, however, fall into this group.

Most general statistical packages include a forecasting module. A sampling of some of the commercially available packages is given in Table 3.2.

Product	Company	Description
CB Predictor	Decisioneering Inc. Denver, Colo. **www.decisioneering.com**	Spreadsheet add-on for demand forecasting including eight time-series methods and regression
Demand Forecasting	Distinction Software, Inc. Atlanta, Ga. **www.distinction.com**	Supports forecast development at multiple levels of summarization based on actual sales history
Eviews	Quantitative Micro Software Irvine, Calif. **www.eviews.com**	Windows interface with menus, dialogs, cut-and-paste, etc.
Forecaster 2000	American Software Atlanta, Ga. **www.amsoftware.com**	Calculates multi-level forecasts with multiple forecast views
Inventory Analyst Pro	User Solutions, Inc. South Lyon, Mich. **www.usersol.com**	Spreadsheet add-on for demand forecasting
MINITAB	Minitab, Inc. State College, Pa. **www.minitab.com**	General statistics package with a great deal of functionality devoted to business/forecasting applications
PEER Planning for Windows	DELPHUS, Inc. Morristown, N.J. **www.delphus.com**	Suited for large-scale, multi-level sales forecasting
SmartForecasts for Windows	Smart Software Belmont, Mass. **www.smartcorp.com**	Contains an automatic forecasting system that produces forecasts of sales, product demand, and inventory levels

TABLE 3.2

Sample of commercial forecasting software

CHOOSING THE BEST FORECASTING METHOD

Earlier in this chapter we discussed three measures of goodness of fit: mean absolute deviation (MAD), root mean squared error (RMSE), and mean absolute percentage error (MAPE). As we have seen, these measures may be used to decide on parameter values for a given method. They may also be used to choose between forecasting methods for a given time series. One procedure to decide on a forecasting method is as follows:

1. Choose, for each method under consideration, the best parameter values based on a goodness-of-fit measure.
2. Select the method with the best goodness-of-fit measure used in step 1.

We will not present an example of this approach, as it is quite straightforward, but encourage you to do the following exercise.

Excel Exercise 3.7

Consider the Atlas cleaners sweater data used in Examples 7 and 8 (found in the worksheets *Atlas Sweaters Additive Seasonal.xls* and *Atlas Sweaters Multiplicative Seasonal.xls*). Use two-way data tables to find the best values of α and γ based on MAPE. Which method is best based on MAPE?

You might suspect that some forecasting software exists that will accomplish this task automatically. Indeed, CB Predictor, a companion product to Crystal Ball, has this capability. The student version of CB Predictor is included with this text. It includes the eight time series approaches previously discussed. The student version has all of the power of the professional version, with the exception that it has no regression option. This spreadsheet add-in falls into the category of automatic software, that is, it finds the best parameter values and method for a given set of data.

FORECASTING WITH CB PREDICTOR

CB Predictor is a product of Decisioneering, Inc. It is part of the Crystal Ball Pro suite of Excel add-ins. It is an easy-to-use forecasting add-in for Excel. Once loaded into Excel as an add-in, CB Predictor is started by selecting CB Predictor from the *Tools* menu. This opens up the CB Predictor dialog box, which contains four tabs:

- *Input Data*—allows you to enter the spreadsheet range where the data to be analyzed reside, specify if the data are in rows or columns, specify if row and/or column headers exist, and view a graph of the data.
- *Data Attributes*—allows you to specify the type of data and its seasonality.
- *Methods Gallery*—allows you to choose one of eight available forecasting techniques depending on whether or not your data are seasonal and whether or not they exhibit a trend. The gallery is shown in Figure 3.23. By double-clicking on a given method, you may set the values of any parameters, or you may choose to

have the values chosen automatically, according to a prespecified fit criterion. The fit criterion is selected by clicking on the *Advanced* button in the *Methods Gallery* (you may choose between MAD, RMSE, and MAPE). The *Select All* option in the *Methods Gallery* will find the best method based on the selected fit criterion.

- *Results*—allows you to specify the number of periods to forecast, confidence interval level, and a variety of report options, including result charts and tables. The confidence interval defines with some probability the range above and below the forecasted value. For example, a confidence interval of 5 percent and 95 percent gives two points for each forecast, where the lower value is the 5th percentile and the higher point the 95th percentile.

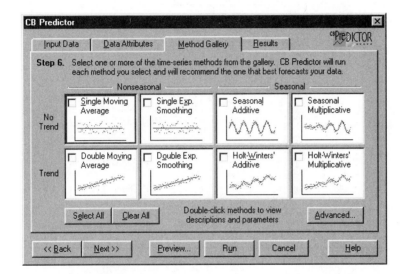

FIGURE 3.23
CD Predictor
Methods Gallery

Example 12: Using CB Predictor for the Atlas Shirt Data

We illustrate here the use of CB Predictor with user-specified method and parameters. Recall the Atlas shirt data introduced in Example 3 (see Figure 3.6). Suppose we would like to forecast for period 11 using *Simple Moving Average* with 3 periods. We follow the following steps:

1. Select the data (cells A3 through B13) and then choose CB Predictor from the *Tools* menu. From the *Input Data* tab, specify that the data are in columns, that the first column has dates, and that the first row has headers. (Note: rather than first selecting the data locations before invoking CB Predictor, you may type the location of the data in the range box.)
2. Click the *Data Attributes* tab and specify that the data are in weeks and that there is no seasonality.
3. Click the *Methods Gallery* tab. Select *Single Moving Average* and then double-click. Select *User Defined Parameters* and set periods to 3. Click *OK*.
4. Click the *Advanced* button and choose *MAD* as the fit criterion.
5. Click the *Results* tab, set number of periods to forecast to 1, and select *none* for *Select a confidence interval*. Select *Report* and *Results Table* as output options.
6. Finally, click the *Run* button at the bottom of the dialog box.

The *Results Table* and *Report* are shown in Figures 3.24 and 3.25. Note that the *Results Table* displays the same results we found in Example 3 (compare Figures 3.7 and 3.24). The *Report Table* shows information on the method used and indicates that the Error = 10.48 (this is the MAD measure we chose), and the forecast for period 11 is 1,061.67. Both of these agree with the results in Example 3.

FIGURE 3.24
CB Predictor results table for Atlas Shirt data

	A	B	C
1	Results Table for ATLAS		
2	Created: 3/25/99 at 8:09:55 AM		
3			
16	Series	# Shirts	
17			
18		Data	
19	Date	Historical Data	Fit & Forecast
20	1	1071	
21	2	1070	
22	3	1054	
23	4	1075	1065
24	5	1055	1066.333333
25	6	1065	1061.333333
26	7	1054	1065
27	8	1074	1058
28	9	1049	1064.333333
29	10	1062	1059
30	11		1061.666667

FIGURE 3.25
CB Predictor report for Atlas Shirt data

(a)

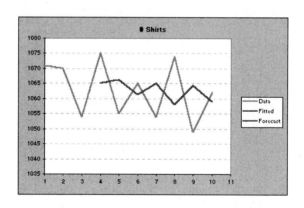

Method Errors:

Method	RMSE	MAD	MAPE
Best: Single Moving Average	11.103	10.048	0.95%

Method Statistics:

Method	Durbin-Watson	Theil's U
Best: Single Moving Average	3.409	0.649

Method Parameters:

Method	Parameter	Value
Best: Single Moving Average	Periods*	3

* = user defined

FIGURE 3.25 CB Predictor report for Atlas Shirt data, continued

(b)

Report for ATLAS
Created: 3/25/99 at 8:09:51 AM

Summary:
    Number of series: 1
    Periods to forecast: 1
    Seasonality: none
    Error Measure: MAD

Series: # Shirts

    Method: Single Moving Average
    Parameters: (user defined)
        Periods: 3
    Error: 10.048

Forecast:

Range: B4:B13

	Series: # Shirts				
Date	Data	Fitted	Forecast	Upper 100%	Lower 0%
1	1071				
2	1070				
3	1054				
4	1075	1065			
5	1055	1066.333333			
6	1065	1061.333333			
7	1054	1065			
8	1074	1058			
9	1049	1064.333333			
10	1062	1059			
11			1061.666667		

Date	Forecast
11	1061.666667

CB Predictor Exercise

Open the worksheet *Atlas.xls*. Use CB Predictor to forecast the number of shirts for period 11, using *Single Exponential Smoothing*, with $\alpha = .7$. Check your results with those given in Example 4.

As previously mentioned, we may also use CB Predictor to automatically find the best method and parameter values for a given data set, based on a chosen measure of fit.

Example 13: Finding the Best Forecasting Method Using CB Predictor

Let us consider again the Atlas sweater data found in *Atlas Sweater.xls*, as shown in Figure 3.12. We will show how to use CB Predictor to find the best forecasting method. First select the data, including dates and row headers, start CB Predictor, and indicate that the first column is dates and includes row headers. From the *Data Attributes* tab, specify that the data are in quarters and seasonal over 4 periods. From the *Methods Gallery*, choose *Select All*, and click the *Advanced* button to set the measure of fit to MAD. From the *Results* tab, specify the forecast for 4 periods, a confidence interval of 5 percent and 95 percent, and a title of "Atlas Sweater Data." Then choose *Report, Charts, Results Table*, and *Methods Table*. Next, select *Run* to generate the outputs. The *Report, Results Table*, and *Methods Table* are shown in Figures 3.26 through 3.28. From Figure 3.26 we see that the Holt-Winters Additive model provides the best MAD. We discuss these outputs more fully in the next section.

FIGURE 3.26
CB Predictor
report for Atlas
Sweater data

(a)

Report for ATLAS Sweaters
Created: 4/16/99 at 10:08:15 AM

Summary:
Number of series: 1
Periods to forecast: 4
Seasonality: 4 quarters
Error Measure: MAD

Series: # Sweaters   **Range: B9:B28**

Method: Holt-Winters' Additive
Parameters:
   Alpha: 0.489
   Beta: 0.482
   Gamma: 0.999
Error: 10.155

Forecast:

Date	Lower 5%	Forecast	Upper 95%
21	286.768331	310.007054	333.245778
22	455.148008	479.677772	504.207535
23	287.704401	313.677091	339.649782
24	171.626696	199.22268	226.818664

(b)

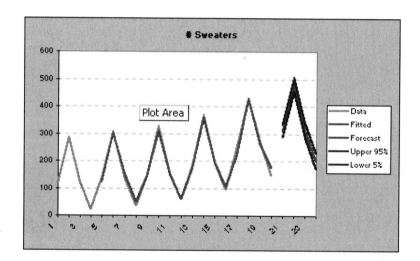

Method Errors:

	Method	RMSE	MAD	MAPE
Best:	Holt-Winters' Additive	13.421	10.155	7.36%
2nd:	Seasonal Additive	18.615	14.072	6.80%
3rd:	Holt-Winters' Multiplicative	22.087	16.852	12.20%
4th:	Seasonal Multiplicative	25.644	20.003	14.78%
5th:	Single Moving Average	106.93	79.609	59.68%
6th:	Double Moving Average	109.33	85.471	75.57%
7th:	Single Exponential Smoothi	118.76	91.423	81.24%
8th:	Double Exponential Smooth	185.88	148.63	193.51%

Method Statistics:

	Method	Durbin-Watson	Theil's U
Best:	Holt-Winters' Additive	1.957	0.082
2nd:	Seasonal Additive	1.125	0.139
3rd:	Holt-Winters' Multiplicative	0.802	0.132
4th:	Seasonal Multiplicative	0.68	0.148
5th:	Single Moving Average	1.865	0.513
6th:	Double Moving Average	1.765	0.434
7th:	Single Exponential Smoothi	1.714	0.465
8th:	Double Exponential Smooth	1.121	1.24

Method Parameters:

	Method	Parameter	Value
Best:	Holt-Winters' Additive	Alpha	0.489
		Beta	0.482
		Gamma	0.999
2nd:	Seasonal Additive	Alpha	0.682
		Gamma	0.999
3rd:	Holt-Winters' Multiplicative	Alpha	0.011
		Beta	0.999
		Gamma	0.999
4th:	Seasonal Multiplicative	Alpha	0.159
		Gamma	0.999
5th:	Single Moving Average	Periods	4
6th:	Double Moving Average	Periods	4
7th:	Single Exponential Smoothi	Alpha	0.11
8th:	Double Exponential Smooth	Alpha	0.273
		Beta	0.999

Series: # Sweaters					
Date	Data	Fitted	Forecast	Upper 95%	Lower 5%
1	120				
2	284				
3	123				
4	22				
5	136	124.125			
6	306	300.8608359			
7	138	150.5119058			
8	35	48.57534184			
9	147	147.9756051			
10	327	309.8181154			
11	155	154.0289888			
12	59	58.99727258			
13	183	175.5198146			
14	370	356.8064719			
15	197	195.8995164			
16	95	105.5711547			
17	254	223.3746455			
18	427	427.0000023			
19	270	258.4548686			
20	149	174.733494			
21			310.0070544	333.2457777	286.768331
22			479.6777718	504.2075353	455.1480083
23			313.6770914	339.6497822	287.7044006
24			199.2226796	226.8186635	171.6266956

FIGURE 3.26
CB Predictor
report for Atlas
Sweater data,
continued

(c)

(d)

FIGURE 3.27
CB Predictor
results table for
Atlas Sweater
data

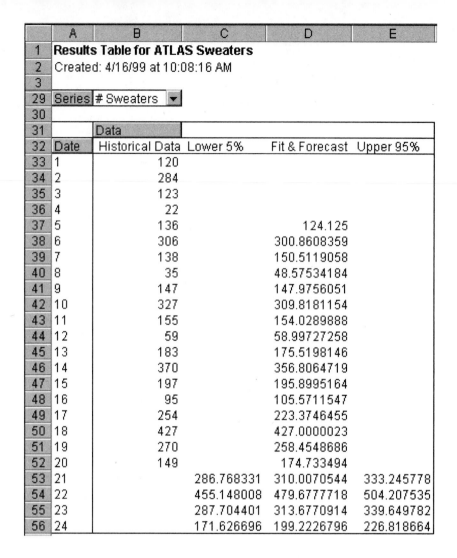

	A	B	C	D	E
1	**Results Table for ATLAS Sweaters**				
2	Created: 4/16/99 at 10:08:16 AM				
3					
29	Series	# Sweaters ▼			
30					
31		Data			
32	Date	Historical Data	Lower 5%	Fit & Forecast	Upper 95%
33	1	120			
34	2	284			
35	3	123			
36	4	22			
37	5	136		124.125	
38	6	306		300.8608359	
39	7	138		150.5119058	
40	8	35		48.57534184	
41	9	147		147.9756051	
42	10	327		309.8181154	
43	11	155		154.0289888	
44	12	59		58.99727258	
45	13	183		175.5198146	
46	14	370		356.8064719	
47	15	197		195.8995164	
48	16	95		105.5711547	
49	17	254		223.3746455	
50	18	427		427.0000023	
51	19	270		258.4548686	
52	20	149		174.733494	
53	21		286.768331	310.0070544	333.245778
54	22		455.148008	479.6777718	504.207535
55	23		287.704401	313.6770914	339.649782
56	24		171.626696	199.2226796	226.818664

FIGURE 3.28 CB Predictor methods table for Atlas Sweater data

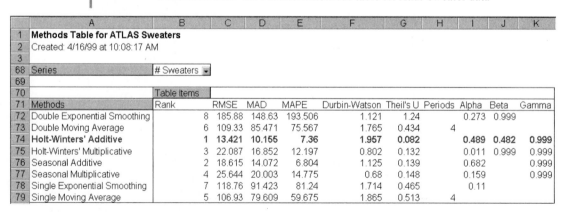

	A	B	C	D	E	F	G	H	I	J	K
1	**Methods Table for ATLAS Sweaters**										
2	Created: 4/16/99 at 10:08:17 AM										
3											
68	Series	# Sweaters ▼									
69											
70		Table Items									
71	Methods	Rank	RMSE	MAD	MAPE	Durbin-Watson	Theil's U	Periods	Alpha	Beta	Gamma
72	Double Exponential Smoothing	8	185.88	148.63	193.506	1.121	1.24		0.273	0.999	
73	Double Moving Average	6	109.33	85.471	75.567	1.765	0.434	4			
74	**Holt-Winters' Additive**	**1**	**13.421**	**10.155**	**7.36**	**1.957**	**0.082**		**0.489**	**0.482**	**0.999**
75	Holt-Winters' Multiplicative	3	22.087	16.852	12.197	0.802	0.132		0.011	0.999	0.999
76	Seasonal Additive	2	18.615	14.072	6.804	1.125	0.139		0.682		0.999
77	Seasonal Multiplicative	4	25.644	20.003	14.775	0.68	0.148		0.159		0.999
78	Single Exponential Smoothing	7	118.76	91.423	81.24	1.714	0.465		0.11		
79	Single Moving Average	5	106.93	79.609	59.675	1.865	0.513	4			

Interpreting CB Predictor Outputs

CB Predictor generates three basic output reports: *Report*, *Results Table*, and *Methods Table*. The latter two are actually pivot tables of results. The *Report* output for Example 13 is shown in Figure 3.26. The vast majority of the first section of the report is self-explanatory. Note that from the *Method Errors* section, the Holt-Winters Additive model is best based on MAD (note that of the methods tested, it is also best based on RMSE, but Seasonal Additive is actually best based on MAPE). The *Methods Statistics* section gives two statistics for each method: the Durban-Watson statistic and Theil's U statistic.

The Durban-Watson (DW) statistic tests autocorrelation in the data, that is, the degree to which a previous time series data point is correlated with the next. The DW statistic ranges from 0 to 4. In general, a DW value less than 1 means the values are positively correlated, values close to 2 mean no correlation, and values greater than 3 mean there is negative correlation.

Theil's U statistic measures how well a method does compared to a naïve forecast (for example a naïve approach would be to use the last period's value to be the estimate for the next period). A Theil's U of less than 1 means essentially that the method is better than a naïve forecast. A value close to 1 implies that the method is about the same as a naïve forecast. A value more than 1 means that the method is actually worse than a naïve forecast. Notice in Figure 3.26 that the Theil's U is fairly aligned with the rankings based on MAD. Finally, the last section of the *Report* contains the best parameter values for each method based on MAD.

The *Results Table* output as shown in Figure 3.27 shows the fitted and forecasts values, along with the confidence interval points for the forecasted value. These data are also shown on the chart (although it is difficult to appreciate in black and white). The *Results Table* also gives the fitted values, along with confidence interval values. The *Methods Table* output, as shown in Figure 3.28, gives a subset of the method information found in the *Report* output.

APPLICATIONS OF FORECASTING TO MODEL BUILDING

Forecasts and, in general, functional forms derived from data are often used as inputs to decision models. For example, demand forecasts are needed for scheduling and resource allocation models, functions estimating manufacturing yields and increased productivity based on learning are needed for scheduling, and exchange and interest rate forecasts are needed for financial modeling. The techniques discussed in this chapter are the methodologies used to provide these important inputs to the models we will discuss in the remainder of this book.

Excel contains several *statistical functions* that perform some of the techniques discussed in this chapter (search for help on statistical functions to see a listing and explanation of these). As functions, they can be incorporated directly into larger optimization and simulation models. One of these functions is the TREND function. The form for this function is

$$=\text{TREND(known\_y's, known\_x's, new\_x's, const)}$$

where

known_y's is the set of *y* values you already know in the relationship $y = mx + b$.

known_x's is an optional set of *x* values that you already know in the relationship $y = mx + b$.

new_x's are new *x* values for which you want TREND to return corresponding *y* values.

const is a logical value specifying whether to force the constant *b* to equal 0. If *const* is TRUE or omitted, *b* is calculated normally. If *const* is FALSE, *b* is set equal to 0 (zero), and the *m*-values are adjusted so that $y = mx$.

We close this chapter with an example of how the TREND function may be used directly in an optimization model. Optimization models, which we introduced in Chapter 1, are the focus of the next two chapters (Chapters 4 and 5). Many of these types of models deal with planning for the future. Good forecasts are needed as inputs to these models.

Example 14: Using a Forecasting Function in an Optimization Model

Optcontrol, Inc., produces scanning devices used in courtrooms and airports to detect metal weapons. The manager is currently developing a production schedule for next year, and would like to determine the optimal production quantities and inventory levels for each quarter. Production in the current quarter will be 205 units, with 10 units of inventory. The cost to hold inventory for one quarter is $45 per unit. The production cost per unit will vary over the year and is projected to be $590, $525, $505, and $575 over the four quarters. Finally, sales for the next four quarters (quarters 9 through 12) must be estimated from the last two years' data. These are shown in Figure 3.29. Since the data appear linear, linear regression may be used to predict the sales volumes for the next four quarters.

FIGURE 3.29 Historical data and trend forecasts for Optcontrol, Inc.

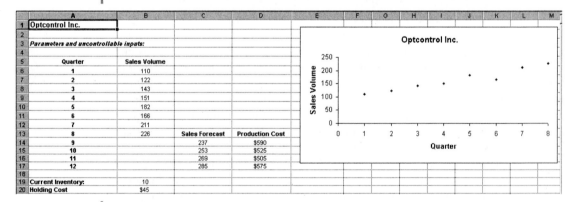

To develop a model for finding an optimal production plan, we first define the decision variables. Let

$$P_t = \text{number of units to produce in quarter } t, \, t = 9, 10, 11, 12$$

$$I_t = \text{number of units to hold in inventory in quarter } t, \, t = 9, 10, 11, 12$$

We seek to minimize the total cost of holding inventory and producing the product:

$$\text{Minimize } \{45I_9 + 45I_{10} + 45I_{11} + 5I_{12} + 590P_9 + 525P_{10} + 505P_{11} + 575P_{12}\}$$

To ensure that forecasted demand will be met, we must satisfy the following equation for each quarter:

$$\text{Inventory held from prior quarter} + \text{production} - \text{sales}$$
$$= \text{inventory held to next quarter}$$

Using the variables defined above and letting s_9, s_{10}, s_{11}, and s_{12} be the forecasted sales for quarters 9 through 12, the constraints of the model are

$$10 + P_9 - s_9 = I_9$$
$$I_9 + P_{10} - s_{10} = I_{10}$$
$$I_{10} + P_{11} - s_{11} = I_{11}$$
$$I_{11} + P_{12} - s_{12} = I_{12}$$

In addition, we must ensure that all variables cannot be negative:

$$I_9, I_{10}, I_{11}, I_{12}, P_9, P_{10}, P_{11}, P_{12} \geq 0$$

An Excel spreadsheet model for implementing this combined forecasting and optimization model is shown in Figure 3.30 and is available in the worksheet *Optcontrol.xls*. The TREND function is used in cells C14 through C17 to forecast sales for the next four quarters. We also use the ROUND function to round the forecast to the nearest integer. For example, cell C14 is

= ROUND(TREND($B$6:$B$13,$A$6:$A$13,A14),0)

By using Excel Solver (which will be discussed fully in the next chapter), we may find the cost-minimizing plan shown in Figure 3.30. In quarters 9 and 10 we should produce the forecasted sales volume and hold no inventory. In quarter 11 we produce the combined forecasted sales for quarters 11 and 12, holding quarter 12 sales in inventory for one quarter. This plan costs $559,350: $546,525 in production cost and $12,825 in holding cost. The plan is intuitive in that production cost is lower in quarters 10 and 11.

	A	B	C	D	E
1	**Optcontrol Inc.**		Key cell formulas		
2					
3	***Parameters and uncontrollable inputs:***				
4					
5	**Quarter**	**Sales Volume**			
6	1	110			
7	2	122			
8	3	143			
9	4	151			
10	5	182			
11	6	166			
12	7	211			
13	8	226	**Sales Forecast**	**Production Cost**	
14	9		237	$590	
15	10		253	$525	
16	11		269	$505	
17	12		285	$575	
18					
19	**Current Inventory:**	10			
20	**Holding Cost**	$45			
21					
22	***Model***				
23					
24	**Quarter**	**Production**	**Inventory**	**Production Cost**	**Inventory Cost**
25	9	227	0	$133,930	$0
26	10	253	0	$132,825	$0
27	11	554	285	$279,770	$12,825
28	12	0	0	$0	$0
29				$546,525	$12,825
30					
31				**Total cost**	$559,350

FIGURE 3.30 Spreadsheet for Optcontrol production and inventory planning model

PROBLEMS

1. Open the worksheet *Components.xls*. This file contains four columns of data (Month, Trend, Seasonality, and Irregular). These are the components of an actual data set on a monthly gas and electric bill. Month 1 is January of the first year.
   a. Display graphically each of the components over time.
   b. Construct a data set of the form $Y_t = T_t \times I_t$. That is, create a data set by multiplying columns B and D together, and display this time series graphically. Which time series methods do you think will provide the best fit to this data?
   c. Construct a data set of the form $Y_t = T_t \times S_t$ (columns B and C). Display this time series graphically. Which time series methods do you think will provide the best fit to this data?
   d. Construct a data set of the form $Y_t = S_t \times I_t$ (columns C and D). Display this time series graphically. Which time series methods do you think will provide the best fit to this data?
   e. Finally, construct a data set of the form $Y_t = T_t \times S_t \times I_t$. Display this time series graphically. Save this data in a worksheet titled *G&E.xls*. Which time series methods do you think will provide the best fit to this data?

2. Open the worksheet *Quarterly Sales.xls*. This worksheet contains data for 12 quarters. Apply the simple moving average method with four periods to forecast sales for the next quarter.

3. Consider the sales data from Problem 2.
   a. Forecast next month's sales using an exponential smoothing model with $\alpha = .2$
   b. Forecast next month's sales using an exponential smoothing model with $\alpha = .3$
   c. Compare the fit of the models used in parts a and b using MAD. Which appears to fit the data better?

4. Use CB Predictor to find the method that gives the best fit to the data in Problem 2 based on MAD. Be sure to specify in the *Data Attributes* tab that the data are in quarters. What is the best method based on MAPE? RMSE?

5. Review the data in *Quarterly Sales.xls*. From the graph, it appears from the pattern of the data that a "season" might involve *six* quarters. Use CB Predictor with the specification of quarterly data with seasonality of 6 quarters in the *Data Attributes* tab. How do these results compare to those of Problem 4?

6. Alan is a marketing manager. He is about to make a presentation to top management concerning the need for corrective action in production planning because of the increase in the number of stockouts. Data over the last 12 months appear in the worksheet *Stockout.xls*. Forecast the number of stockouts for the next three months, using the *Add Trendline* option on the graph.

7. Consider the stock price data in the worksheet *Monthly Stock.xls*. The stock price given is the price of the stock at the end of each month.
   a. Forecast the future stock price using one of the methods discussed in this chapter for CB Predictor. Justify using the method you choose.
   b. Assuming you've bought the stock at the last price in the data set, how long will you have to hold the stock to realize a return of 10 percent?

8. Open the worksheet *Cycle.xls*. This file contains the data discussed in Example 2 from this chapter. Choose one of the methods discussed in this chapter to forecast the next quarter (21), using CB Predictor. Justify your choice of method.

9. Jennifer's Doll Shop specializes in handmade toy dolls. The worksheet *Jennifer.xls* contains five years of quarterly sales volume. Choose one of the methods discussed in this chapter to forecast the next quarter (21) using CB Predictor. Justify your choice of method.

10. Use CB Predictor to find the method providing the best MAD fit to the data generated in Problem 1(e) (gas and electric bill by month over 5 years). Is the best method the one you suggested in Problem 1(e)? Forecast the gas and electric bill for the next year.

11. Refer to the data in *Cycle.xls* (Problem 8).
    a.   Use CB Predictor to find the method providing the best RMSE fit for the data in *Cycle.xls*.
    b.   Use *Add Trendline* to fit a linear regression trend line to these data. Calculate the RMSE for the fit based on this trendline.
    c.   How does this fit compare to that found by CB Predictor? How much do the forecasts for the next period differ between these the two methods?

12. For the monthly stock price data from Problem 7 (worksheet *Monthly Stock.xls*):
    a.   Use CB Predictor to find the best method based on MAPE. What are the best methods based on MAD and RMSE?
    b.   Forecast the stock price for the next six months based on the best methods and compare.

13. Use CB Predictor to find the best method based on MAD for the toy doll data from Problem 9 (worksheet *Jennifer.xls*). Forecast the volume for the next two years with 5 percent and 95 percent confidence endpoints.

14. Many labor-intensive production processes behave according to a learning curve. Learning curve theory states that the time to produce a unit is a function of the unit number (that is, time to produce a unit of production declines as total cumulative production volume increases). The worksheet *Learning Curve Data.xls* contains labor hours per unit for the first 28 units of production of a product. Right-click on the data in the graph and use the *Add Trendline* option to find a good fit for these data (display R^2 as your measure of goodness-of-fit). Forecast the labor hours to produce the units 29 through 35.

15. Bullet Fresh Pasta has been in operation for 10 years. The annual revenue for the last 10 years is in the worksheet *Pasta.xls*. Use the *Add Trendline* option by right-clicking on the data in the graph. Select an appropriate regression type for these data, display the fitted equation, the R^2, and forecasts for the next three years.

16. Right-click on the data in the chart in the worksheet *Quarterly Sales.xls* (Problem 2) and choose *Add Trendline*. Fit a polynomial of order 4 to these data (be sure to select *Show equation* and *Show R^2* options).
    a.   Give the fitted equation and R^2 for this 4th order polynomial. From the graph, does the fit seem reasonable?
    b.   Again, right-click on the data and, using the same choices as the fit in part (a), forecast 4 periods out. Does this polynomial approach still seem reasonable as a forecast?

17. Schneider's Sweets specializes in homemade candy and ice cream. Sales have steadily increased over the last few years. The worksheet *Schneider.xls* contains sales data for the last 29 months. In addition to the steady increase in sales, it is also apparent that sales are related to two other factors: Christmas and Easter. Christmas always occurs during December, so that December sales always show an extraordinary increase. Easter can occur in March or April depending on the year. The month in which Easter occurs also shows an extraordinary increase.
    a.   Use the Excel Regression tool to fit the following model:

$$\text{SALES}_t = a + b\,t + c\,\text{XMAS}_t + d\,\text{EASTER}_t$$

where

$$\text{SALES}_t = \text{sales (\$000) in month } t$$

$$\text{XMAS}_t = 1 \text{ if month } t \text{ is December, 0 if not}$$

$$\text{EASTER}_t = 1 \text{ if Easter occurs in month } t, \text{ 0 if not}$$

b. Forecast sales for periods 30 through 36 and plot these with the historical data.

18. For staff planning, hospitals must attempt to forecast their emergency room demand. The worksheet *Emergency Room.xls* contains a sample of 20 observations for the night shift of a hospital during the summer months. It is believed that the average demand is increased during a full moon and on Friday and Saturday nights. These factors are indicated by 0–1 variables in the data set. For example, the first observation corresponds to a night that is neither a Friday nor a Saturday, but does have a full moon.

a. Use the Excel Regression tool to fit the following model:

$$\text{DEMAND} = a + b(\text{FRIDAY}) + c(\text{SATURDAY}) + d(\text{FULL})$$

where

$$\text{DEMAND} = \text{number of patients}$$

$$\text{FRIDAY} = 1 \text{ if Friday, 0 if not}$$

$$\text{SATURDAY} = 1 \text{ if Saturday, 0 if not}$$

$$\text{FULL} = 1 \text{ if there is a full moon, 0 if not}$$

Give the estimated values of a, b, c, d, and the R^2 for the model.

b. Forecast the demand for a summer Friday night with a full moon.

19. As we have seen, seasonal data can be forecasted using additive and multiplicative time series models. Another approach to dealing with seasonality is to use indicator variables in a linear regression. Consider the toy doll data from Problem 13 (worksheet *Jennifer.xls*). We may use two indicator variables to define the quarters in a regression model (and hence capture the effect of the quarters on sales volume). Define these to variables as follows:

$$X1_t = 1 \text{ if } t = 1, 5, 9, 13, 17 \qquad 0 \text{ otherwise}$$

$$X2_t = 1 \text{ if } t = 2, 6, 10, 14, 18 \qquad 0 \text{ otherwise}$$

$$X3_t = 1 \text{ if } t = 3, 7, 11, 15, 19 \qquad 0 \text{ otherwise}$$

Let

$$Y_t = \text{the sales volume of dolls in quarter } t$$

Use regression to fit the following model:

$$Y_t = a + bt + cX1_t + dX2_t + eX3_t$$

In this model, c, d and e are estimated to adjust for the quarter effect (note that the fourth quarter each year is captured by the terms $a + bt$ (all x variables are zero for the fourth quarter). This indicator information is contained in the worksheet *Jennifert01.xls*. Estimate the regression model using the Excel Regression tool. Compare your results to those found in Problems 9 and 13.

20. Return to the Bullet Fresh Pasta data from Problem 15 (file *Pastsa.xls*). Use CB Predictor to find the best method based on RMSE. Forecast for the next three years. Compare these results to those you found in Problem 15. What method do you recommend for forecasting?

NOTE

1. Adapted from "Forecasting That Fits," and accompanying software survey, *OR/MS Today*, February 1998, 42–55.

BIBLIOGRAPHY

Andrews, B. H., and S. M. Cunningham. 1995. "L. L. Bean Improves Call-Center Forecasting." *Interfaces*, Vol. 25, No. 6, November–December, pp. 1–13.

Clements, D. W., and R. A. Reid. 1994. "Analytical MS/OR Tools Applied to a Plant Closure." *Interfaces*, Vol. 24, No. 2, March–April, pp. 1–12.

Cohen, M., P. V. Kamesam, H. Lee, and A. Tekerian. 1990. "Optimizer: IBM's Multi-Echelon Inventory System for Managing Service Logistics." *Interfaces*, Vol. 20, No. 1, January–February, pp. 65–82.

Makridakis, S., S. C. Wheelwright, and R. J. Hyndman. 1998. *Forecasting: Methods and Applications*, 3d ed. New York: John Wiley and Sons.

Linear and Multiobjective Optimization Models

O U T L I N E

The Structure of Linear and Multiobjective Optimization Models

Applications of Linear and Multiobjective Programming Models
Land Management at the National Forest Service
Credit Collection
Research and Development Funding
Site Selection

Modeling Optimization Problems in Excel

Building Linear Programming Models
Dimensionality Checks
LP Modeling Examples
Standard Form
Management Science Practice: Using Transportation Models for Procter & Gamble's Product Sourcing Decisions

Solving Linear Programming Models
Intuition and LP Solutions
When Intuition Can Fail
Using Excel Solver

Interpreting Solver Reports and Sensitivity Analysis
Answer and Limit Reports
Sensitivity Report
Scenario Analysis

Solving Multiobjective Models
Weighted Objective Approach
Absolute Priorities Approach
Goal-Programming Approach
Management Science Practice: Goal Programming for Site Location at Truck Transport Corporation

Technology For Linear Optimization

Problems

Appendix: Using Premium Solver to Solve Linear Programs

$\mathbf{A}$ll organizations are faced at one time or another with making decisions involving limited resources or ensuring that certain requirements must be satisfied. We often formulate these decision problems as *optimization models* that seek to maximize or minimize some objective function while satisfying a set of constraints. An important category of optimization models is called **mathematical programming**. In a mathematical programming model, constraints are expressed mathematically using inequalities ($\leq$ or $\geq$) and equalities ($=$). The term "programming" was coined because these models find the best "program," or course of action, to follow. We will deal with four important types of mathematical programming models:

1. *Linear models*—those in which all constraints are linear functions of the decision variables, all of which can assume any continuous value;
2. *Multiobjective models*—linear models that have more than one objective to meet or conflicting goals that must be resolved;
3. *Integer models*—linear models for which some or all of the decision variables must have integer (whole number) values for an optimal solution to be realistic; and
4. *Nonlinear models*—those for which the objective function and/or constraint functions are not linear.

In this chapter we address exclusively linear and multiobjective models; in the next chapter we will discuss integer and nonlinear models. Our focus will be on understanding how to formulate mathematical programming models for practical problems, implement them on spreadsheets, solve them within Excel, and interpret the results.

THE STRUCTURE OF LINEAR AND MULTIOBJECTIVE OPTIMIZATION MODELS

We have already seen an example of a linear programming model in Chapter 1 (Example 2). Recall that the Santa Cruz MicroProducts example concerned the question of how many LaserStop and SpeedBuster radar detectors to make in order to maximize profit. We defined the decision variables of the model as LS = the number of units of LaserStop to produce, and SB = the number of units of SpeedBuster to produce. Each LaserStop yields a profit of \$24, and each SpeedBuster gives a profit of \$40. However, profit maximization is restricted by the fact that there are limited quantities of the two components that both products require. Only 6,000 units of component A and 3,500 units of component B are available. Each LaserStop requires 12 units of component A and six units of component B. Each SpeedBuster requires 12 units of component A and 10 units of B. The model we developed in Chapter 1 is:

Maximize $24LS + 40SB$

subject to

$$12LS + 12SB \leq 6,000 \qquad \text{(Component A limitation)}$$

$$6LS + 10SB \leq 3,500 \qquad \text{(Component B limitation)}$$

$$LS, SB \geq 0 \qquad \text{(non-negativity)}$$

We refer to the unit profits as *objective function coefficients*, the per-unit requirements for the limited components of each product as *constraint coefficients*, and the available quantities of components A and B as the *right-hand side values*. The left-hand sides of the constraints are called *constraint functions*. Thus, for the component A limitation, $12LS + 12SB$ is the constraint function with constraint coefficients of 12 and 12, and 6,000 is the right-hand side value.

This type of model is called a **linear programming (LP) model**. To be a *linear* program, the following three conditions must be met:

1. The objective and constraints must be represented using *linear functions* of the decision variables. This means that all decision variables can be raised *only* to the first power and can be multiplied *only* by a constant term. Therefore, nonlinear terms such as x^2, xy, and 2^x are prohibited. The linearity assumption implies that with respect to the decision variables, the rate of change of the objective function and constraints is always a constant. For example, in the product mix example, the unit profit of a product is the same for every unit, regardless of volume.

2. Constraints must be of a $\le$, $\ge$, or $=$ type. A constraint using a *strict inequality* ($<$ or $>$) is not permitted. To see why, suppose we have the problem

Maximize x

subject to

$$x + 4y < 5$$

You can see that $x = 5$ is not feasible. Can you find any value of x that maximizes the objective and is feasible? For any value of x strictly less than 5, we can always find a number closer to 5 that gives a larger value for the objective function. Therefore, it is impossible to find an absolute numerical solution.

3. Variables can assume any fractional numerical value; that is, they are continuous. In many problems, such as those involving the allocation of time or money, or decisions involving large production quantities, this is a reasonable assumption. However, suppose that we are trying to decide the number of different models of high-cost machine tools to purchase. In this case, fractional values do not make sense. In such cases, we would require a model and solution method that *forces* the variables to assume integer values; these are discussed in the next chapter.

Multiobjective linear programs have the same assumptions as ordinary LPs: the objective and constraint functions must be linear, constraints must be of the $=$, $\le$, or $\ge$ type, and fractional values for the decision variables are allowed. The major difference is that we cannot define the problem using a single objective. Many practical problems have multiple objectives. Examples include minimizing cost vs. maximizing customer service, minimizing risk vs. maximizing profit, maximizing exposure vs minimizing advertising cost. When multiple objectives exist, they are usually in conflict with one another; thus, we need an approach for modeling and evaluating the tradeoffs among the conflicting objectives. We illustrate a multiobjective LP model with the following example.

Example 1: A Multiobjective Linear Program

Mary Trump has $1,000 she would like to allocate to two investments. Investment 1 has an expected return of 6 percent. The risk associated with this investment is measured by a risk factor—a measure of possible variation in the return—which is 4 percent. (Roughly speaking, this means that the true return might be anywhere between 2 percent and 10 percent.) Investment 2 is risk free and has a fixed return of 3 percent. Mary knows that it is not a good idea to put all of her eggs in one basket, so she has decided to invest at least $200, but no more than $700, in each of the two investments. She would like to maximize return, but would also like to keep the risk at a minimum level. Mary's decision variables are:

Investment1 = the amount (in dollars) to invest in investment 1

Investment2 = the amount (in dollars) to invest in investment 2

She has two objectives. First, maximize the total return:

$$Maximize\ .06\ Investment1 + .03\ Investment2$$

and second, minimize risk:

$$Minimize\ .04\ Investment1 + 0\ Investment2$$

These objectives are subject to the constraints:

$$Investment1 + Investment2 \leq 1,000 \qquad \text{(available funds)}$$

$$200 \leq Investment1 \leq 700 \qquad \text{(limits on investment 1)}$$

$$200 \leq Investment2 \leq 700 \qquad \text{(limits on investment 2)}$$

To understand the conflicts that Mary faces, suppose that she decides to invest $700 in one investment and $300 in the other. These potential solutions have the following returns and risks:

Plan	Investment 1	Investment 2	Expected return	Risk factor
A	$700	$300	$51	28
B	$300	$700	$39	12

Which plan is better? There is no definitive answer, because the objectives conflict. Later in this chapter we will see how to address this problem.

APPLICATIONS OF LINEAR AND MULTIOBJECTIVE PROGRAMMING MODELS

Land Management at the National Forest Service

The National Forest Service is responsible for the management of over 191 million acres of national forest in 154 designated national forests. The National Forest Management Act of 1976 mandated the development of a comprehensive plan to guide the management of each national forest.

Decisions need to be made regarding the use of land in each of the national forests. The overall management objective is "to provide multiple use and the sustained yield of goods and services from the National Forest System in a way that maximizes long-term net public benefits in an environmentally sound manner." Linear programming models are used to decide the number of acres of various types in each forest to be used for the many possible usage strategies. For example, the decision variables of the model include the number of acres in a particular forest to be used for timber, the number of acres to be used for recreational purposes, and the number of acres to be left undisturbed.

The model has as its objective the maximization of the net discounted value of the forest over the planning time horizon. Constraints include the availability of land of different types, bounds on the amount of land dedicated to certain uses, and resources available to manage land under different strategies (for example, recreational areas must be staffed; harvesting the land requires a budget for labor, the transport of goods, and replanting). These LP planning models are quite large. Some have over 6,000 constraints and over 120,000 decision variables! The data requirements for these models are quite intense, requiring the work of literally hundreds of agricultural resource specialists to estimate reasonable values for the inputs to the model. These specialists are also used to ensure that the solutions make sense operationally (Field, 1984).

Credit Collection

GE Capital provides credit card services for a consumer credit business that exceeds $12 billion in total outstanding dollars. Managing delinquent accounts is a problem of paramount importance to the profitability of the company. Different collection strategies may be used for delinquent accounts, including mailed letters, interactive taped telephone messages, live telephone calls, and legal procedures. Accounts are categorized by outstanding balance and expected payment performance, which is based on factors like customer demographics and payment history. GE uses a multiple time period linear programming model to determine the most effective collection strategy to apply to its various categories of delinquent accounts. This approach has reduced annual losses due to delinquency by $37 million (Makuch et al., 1992).

Research and Development Funding

The allocation of funds to competing research and development (R&D) projects can have a dramatic impact on the future growth of a company. Top management at Lord Corporation used multiobjective programming to analyze the alternatives in allocating funds to 25 proposed R&D projects. Management established a variety of objectives to accomplish, including the following: (1) No program should consume more than 10 percent of the R&D budget; (2) sales growth should exceed 15 percent; (3) the discounted cash flow rate of return should exceed 30 percent; (4) projects should have five-year capital limits; and (5) the company should promote constructive change in the industry and be involved in the development of new technology. The need for multiobjective programming techniques arose from the fact that these objectives are in conflict with one another, making it impossible to satisfy all of them simultaneously. The multiobjective programming approach allowed management to assess these conflicts between the company's financial goals and those relating more to corporate purpose, and to resolve them to reach an allocation of R&D funds in line with the stated objectives of the company (Salvia and Ludwig, 1979).

Site Selection

Investing in a trucking terminal is one of the most crucial decisions a trucking firm faces. An important part of this decision is site selection. A poor location decision for a terminal can result in a loss of valued customers because of low quality service. If the company uses independent truckers (who own their equipment), poor site selection could also lead to a loss of reliable company workers, since the cost of traveling from home to the terminal and back is borne by the truckers. The liquid food division of Truck Transport Corporation used multiobjective programming models to evaluate potential sites for the relocation of their East St. Louis terminal (one of five terminals the division operates). The model was used to evaluate each site with respect to customer and driver satisfaction, transportation cost to the major customers, and the transportation cost to the homes of its independent truckers. The recommendation obtained from the model was implemented with minimal driver and customer turnover and no significant increase in transportation cost to the drivers (Schneiderjans, *et al.*, 1982).

MODELING OPTIMIZATION PROBLEMS IN EXCEL

Some general guidelines for creating good spreadsheets for mathematical programming models include the following:

- Put the objective function coefficients, constraint coefficients, and right-hand side values in a logical format in the spreadsheet. For example, you might assign the decision variables to columns and the constraints as rows, much like the mathematical formulation of the model, and input the model parameters in a matrix. If you have many more variables than constraints, it might make sense to use rows for the variables and columns for the constraints.
- Define a set of cells (either rows or columns) for the values of the decision variables. In some models, it may be necessary to define a matrix to represent the decision variables; we will see an example of this later. The names of the decision variables should be listed directly above or immediately to the left of the decision variable cells.
- Define separate cells for the objective function and each constraint function (the left-hand side of a constraint). Use descriptive labels either immediately to the left or directly above these cells.
- Use different fonts, borders, colors or shading to clearly delineate your parameters, decision variables, and objective/constraint functions.

Following these guidelines will not only make your model more readable and more manageable, but also facilitates the application of the Excel-based solution process that we will describe later in this chapter. We will follow the general format described in Chapter 1. That is, data inputs will be placed in a separate section, followed by a section for model relationships. The decision variable cells will be bordered and shaded.

Example 2: A Spreadsheet Model for a Linear Program

Figure 4.1 shows a spreadsheet model for the Santa Cruz linear programming problem. The *Parameters and Uncontrollable Inputs* section provides the objective function coefficients, constraint coefficients, and right-hand sides of the model. In the *Model* section, the number of each product to make is given in the shaded cells B13 and D13. Calculations for the objective function, $24LS + 40SB$, are in cells B16 and D16, which are summed in cell F16. The constraint functions are defined in the following cells:

12*LS* (Units of A used in production of LaserStops, cell B14)

12*SB* (Units of A used in production of SpeedBusters, cell D14)

6*LS* (Units of B used in production of LaserStops, cell B15)

10*SB* (Units of B used in production of SpeedBusters, cell D15)

12*LS* + 12*SB* (Total units of component A used, cell F14)

6*LS* + 10*SB* (Total units of component B used, cell F15)

Note that each term in the model, as well as each constraint function and the objective function, is associated with a cell in the spreadsheet. This facilitates the process of translating a model into a correct spreadsheet implementation. We may write the model in terms of the spreadsheet cells as

*Maximize F*16 = *B*16 + *D*16

subject to

$$F14 = B14 + D14 \le F7 \qquad \text{(component A)}$$

$$F15 = B15 + D15 \le F8 \qquad \text{(component B)}$$

$$B13, D13 \ge 0$$

We recommend this approach as a way of checking your spreadsheet implementations.

FIGURE 4.1
Santa Cruz
MicroProducts
LP model

	A	B	C	D	E	F	G
1	**Santa Cruz MicroProducts**					*Key cell formulas*	
2							
3	*Parameters and Uncontrollable Inputs*						
4							
5	Product	LaserStop		SpeedBuster			
6	Unit Profit	$25.00		$40.00		Availability	
7	Component A	12		12		6000	
8	Component B	6		10		3500	
9							
10	*Model*						
11							
12		LaserStop		SpeedBuster		Total	Unused
13	Quantity Produced	0		0		0	
14	Component A Used	0		0		0	6000
15	Component B Used	0		0		0	3500
16	Profit	$0.00		$0.00		$0.00	

BUILDING LINEAR PROGRAMMING MODELS

Three helpful questions can assist you in formulating a problem as a linear programming model:

1. *What am I trying to decide?* Identify the decisions needed to solve the problem and define appropriate variables that represent them. For instance, in the Santa Cruz MicroProducts example, we needed to know how many LaserStops and Speed-

Busters to produce. This led us to the decision variables LS = number of Laser-Stops to produce and SB = number of SpeedBusters to produce.

2. *What is the objective to be maximized or minimized?* Determine the objective and express it as a linear function of the decision variables. Because we were given the per-unit profit for each product in the radar detector example, it is clear that the objective function is to maximize profit, and the objective function is easy to define.

   In problems involving accounting data, it is important to fully understand how profits or costs are defined. When building a linear programming model, *only those costs that are relevant to the decision should be included.* Sunk (or fixed) costs, which have already been incurred or have to be incurred regardless of the decision made, should not be included. For example, in the radar detector problem, if the component inventory is already on hand, then the cost of these components is a sunk cost, and will not affect the decision of how many to make. We are simply concerned with using the components in an optimal fashion. Thus, in accounting terms, the objective function coefficients of $24 and $40 are really *per-unit contribution margins*, namely, the contribution of each unit of product toward the payoff of fixed costs and profit.

3. *What limitations or requirements restrict the values of the decision variables?* Identify and write the constraints as linear functions of the decision variables. Constraints generally fall into one of the following categories:

   - *Limitations*—"The amount of material used in production cannot exceed the amount available in inventory," "Production should not exceed the market potential of the product," or "The amount shipped from the Austin plant in July cannot exceed the plant's capacity." Problems with limitation constraints usually involve the allocation of scarce resources.
   - *Requirements*—"Enough cash must be available in February to meet financial obligations," "Production must be sufficient to meet promised customer orders," or "The marketing plan should ensure that at least 400 customers are contacted each month." Problems with requirements involve the specification of minimum levels of performance.
   - *Simple bounds*—"No more than $5000 can be invested in any single stock," or "We must produce at least 350 units of product Y to meet customer commitments this month." Simple bounds constrain the value of a single variable.
   - *Proportional relationships*—"The amount invested in aggressive growth stocks cannot be more than twice the amount invested in equity-income funds," "The octane rating of gasoline obtained from mixing different crude blends must be at least 89," or "At least 40 percent of client contacts must have an income level above $50,000." Proportional relationships are often found in problems involving mixtures or blends of materials or strategies.
   - *Balance constraints*—"Production in June plus any available inventory must equal June's demand plus inventory held to July," "The total amount shipped to a distribution center from all plants must equal the amount shipped from the distribution center to all customers," or "The total amount of money invested or saved in March must equal the amount of money available at the end of February." Balance constraints essentially state that "input = output" and ensure that the flow of material or money is accounted for at locations or between time periods.

Constraints in nearly all linear programming models are some combination of constraints from these categories. When reading a problem, you should think about these categories. Problem data or verbal clues in a problem statement often help you identify the appropriate constraint. For example, in the Santa Cruz MicroProducts problem, we are given *limited amounts* of each component. This suggests that the number of components used cannot exceed the amount available. We need only translate the "number of components used" into a constraint function or spreadsheet formula. As described in Chapter 1, influence diagrams can be helpful in visualizing and organizing these relationships.

In many situations, all constraints may not be explicitly stated, but are required for the model to represent the real problem accurately. A simple example of implicit constraints is *nonnegativity restrictions* on the decision variables. It is obvious (to us, but not necessarily to a computer!) that it makes no sense to recommend producing a negative number of LaserStops or SpeedBusters.

Dimensionality Checks

A quick way to provide some confidence that a linear programming model is correct is to check the dimensions of all model components. The **dimension** of a term in the model is its unit of measure. Inconsistencies in dimensions within an objective or constraint function, or between a constraint function and its right-hand side, indicate that an error has been made in constructing the model.

We will illustrate this with the radar detector example. Consider the objective function term $24LS$. This dimension of this term is (dollars/LaserStop)(quantity of LaserStops) = dollars. The term $40SB$ is likewise in dollars, so that the dimension of each term in the objective function is consistent. The constraint functions, which represent the number of units of each component used, should be consistent with the right-hand sides. For example, the dimension of $12LS + 12SB$ is (Units of Component A/LaserStop) $\times$ (Number of LaserStops made) + (Units of Component A/Speed-Buster) $\times$ (Number of SpeedBusters made) = Units of Component A. Note that the right-hand side of the constraint is expressed in the same dimensions. Passing the dimensionality check does not necessarily imply that your model is correct; however, if your model fails the dimensionality check, you can be sure that your model is not correct. Therefore, we encourage you to do this for every model you develop.

LP Modeling Examples

The best way to learn to develop LP models is to study many examples and practice developing models on your own! We will present several examples of LP models to help you.

Example 3: Investment Allocation

Kathryn, a college student who is entering her sophomore year, has identified four potential stocks in which to invest. Kathryn has $2,000 to invest from her summer job, and wants to maximize her total return upon graduation. The cost per share of each stock and the expected returns per share over the next three years are shown in Table 4.1. How much should she invest in each stock to maximize her expected return?

Stock	1	2	3	4	
Price/share	$34.00	$20.00	$16.00	$50.00	**TABLE 4.1** Data for Kathryn's investment problem
Return/share	$19.04	$10.20	$ 9.28	$28.00	

Let us use the three basic questions previously discussed to guide the formulation process.

What does Kathryn need to decide? Clearly, this is the number of shares of each stock to purchase. The decision variables are

$$STOCK1 = \text{number of shares of stock 1 to purchase}$$

$$STOCK2 = \text{number of shares of stock 2 to purchase}$$

$$STOCK3 = \text{number of shares of stock 3 to purchase}$$

$$STOCK4 = \text{number of shares of stock 4 to purchase}$$

What is the objective to be maximized or minimized? Kathryn's objective is to maximize expected return. The return per share multiplied by the number of shares invested gives the return for each stock in dollars [($ per share)(number of shares) = $]. Summing over all four stocks gives total return.

$$\text{Total return} = 19.04STOCK1 + 10.20STOCK2 + 9.28STOCK3 + 28STOCK4$$

How is the problem restricted? Kathryn can invest at most $2,000. Thus, this constraint can be expressed as

$$\text{Total amount invested} \leq \$2,000$$

As a function of the decision variables, the total amount invested is calculated by multiplying the price per share of each stock by the number of shares bought [($/share)(number of shares) = $]. Thus, the investment constraint is

$$34STOCK1 + 20STOCK2 + 16STOCK3 + 50STOCK4 \leq 2,000$$

Since it is impossible to invest a negative amount, the decision variables must also be restricted to be nonnegative.

The complete LP model is

Maximize $19.04STOCK1 + 10.20STOCK2 + 9.28STOCK3 + 28STOCK4$

subject to

$$34STOCK1 + 20STOCK2 + 16STOCK3 + 50STOCK4 \leq 2,000$$

$$STOCK1, STOCK2, STOCK3, STOCK4 \geq 0$$

Figure 4.2 shows an Excel spreadsheet model. Row 15 defines the objective function, and row 16 defines the constraint function.

FIGURE 4.2

Spreadsheet model for Kathryn's investment problem

	A	B	C	D	E	F
1	Kathryn's Investment Problem					
2						
3	*Parameters and uncontrollable inputs*					
4						
5	Stock	Stock 1	Stock 2	Stock 3	Stock 4	
6	Price/share	$34.00	$20.00	$16.00	$50.00	
7	Return/share	$19.04	$10.20	$9.28	$28.00	
8						
9	Amount available	$2,000.00				
10						
11	*Model*					
12						
13	Stock	Stock 1	Stock 2	Stock 3	Stock 4	
14	Number of shares	0	0	0	0	Total
15	Return	$0.00	$0.00	$0.00	$0.00	$0.00
16	Investment	$0.00	$0.00	$0.00	$0.00	$0.00
17						
18	*Key cell formulas*				Unused Funds	$2,000.00

Excel Exercise 4.1

Open the worksheet *Kathryn.xls*. Verify that the appropriate cells in the spreadsheet correctly represent the terms of the model, and write the model in terms of the spreadsheet cells. In the spreadsheet, we use a simple summation for the total investment and return. Simplify these formulas using the SUMPRODUCT function.

Example 4: Allocation of Marketing Effort[1]

Phillips, Inc., sells two products and uses two marketing instruments—advertising and selling—to stimulate demand for these products. The two products do not interact, in the sense that neither the price, cost, nor demand associated with one product will affect those of the other product. The company must determine the amount of selling and advertising resources to devote to each of the products in the coming quarter. Phillips has budgeted $25,000 for advertising and has allotted a total of 5,000 hours of labor to selling for the quarter. Based on past data and managerial judgment, Phillips has estimated that a dollar of advertising will yield $8 profit when spent on product 1 and $15 when spent on product 2. These estimates hold over the entire range of the advertising budget. Management has likewise estimated the productivity of personal selling. An hour of personal selling devoted to product 1 will generate a profit of $25, and an hour devoted to product 2 will generate a profit of $45. These estimates are assumed to hold over the ranges of personal selling hours allowable by the company. Company policy stipulates that at least 2,000, but no more than 3,500 hours of labor can be used on each product. Company policy also dictates that at least $5,000 but no more than $20,000 can be spent on advertising for each product and that advertising dollars allocated to any product can be no more than twice that allocated to any other product.

What does Phillips need to decide? Phillips needs to decide how many labor hours for selling and how much money for advertising to allocate to each of its two products. We therefore need to define a set of decision variables that accurately represent these decisions:

A_i = amount of advertising dollars to spend on product i, $i = 1, 2$

S_i = number of hours of selling to allocate to product i, $i = 1, 2$

What is the objective to be maximized or minimized? Phillips would like to maximize the next quarter's profit contribution realized from the allocation of its advertising budget and its selling hours allotment. This can be expressed as

$$Maximize\ 8A_1 + 15A_2 + 25S_1 + 45S_2$$

Note that the dimensions of the first two terms are (profit/advertising dollar) (advertising dollars spent), while the dimensions of the second two terms are (profit/allocated sales hours)(number of allocated sales hours). In both cases, the final dimension is profit.

How is the problem restricted? First, we must ensure that the total dollars expended on advertising does not exceed the budget. This can be expressed as

$$A_1 + A_2 \leq 25{,}000$$

Likewise, the total number of hours allocated to selling for the two products cannot exceed 5,000 hours:

$$S_1 + S_2 \leq 5{,}000$$

Furthermore, we have simple upper and lower bounds on the amount of advertising dollars (between \$5,000 and \$20,000) and selling hours (between 2,000 and 3,500 hours):

$$A_1 \geq 5{,}000$$
$$A_1 \leq 20{,}000$$
$$A_2 \geq 5{,}000$$
$$A_2 \leq 20{,}000$$
$$S_1 \geq 2{,}000$$
$$S_1 \leq 3{,}500$$
$$S_2 \geq 2{,}000$$
$$S_2 \leq 3{,}500$$

Finally, the advertising dollars allocated for either product cannot be more than twice that of the other.

$$A_1 \leq 2A_2$$
$$A_2 \leq 2A_1$$

The entire model can be stated as

$$Maximize\ 8A_1 + 15A_2 + 25S_1 + 45S_2$$

subject to

$$A_1 + A_2 \le 25{,}000$$

$$S_1 + S_2 \le 5{,}000$$

$$5{,}000 \le A_1 \le 20{,}000$$

$$5{,}000 \le A_2 \le 20{,}000$$

$$2{,}000 \le S_1 \le 3{,}500$$

$$2{,}000 \le S_2 \le 3{,}500$$

$$A_1 \le 2A_2$$

$$A_2 \le 2A_1$$

Notice that nonnegativity is satisfied implicitly by the lower bound constraints.

Excel Exercise 4.2

Construct a spreadsheet model for Example 4. Verify your implementation by writing the model in terms of the spreadsheet cells and compare it to the mathematical formulation.

Example 5: A Blending Problem

The Colorado Cattle Company (CCC) can purchase three types of raw feed ingredients from a wholesale distributor. The company's cattle have certain nutritional requirements for fat, protein, calcium, and iron. Each cow requires at least 10 units of calcium, at least 12 units of iron, at least 15 units of protein, and no more than 7.5 units of fat per day. Table 4.2 shows the amount of calcium, iron, protein, and fat in each pound of the three feed ingredients. Grade 1 feed costs $.25 per pound, grade 2, $.10, and grade 3, $.08. The cattle can be fed a mixture of the three types of raw feed, and CCC would like to feed its herd as cheaply as possible.

TABLE 4.2
Data for the Colorado Cattle Company example

Grade	Grade 1	Grade 2	Grade 3
Cost/pound	$0.25	$0.10	$0.08
Calcium/pound	0.70	0.80	–
Iron/pound	0.90	0.80	0.80
Protein/pound	0.80	1.50	0.90
Fat/pound	0.50	0.60	0.40

What is the Colorado Cattle Company trying to decide? We must determine how to blend the raw feed to meet the nutritional requirements. We can do this for a single cow, and then simply order the amounts required per cow times the number of cattle in the herd. Therefore, we define the decision variables as

GRADE1 = amount (in lbs.) of grade 1 to use per day in feeding a cow

GRADE2 = amount (in lbs.) of grade 2 to use per day in feeding a cow

GRADE3 = amount (in lbs.) of grade 3 to use per day in feeding a cow.

What is the objective to be maximized or minimized? The stated objective is to minimize the daily cost of feeding a cow. The cost of any grade feed is simply the cost per pound multiplied by the amount of each feed purchased [($/lb.)(lbs.) = $]. Thus, the total daily cost for feeding a cow can be expressed as

$$\text{Total daily cost} = .25GRADE1 + .10GRADE2 + .08GRADE3$$

How is the problem restricted? The constraints are a combination of requirements (at least 10 units calcium, 12 units of iron, and 15 units of protein) and restrictions (no more than 7.5 units of fat) relating to the daily nutritional needs of the cattle. The constraint functions define the amount of nutritional ingredient provided each day by the feed mix. For example, each pound of grade 1 provides 0.7 units of calcium, each pound of grade 2 provides 0.8 units, and grade 3 does not provide any. Thus, the amount of calcium in a feed mixture is $0.7GRADE1 + 0.8GRADE2 + 0GRADE3$. Since the mix must provide at least 10 units/day, the appropriate constraint is

$$0.7GRADE1 + 0.8GRADE2 + 0GRADE3 \geq 10$$

Note the dimensions [(units of calcium/lb.)(lbs./day) = units of calcium/day]. Similar constraints can be developed for the other nutritional ingredients. The complete linear program can be stated as

Minimize $.25GRADE1 + .10GRADE2 + .08GRADE3$

subject to

$$
\begin{array}{ll}
0.7GRADE1 + 0.8GRADE2 + 0GRADE3 \geq 10 & \text{(Calcium)} \\
0.9GRADE1 + 0.8GRADE2 + 0.8GRADE3 \geq 12 & \text{(Iron)} \\
0.8GRADE1 + 1.5GRADE2 + 0.9GRADE3 \geq 15 & \text{(Protein)} \\
0.5GRADE1 + 0.6GRADE2 + 0.4GRADE3 \leq 7.5 & \text{(Fat)} \\
GRADE1, GRADE2, GRADE3 \geq 0 &
\end{array}
$$

This problem is also known as the *diet problem*. Similar models are used in hospitals to plan diets for patients requiring specific nutritional needs. Figure 4.3 shows a spreadsheet model for the CCC problem.

	A	B	C	D	E	F	G
1	Colorado Cattle Company Model				Key cell formulas		
2							
3	*Parameters and uncontrollable inputs*						
4							
5	Grade	Grade 1	Grade 2	Grade 3			
6	Cost / pound	$0.25	$0.10	$0.08		Minimum	Maximum
7	Calcium / pound	0.7	0.8	0	Calcium	10	
8	Iron / pound	0.9	0.8	0.8	Iron	12	
9	Protein / pound	0.8	1.5	0.9	Protein	15	
10	Fat / pound	0.5	0.6	0.4	Fat		7.5
11							
12	*Model*						
13							
14		Grade 1	Grade 2	Grade 3			
15	Quantity	0.000	0.000	0.000	Total		
16	Cost	$0.00	$0.00	$0.00	$0.00000		
17	Calcium	0.000	0.000	0.000	0.000		
18	Iron	0.000	0.000	0.000	0.000		
19	Protein	0.000	0.000	0.000	0.000		
20	Fat	0.000	0.000	0.000	0.000		

FIGURE 4.3
Spreadsheet model for Colorado Cattle Company

Standard Form

Computer programs that solve linear programs preprocess the model and convert it into what is called **standard form**—a formulation of the model that has four properties: (1) all variables appear on the left-hand side of each constraint; (2) all constants appear on the right-hand side of the constraints; (3) all constraints (excluding the non-negativity constraints) are expressed as *equalities*; and (4) all variables are restricted to be nonnegative.

The first two properties are easy to satisfy using simple algebra. To satisfy property (3), we need a way to convert inequality ($\leq$ and $\geq$) constraints to equalities ($=$). First consider the calcium constraint. Because it is a greater-than-or-equal-to constraint, the left-hand side, which represents the amount of calcium in the mix, can be *equal to* or *greater than* 10. If the left-hand side is strictly greater than 10, then we have included more calcium than required. We represent this excess amount by a **slack variable** s_1.[2] Thus, the amount of calcium in the mix minus the excess must equal the requirement. Written mathematically, we have

$$0.7GRADE1 + 0.8GRADE2 + 0GRADE3 - s_1 = 10 \qquad \text{(Calcium)}$$

Using similar logic, we define s_2 and s_3 as the amounts exceeding the minimum requirement for iron and protein, respectively. These constraints, written in standard form, are

$$0.9GRADE1 + 0.8GRADE2 + 0.8GRADE3 - s_2 = 12 \qquad \text{(Iron)}$$

$$0.8GRADE1 + 1.5GRADE2 + 0.9GRADE3 - s_3 = 15 \qquad \text{(Protein)}$$

The fat constraint requires that the amount of fat must be equal to or less than 7.5. Defining a slack variable s_4 as the amount by which the left-hand side is *less than* 7.5, then, the amount of fat in the mix plus s_4 must total 7.5:

$$0.5GRADE1 + 0.6GRADE2 + 0.4GRADE3 + s_4 = 7.5 \qquad \text{(Fat)}$$

Note that all slack variables must be nonnegative for the solution to be feasible.

The complete model for the Colorado Cattle Company in standard form is

Minimize $.25GRADE1 + .10GRADE2 + .08GRADE3$

subject to

$$0.7GRADE1 + 0.8GRADE2 + 0GRADE3 - s_1 = 10 \qquad \text{(Calcium)}$$

$$0.9GRADE1 + 0.8GRADE2 + 0.8GRADE3 - s_2 = 12 \qquad \text{(Iron)}$$

$$0.8GRADE1 + 1.5GRADE2 + 0.9GRADE3 - s_3 = 15 \qquad \text{(Protein)}$$

$$0.5GRADE1 + 0.6GRADE2 + 0.4GRADE3 + s_4 = 7.5 \qquad \text{(Fat)}$$

$$GRADE1, GRADE2, GRADE3, s_1, s_2, s_3, s_4 \geq 0$$

Although you will not have to formulate problems in standard form to solve them on a computer, it is important to understand the meaning of slack variables because they are used in the output reports generated by computer programs that solve LPs.

Excel Exercise 4.3

Open the Colorado Cattle Company spreadsheet model *ccc.xls*. Modify the spreadsheet to compute the slack variables for each constraint. Enter the solution 3, 4, and 8 for Grades 1, 2, and 3, respectively. What do the slack variables tell you about feasibility?

Example 6: Multiperiod Production Planning

Suzie's Sweatshirts is a home-based company that makes upscale, hand-painted sweatshirts for children. Forecasts of sales for the next year are 125 in the autumn, 350 in the winter, and 75 in the spring. Shirts are purchased for $15. The cost of capital is 24 percent per year (or 6 percent per quarter); thus, the holding cost per shirt is .06(15) = $.90 per quarter. Suzie hires part-time workers to paint the shirts during the autumn, and they earn $4.50 per hour. Because of the high demand for part-time help during the winter holiday season, labor rates are higher in the winter, and Suzie must pay the workers $6.00 per hour. In the spring, labor is more difficult to keep, and Suzie finds that she must pay $5.50 per hour to get qualified help. Each shirt takes 1.5 hours to complete. How should Suzie plan production over the three quarters to minimize the combined production and inventory holding costs?

What is Suzie trying to decide? Suzie's primary decision variables are the numbers of shirts to produce during each of the three quarters. However, it may be advantageous to produce more than the demand during some quarters and carry the shirts in inventory, thereby letting lower labor rates offset the carrying costs. Therefore, we must also define decision variables for the number of units to hold in inventory at the end of each quarter. The decision variables are

NA = amount to produce in quarter 1 (autumn)

NW = amount to produce in quarter 2 (winter)

NS = amount to produce in quarter 3 (spring)

$INVA$ = inventory held at the end of autumn

$INVW$ = inventory held at the end of winter

$INVS$ = inventory held at the end of spring

When building larger and more complicated LP models, it is convenient to define variables using a simple letter symbol and a range of indexes written as subscripts. For example, in this problem we might let the variable Q_i represent production in quarter i, and I_i represent the inventory held at the end of quarter i (where $i = 1$ represents autumn, etc.). Although a more descriptive name should be used when formulating the problems on a spreadsheet to enhance readability, this style enables us to write large LPs in a more compact from and simplify the presentation of the model. Therefore, we will use these variable definitions:

Q_i = quantity to produce in quarter i, for $i = 1, 2, 3$

I_i = inventory held from quarter i to quarter $i + 1$, for $i = 1, 2, 3$

What is the objective to be maximized or minimized? The production cost per shirt is computed by multiplying the labor rate by the number of hours required to produce a shirt. Thus, the unit cost in the autumn is ($4.50)(1.5) = $6.75; in the winter, ($6.00)(1.5) = $9.00; and in the spring, ($5.50)(1.5) = $8.25. The objective function is to minimize the total cost of production and inventory. (Since the cost of the shirts themselves is constant, it is not relevant to the problem we are addressing.) The objective function is therefore

$$\text{Minimize } 6.75Q_1 + 9.00Q_2 + 8.25Q_3 + 0.90I_1 + 0.90I_2 + 0.90I_3$$

How is the problem restricted? The only explicit restriction in this problem is that demand must be satisfied. Note that both the production in a quarter as well as the inventory held from the *previous* quarter can be used to satisfy demand. In addition, any amount in excess of the demand is held to the next quarter. Therefore, the constraints take the form of *inventory balance equations* that essentially say, "What is available in any time period must be accounted for somewhere." More formally,

Production + inventory from the previous quarter = demand + inventory held to the next quarter

This can be represented visually using the diagram in Figure 4.4. For each quarter, the sum of the variables coming in must equal the sum of the variables going out. Drawing such a figure is very useful for any type of multi–time period problem. This results in the constraint set:

$$Q_1 + 0 = 125 + I_1$$
$$Q_2 + I_1 = 350 + I_2$$
$$Q_3 + I_2 = 75 + I_3$$

The full model, after putting the constraints in standard form and including nonnegativity constraints, is

$$\text{Minimize } 6.75Q_1 + 9.00Q_2 + 8.25Q_3 + 0.90I_1 + 0.90I_2 + 0.90I_3$$

subject to

$$Q_1 - I_1 = 125$$
$$Q_2 + I_1 - I_2 = 350$$
$$Q_3 + I_2 - I_3 = 75$$
$$Q_i \geq 0, \text{ for all } i$$
$$I_i \geq 0, \text{ for all } i$$

A spreadsheet model is shown in Figure 4.5. Inventory levels are calculated from the production quantities. For example, if we choose to make 150 sweatshirts in the Autumn quarter, we know that $I_1 = 25$. Knowing I_1 and Q_2, we could easily compute the value of I_2 from the second constraint, and so on. Thus, only the production quantities need be defined as decision variables. The plan shown in the spreadsheet produces all demand in the autumn quarter. However, this is not an optimal solution.

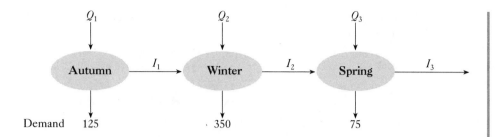

FIGURE 4.4
Graph illustrating material balance constraints

FIGURE 4.5
Spreadsheet model for Suzie's Sweatshirts

	A	B	C	D	E
1	Suzie's Sweatshirts	Key cell formulas			
2					
3	*Parameters and uncontrollable inputs*				
4					
5	Cost of Capital	6%			
6	Cost per Shirt	$15.00			
7	Hours per Shirt	1.5			
8	Autumn Beginning Inventory	0			
9					
10		Autumn	Winter	Spring	
11	Hourly Wage	$4.50	$6.00	$5.50	
12	Demand	125	350	75	
13					
14					
15	*Model*				
16					
17		Autumn	Winter	Spring	
18	Inventory cost / shirt	$0.90	$0.90	$0.90	
19	Production cost / shirt	$6.75	$9.00	$8.25	
20	Production Quantity	550	0	0	
21	Beginning Inventory	0.0	425.0	75.0	
22	Ending Inventory	425.0	75.0	0.0	Total
23	Inventory cost	$382.50	$67.50	$0.00	$450.00
24	Production cost	$3,712.50	$0.00	$0.00	$3,712.50
25				Total	$4,162.50

Developing models is more an art than a science; consequently, there is often more than one way to model a particular problem. Using the ideas presented in this example, we may construct an alternative model involving only the variables Q_j. We simply have to make sure that demand is satisfied. We can do this by ensuring that the cumulative production in each quarter is at least as great as the cumulative demand. This is expressed by the following constraints:

$$Q_1 \geq 125$$

$$Q_1 + Q_2 \geq 475$$

$$Q_1 + Q_2 + Q_3 \geq 550$$

$$Q_1, Q_2, Q_3 \geq 0$$

The differences between the left- and right-hand sides of these constraints are the ending inventories for each period—and we need to keep track of these amounts because inventory has a cost associated with it. Thus, we use the following objective function:

$$\textit{Minimize } 6.75Q_1 + 9.0Q_2 + 8.25Q_3 + .9(Q_1 - 125) + .9(Q_1 + Q_2 - 475)$$
$$+ .9(Q_1 + Q_2 + Q_3 - 550)$$

Of course, this model can be simplified algebraically by combining like terms. The complete alternative model is

Minimize $9.45Q_1 + 10.8Q_2 + 9.15Q_3 - 1035$

subject to

$$Q_1 \geq 125$$
$$Q_1 + Q_2 \geq 475$$
$$Q_1 + Q_2 + Q_3 \geq 550$$
$$Q_1, Q_2, Q_3 \geq 0$$

Although these two models look very different, they are equivalent and will produce the same solution. However, the information provided from the computer solution reports will be quite different. There is no one "right" model, there are only *correct* models!

Excel Exercise 4.4

Open the worksheet for Suzie's Sweatshirts, *Suzie.xls*. Examine the spreadsheet formulas to see how inventory values are calculated. How does this spreadsheet compare to the two models previously discussed? Try to find a better solution than the one shown in Figure 4.5 by changing the values for the production variable cells.

Example 7: Working Capital Management

The financial manager of Vohio, Inc., is charged with managing the company's cash flow. She must manage short-term investments to maximize interest income, while making sure that funds are available to pay company expenditures. She is considering three investment options over the next six months: A, a 1-month investment that pays 0.5 percent; B, a 3-month investment that pays 2.1 percent; and C, a 6-month CD that pays 3.5 percent. The net expenditures for the next six months are as follows: $50,000, ($12,000), $23,000, ($20,000), $41,000, ($13,000). Note that dollar amounts in parentheses indicate a net inflow of cash (for example, at the start of the month 2, $12,000 will be added to the company accounts). The company currently has $300,000 in cash.

What is it that the financial manager is trying to decide? The decision variables are how much to invest each month in each alternative. We may define them as

A_i = amount ($) to invest in a 1-month CD at the start of month i

B_i = amount ($) to invest in a 3-month CD at the start of month i

C_i = amount ($) to invest in a 6-month CD at the start of month i

Because the time horizons on these alternatives vary, it is helpful to draw a picture to represent the investments and returns for each year, as shown in Figure 4.6. Each circle represents the beginning of a month. Arrows represent the investments and times to maturity and show expenditures (an arrow out is an expenditure, an arrow in is an inflow). For example, investing in a 3-month CD at the start of month 1 (B_1) matures at the beginning of month 4; hence the arrow from month 1 to month 4. Note that alternative A is available every month, alternative B is available in months 1, 2, 3, and 4; and alternative C is available only at the start of month 1. We assume that all available funds are invested.

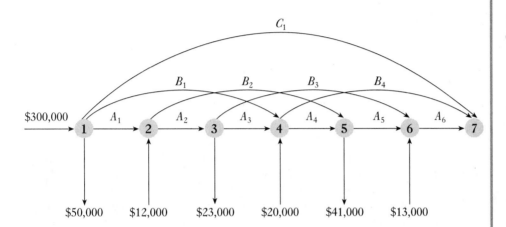

FIGURE 4.6
Graphical illustration of cash flows and investments

What is the objective to be maximized or minimized? The financial manager wishes to maximize the amount of cash on hand at the end of the planning period. From Figure 4.6, we see that investments A_6, B_4, and C_6 will mature at the beginning of month 7 (end of month 6). Because a dollar invested in A will return \$1.005, a dollar invested in B will return \$1.021, and a dollar invested in C will return \$1.035, the objective function is

$$Maximize \ 1.005A_6 + 1.021B_4 + 1.035C_1$$

How is the problem restricted? Only one type of constraint—a cash balance equation—is necessary. For each month, the amount of money available for investing must equal the amount invested plus expenditure. Consider the start of the first month. From Figure 4.6, alternatives A, B, and C are available, and a total of \$300,000 is on hand and there is an expenditure in month 1 of \$50,000. Therefore, we must have

$$300,000 = (A_1 + B_1 + C_1 + 50,000)$$

For the start of the second month, only those funds obtained from investing in alternative A_1 $(1.005A_1)$ plus the inflow of \$12,000 will be available to reinvest in options A_2 and B_2. Thus, the cash balance constraint is

$$1.005A_1 + 12,000 = (A_2 + B_2)$$

The constraints for the remaining months are similar; the full model for this investment problem after moving all decision variables to the left-hand side is

Maximize $1.005A_6 + 1.021B_4 + 1.035C_1$

subject to

$$300,000 - (A_1 + B_1 + C_1 + 50,000) = 0 \qquad \text{(Month 1)}$$

$$1.005A_1 + 12,000 - (A_2 + B_2) = 0 \qquad \text{(Month 2)}$$

$$1.005A_2 - (A_3 + B_3 + 23,000) = 0 \qquad \text{(Month 3)}$$

$$1.005A_3 + 1.021B_1 + 20,000 - (A_4 + B_4) = 0 \qquad \text{(Month 4)}$$

$$1.005A_4 + 1.021B_2 - (A_5 + 41,000) = 0 \qquad \text{(Month 5)}$$

$$1.005A_5 + 1.021B_3 + 13,000 - A_6 = 0 \qquad \text{(Month 6)}$$

$$A_i, B_i, C_i \geq 0, \text{ for all } i$$

Example 8: A Transportation Problem

In Chapter 2, we described the problem of locating a new production facility for Hanover, Inc. We will develop a prototype model to find a minimal cost sourcing plan for a single product for a given new plant location. For simplicity we will consider only board A. Currently, Paris has a capacity of 2,600,000 units and Austin has a capacity of 4,000,000 units. Suppose we decided to establish a new facility in Spain with 4,000,000 units of capacity. What is the most cost-efficient way of supplying the customers?

What is it that we need to decide? We need to determine how many of board type A to produce and ship from each plant to each customer. Suppose we number the three plants in Paris, Austin, and Spain 1, 2, and 3, respectively. Likewise, number the customers in the order in which they appear in the Hanover database (that is, Malaysia, China, France, Brazil, U.S. Northeast, U.S. Southeast, U.S. Midwest, U.S. West as 1, 2,....8). We may then define the decision variables as

$$X_{ij} = \text{number of units of board A to produce and ship}$$
$$\text{from plant } i \text{ to customer } j$$

What is the objective to be maximized or minimized? Hanover would like to minimize the cost of producing and delivering circuit boards to its customers. Define C_{ij} to be the per-unit cost of supplying board A from plant i to customer j. How is this cost determined? The fixed costs associated with any plant are irrelevant (that is, our sourcing decisions do not impact fixed costs). The costs that *are* relevant are those that are product-, source-, or destination-dependent. These include variable production cost, freight & warehousing cost, and duty cost associated with shipping a product from a particular plant to a particular customer, all of which are available in the Hanover database. For example, the per-unit cost of producing and shipping board A from Paris to Malaysia is the cost of producing in Paris + cost of warehousing and freight from Paris to Malaysia + duty cost from Paris to Malaysia ($C_{11} = \$1.256 + \$1.41 + \$.92 = \$3.586$ or $\$3.59$). Other costs are calculated in similar manner. The total cost of shipping X_{ij} units from plant i to customer j is then $C_{ij}X_{ij}$. The complete objective function is

$$\text{Minimize } C_{11}X_{11} + C_{12}X_{12} + \dots C_{18}X_{18} + C_{21}X_{21} + C_{22}X_{22} + \dots C_{28}X_{28} +$$
$$C_{31}X_{31} + C_{32}X_{32} + \dots C_{38}X_{38}$$

We may use summation notation to express this more efficiently as

$$Minimize \sum_{i=1}^{3} \sum_{j=1}^{8} C_{ij} X_{ij}$$

How is the problem restricted? The problem is restricted in two simple ways. First, we cannot produce more than the available capacity at a given plant. Second, we must satisfy the demand for each customer. We refer to these constraints as *supply constraints* and *demand constraints*, respectively.

First consider the supply constraints. The amount produced and shipped from Paris to all destinations cannot exceed its capacity. This is expressed by the constraint

$$X_{11} + X_{12} + X_{13} + X_{14} + X_{15} + X_{16} + X_{17} + X_{18} \leq 2,600,000$$

or equivalently,

$$\sum_{j=1}^{8} X_{1j} \leq 2,600,000$$

Similarly, for the Austin and Spain plants, the capacity constraints are

$$\sum_{j=1}^{8} X_{2j} \leq 4,000,000$$

$$\sum_{j=1}^{8} X_{3j} \leq 4,000,000$$

The demand constraints ensure that each customer's demand is satisfied. For example, the amount of board A shipped to Malaysia from all plants must equal the demand for board A in Malaysia. This is expressed as

$$X_{11} + X_{21} + X_{31} = 399,000$$

or equivalently,

$$\sum_{i=1}^{3} X_{i1} = 399,000$$

We may write similar constraints for all other destinations.

You can see that for a problem with *m* plants and *n* customers, we will have *m* + *n* supply and demand constraints. Rather than write each individually, we may write the model in a generic form by letting CAP_i be the capacity at plant *i* and DEM_j be the demand for customer *j*. The complete model is

$$Minimize \sum_{i=1}^{3} \sum_{j=1}^{8} C_{ij} X_{ij}$$

subject to

$$\sum_{j=1}^{8} X_{ij} \leq CAP_i \qquad i = 1, 2, 3 \qquad (Plant\ i\ capacity)$$

$$\sum_{i=1}^{3} X_{ij} = DEM_j \qquad j = 1, 2,... 8 \qquad (Customer\ j\ Demand)$$

$$X_{ij} \geq 0 \qquad i = 1, 2, 3; \qquad j = 1, 2,... 8$$

This model is called a *transportation model* (see the next Management Science Practice box). A spreadsheet model for this problem is shown in Figure 4.7. In the model section, the cost coefficients are created from the raw data. Note that the total shipped out of each plant (J41:J43) are constrained by the capacities in E7:E9, and the total shipped into each destination (B44:I44) must equal the demands in row 27.

FIGURE 4.7 Spreadsheet model for Hanover transportation problem

	A	B	C	D	E	F	G	H	I	J
1	**Hanover Inc. Spain Transportation Model**			Key cel formulas						
2										
3	*Parameters and uncontrollable inputs*									
4										
5	Variable Cost									
6		Product A			Capacity					
7	Paris	$ 1.256		Paris	2600000					
8	Austin	$ 1.202		Austin	4000000					
9	Spain	$ 0.480		Spain	4000000					
10										
11	Duty Cost A									
12										
13	Plant	Malaysia	China	France	Brazil	US Northeast	US Southeast	US Midwest	US West	
14	Paris	$0.92	$1.26	$0.00	$0.46	$0.16	$0.16	$0.16	$0.16	
15	Austin	$0.92	$1.26	$0.38	$0.46	$0.00	$0.00	$0.00	$0.00	
16	Spain	$0.92	$1.26	$0.00	$0.46	$0.16	$0.16	$0.16	$0.16	
17										
18	Warehousing & Freight									
19										
20	Plant	Malaysia	China	France	Brazil	US Northeast	US Southeast	US Midwest	US West	
21	Paris	$1.41	$1.38	$1.27	$1.45	$1.43	$1.43	$1.45	$1.51	
22	Austin	$1.39	$1.51	$1.43	$1.54	$1.36	$1.32	$1.31	$1.32	
23	Spain	$1.41	$1.45	$1.37	$1.46	$1.52	$1.52	$1.53	$1.54	
24										
25	Demand									
26		Malaysia	China	France	Brazil	US Northeast	US Southeast	US Midwest	US West	
27	Product A	399000	3158300	1406000	163500	68700	999700	544900	1804000	
28										
29										
30	*Model*									
31										
32	Total Sourcing Cost Per Unit									
33	Product A									
34	Plant	Malaysia	China	France	Brazil	US Northeast	US Southeast	US Midwest	US West	
35	Paris	$3.59	$3.89	$2.53	$3.17	$2.84	$2.84	$2.86	$2.93	
36	Austin	$3.51	$3.97	$3.08	$3.20	$2.56	$2.53	$2.51	$2.52	
37	Spain	$2.81	$3.19	$1.85	$2.40	$2.16	$2.16	$2.17	$2.18	
38										
39	Sourcing Product A									
40	Plant	Malaysia	China	France	Brazil	US Northeast	US Southeast	US Midwest	US West	Total A
41	Paris	0	0	0	0	0	0	0	0	0
42	Austin	0	0	0	0	0	0	0	0	0
43	Spain	0	0	0	0	0	0	0	0	0
44	Total	0	0	0	0	0	0	0	0	0
45										
46	Total Cost	$0								

Management Science Practice:
Using Transportation Models for Procter & Gamble's Product Sourcing Decisions

In 1993, Procter & Gamble began an effort entitled "Strengthening Global Effectiveness (SGE)," to streamline work processes, drive out non–value-added costs, and eliminate duplication. A principal component of SGE was the North American Product Supply Study, designed to reexamine and reengineer P&G's product-sourcing and distribution system for its North American operations,

continued

with an emphasis on plant consolidation. Prior to the study, the North American supply chain consisted of hundreds of suppliers, over 50 product categories, over 60 plants, 15 distribution centers (DCs), and over 1,000 customers. The need to consolidate plants was driven by the move to global brands and common packaging, and by the need to reduce manufacturing expense, improve speed to market, avoid major capital investments, and deliver better consumer value.

One of the key submodels in the overall optimization effort was a simple transportation model, called the *product sourcing model*, for each of 30 product categories. Product-strategy teams used this model to specify plant locations and capacity options and to optimize the flow of product from plants to DCs and customers. Costs included manufacturing, warehousing at the plant, and transportation costs. Transportation costs were estimated using actual negotiated rates or an estimation model that included fixed costs plus a linear function of road distance. Comparing them to the costs of the existing distribution system validated these costs.

The product sourcing model was integrated with a geographic information system to display a map and manipulate data. The strategy teams could easily alter dozens of different aspects of the supply system and then quickly compare the cost of the option to others under consideration. The user would choose the product type, potential plant locations, transportation mode, cost modes, and so on. The system would optimize the model and display the results on a map, as shown in Figure 4.8. This spatial visualization helped the teams to better understand the relationships among manufacturing costs, capacity, and the distribution system design.

FIGURE 4.8 Example of product sourcing output

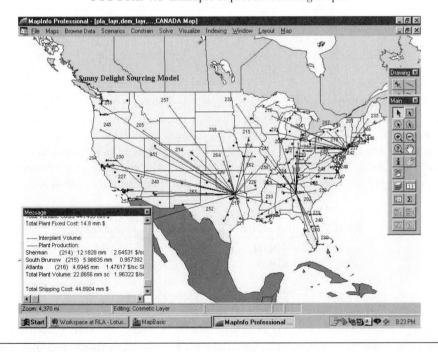

Source: Reprinted by permission, Jeffrey D. Camm, Thomas E. Chorman, Franz A. Dill, James R. Evans, Dennis J. Sweeney, and Glenn W. Wegryn, "Blending OR/MS, Judgment, and GIS: Restructuring P&G's Supply Chain," *Interfaces*, Vol. 27, No. 1, January–February 1997, 128–142. Copyright 1997, the Institute for Operations Research and the Management Sciences (INFORMS), 901 Elkridge Landing Road, Suite 400, Linthicum, MD 21090 USA.

SOLVING LINEAR PROGRAMMING MODELS

In this section we discuss how to solve LP models to optimality. Although solving these within a spreadsheet environment requires little more than point-and-click, we will first try to develop some intuition to help you gain insight into the interactions among variables and the necessary trade-offs between optimizing the objective function and meeting constraints. This will give you a better appreciation of the power of LP models.

Intuition and LP Solutions

As you already know, one of the advantages of using spreadsheets is the ability to change cell values and quickly recalculate the effects. Once an LP model is formulated on a spreadsheet, you can easily adjust values of the decision variables to attempt to find a good solution. Let us use Kathryn's investment problem (Example 3), which, for convenience, is repeated below.

Maximize 19.04*STOCK1* + 10.20*STOCK2* + 9.28*STOCK3* + 28.0*STOCK4*

subject to

$$34STOCK1 + 20STOCK2 + 16STOCK3 + 50STOCK4 \le 2{,}000$$

$$STOCK1, STOCK2, STOCK3, STOCK4 \ge 0$$

Using the spreadsheet in Figure 4.2, you might simply try different values for the commodities, make sure the budget restriction is satisfied, and examine the profit, seeking better and better solutions. Usually, some intuition can lead us in good directions. For example, since we want to maximize return, we might select the stock with the highest return per share, which is *STOCK4*, and purchase as many shares as possible. Because each share costs $50, we can purchase 2000/50 = 40 shares, giving a profit of $28(40) = $1,120. However, this is *not* the best solution! The optimal solution is to purchase 125 shares of stock 3 (and none of stocks 1, 2, and 4), for a total profit of $1,160.

Let us see why this solution makes sense. Even though stock 4 has the largest return per share, it also has the highest price per share, and thus requires fewer shares to reach the budget limitation. Stock 3, on the other hand, costs less than a third as much as stock 4, and therefore Kathryn can purchase more than three times as many shares of stock 3. Even though the return per share is smaller, the *total return* is larger. This is easy to see if we consider the *return per dollar* invested for each stock:

Stock	Return per share	Price per share	Return per dollar
Stock 1	$19.04	$34.00	$.56
Stock 2	$10.20	$20.00	$.51
Stock 3	$ 9.28	$16.00	$.58
Stock 4	$28.00	$50.00	$.56

Clearly Stock 3 has the highest return per dollar invested and consequently is the stock that maximizes *total* return. The optimal LP solution considers simultaneously the return per share and the price per share contribution to the budget constraint. For a linear program with a single constraint and positive coefficients, identifying the variable with the best ratio of objective-to-constraint coefficients will always yield the optimal solution.

When Intuition Can Fail

Let us now consider the Colorado Cattle Company problem (Example 5). The model is

Minimize .25GRADE1 + .10GRADE2 + .08GRADE3

subject to

$$0.7GRADE1 + 0.8GRADE2 + 0GRADE3 \geq 10 \qquad \text{(Calcium)}$$

$$0.9GRADE1 + 0.8GRADE2 + 0.8GRADE3 \geq 12 \qquad \text{(Iron)}$$

$$0.8GRADE1 + 1.5GRADE2 + 0.9GRADE3 \geq 15 \qquad \text{(Protein)}$$

$$0.5GRADE1 + 0.6GRADE2 + 0.4GRADE3 \leq 7.5 \qquad \text{(Fat)}$$

$$GRADE1, GRADE2, GRADE3 \geq 0$$

Is the solution to this LP as obvious as the solution to Kathryn's problem? The presence of multiple constraints makes this problem much more difficult. Using intuition, not only might you have difficulty finding an optimal solution, but you might not even find a *feasible* solution. For instance, if you try to buy all grade 3 feed because it is the cheapest, it is impossible to satisfy the calcium requirement. If we try the two least expensive grades (2 and 3), we find that we need at least 12.5 pounds of grade 2 to meet the calcium requirement ((10 units)/(.8 units/lb) = 12.5 lbs), but that takes us exactly to the limit of fat (.6[12.5] = 7.5). Adding *any* grade 3 (to satisfy the iron restriction) will add fat and consequently violate the fat constraint. Before continuing, we encourage you to experiment with the spreadsheet model to try to identify the best feasible solution you can. You will probably be able to find a good one, but it is unlikely that you will identify an optimal solution very easily.

The optimal solution is to mix 8 pounds of grade 1, 5.5 pounds of grade 2, and .5 pounds of grade 3, for a total cost of $2.59 per cow per day. The solution is shown on the spreadsheet model in Figure 4.9. Somewhat counterintuitively, we are using only .5 lbs. of the cheapest grade. A look at the constraint set shows that grade 3 is relatively comparable to grade 2 in its contribution of fat and iron, but considerably lower in calcium and protein. Also, notice that the iron and calcium minimum requirements are met exactly and that this solution contains the maximum amount of fat allowed.

The power of LP solution approaches is the ability to consider the complex trade-offs among constraints and the objective function, and determine which of the constraints are "bottlenecks" that keep us from doing even better. Microsoft Excel includes an add-in called Solver that allows you to find optimal solutions for spreadsheet-based LP models.

Excel Exercise 4.5

Open the transportation model spreadsheet *Hanover Spain.xls* (Figure 4.7). Find a solution to this problem using your intuition, and explain your reasoning.

FIGURE 4.9
Optional
solution to
Colorado Cattle
Company
model

	A	B	C	D	E	F	G
1	Colorado Cattle Company Model				Key cell formulas		
2							
3	*Parameters and uncontrollable inputs*						
4							
5	Grade	Grade 1	Grade 2	Grade 3			
6	Cost / pound	$0.25	$0.10	$0.08		Minimum	Maximum
7	Calcium / pound	0.7	0.8	0	Calcium	10	
8	Iron / pound	0.9	0.8	0.8	Iron	12	
9	Protein / pound	0.8	1.5	0.9	Protein	15	
10	Fat / pound	0.5	0.6	0.4	Fat		7.5
11							
12	*Model*						
13							
14		Grade 1	Grade 2	Grade 3			
15	Quantity	8.000	5.500	0.500	Total		
16	Cost	$2.00	$0.55	$0.04	$2.59000		
17	Calcium	5.600	4.400	0.000	10.000		
18	Iron	7.200	4.400	0.400	12.000		
19	Protein	6.400	8.250	0.450	15.100		
20	Fat	4.000	3.300	0.200	7.500		

Using Excel Solver[3]

Solver is an add-in included with Excel for solving linear programs and other types of optimization problems. (See Chapter 1 for installation issues). Solver may be used to solve problems with up to 200 decision variables, 100 regular constraints, and 400 simple constraints (lower and upper bounds on the decision variables). Although we will provide sufficient information on using it, you may find additional examples in the *Solvsamp.xls* workbook included in the Excel examples folder. Further information may be found at the web site of Frontline Systems, Inc., the developers of Solver (**www.frontsys.com**).

To begin Solver, select *Tools* from the main menu and then *Solver*. The *Solver Parameters* dialog box will appear, as shown in Figure 4.10. The *Set Target Cell* box should contain the cell location of the objective function. Either *Max* or *Min* should be checked to define the objective. (If *Value* is selected, the Solver will attempt to find a solution with that specific value, but will not optimize the problem.) In the *By Changing Cells* box, you specify the location of the decision variables. Finally, constraints are specified in the *Subject to the Constraints* box by clicking on *Add*. *Change* allows you to modify a constraint already entered, and *Delete* allows you to delete a previously entered constraint. *Reset All* clears the current problem and resets all parameters to their default values. *Options* invokes the Solver options dialog box (to be discussed later). The *Guess* selection is not particularly useful for our purposes and will not be discussed here.

When the *Add* button is clicked, the *Add Constraint* dialog box appears (Figure 4.11). Clicking on the *Cell Reference* box allows you to specify a cell location (usually corresponding to the constraint function). The *constraint type* is set by selecting the drop-down menu and choosing from $<=$, $>=$, $=$, *int*, or *bin* (*int* and *bin* will be discussed in the next chapter). The *Constraint* box may contain a formula, a simple cell reference, or a numerical value. The *Add* button adds the currently specified constraint to the existing model and returns to the *Add Constraint* dialog box. The *OK* button adds the current constraint to the model and returns you to the *Solver Parameters* dialog box.

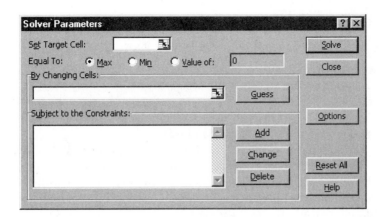

FIGURE 4.10
Solver parameters dialog box

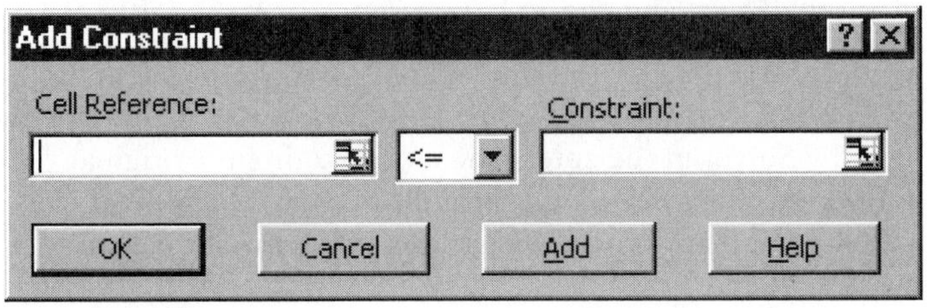

FIGURE 4.11
Solver add constraint dialog box

Solver does *not* assume nonnegativity of the decision variables. This should be specified in the *Options* dialog box by selecting the *Options* button and checking *Assume Non-Negative* (Figure 4.12). (Older versions of Excel do not have this option; in this case, you must add nonnegativity constraints in the *Add Constraint* dialog box.) For linear programs, you must check the *Assume Linear Model* box.

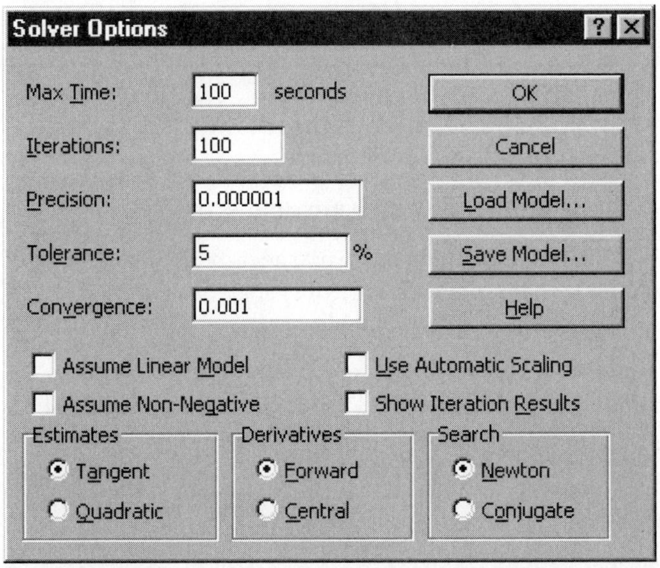

FIGURE 4.12
Solver options dialog box (note the assume linear model and assume non-negative check boxes)

Example 9: Solving the Colorado Cattle Company Model

To solve the CCC model, perform the following steps:

1. Select *Solver* from the Tools menu.
2. Click on the *Set Target Cell* box and type E16.
3. Click on *Min*.
4. Click inside the *By Changing Cells* box and, in the spreadsheet, click and drag the mouse from B15 to D15 (or type B15:D15).
5. Click the *Add* button to invoke the *Add Constraint* dialog box. To enter the minimum requirement constraints: Click inside the *Cell Reference* box and enter E17:E19; then select >= from the drop-down menu; and then click on the *Constraint* box and enter F7:F9. Click on *Add*. To enter the maximum allowed constraint, click inside the *Cell Reference* box and enter E20; select <= from the drop-down menu; and then click on the *Constraint* box and enter G10. Click on the *OK* Button.
6. Finally, select *Options*, and check *Assume Non-Negative* and *Assume Linear Model*.

To solve the model, click the *Solve* button. After the problem has been solved, the *Solver Results* box will appear (Figure 4.13). You have the option of keeping the solution found by Solver in the spreadsheet or restoring the original values. Also, as shown in the *Reports* box, three different reports may be automatically generated. You may select any of these reports by clicking on them. In the next section, each of the reports is described.

FIGURE 4.13
Solver results dialog box (Click on the reports you wish to save)

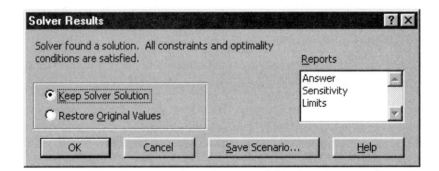

Excel Exercise 4.6

Open the Colorado Cattle Company worksheet *ccc.xls*. Use Solver to verify the optimal solution of 8 lbs. of grade 1, 5.5 lbs. of grade 2, and 0.5 lbs. of grade 3.

Solving a linear programming model can result in four possible outcomes:

1. Unique optimal solution
2. Alternate optimal solutions
3. Unboundedness
4. Infeasibility

When an LP has a **unique optimal solution,** it means that there is exactly one solution that will result in the maximum (or minimum) objective. If an LP has **alternate optimal solutions,** the objective is maximized (or minimized) by more than a single plan. This is good news, because this type of solution gives management more flexibility in achieving the objective. Although any alternate optimal solution is as good as any other, there may be qualitative factors that make one solution more attractive than the others. Unfortunately, computer programs such as Solver typically report only one of the many possible alternate optimal solutions, and it is difficult to identify others.

A problem is **unbounded** if the objective can be maximized to infinity (or minimized to negative infinity). Ordinarily, this means that the real system has not been correctly modeled. Since we live in a world of limited resources, it is impossible to achieve an infinite profit or a negatively infinite cost; thus, an unbounded problem typically means that some constraint or set of constraints has been left out of the model. If you have an unbounded problem, you should check to make sure that all of the restrictions of the problem are in the model and that the model inputs have the appropriate sign. Your problem is unbounded if Solver reports, "The Set Cells do not converge."

Finally, if an LP is **infeasible,** it means that there is no plan that satisfies all of the restrictions of the problem. If this occurs, some of the restrictions originally imposed will have to be relaxed in order to obtain a feasible solution. Infeasible problems *can* occur in practice. For example, a company could be in the enviable position of having increasing demand for its product, but could find that it no longer has the capacity to satisfy this demand. In the infeasible case, Solver will report, "Solver could not find a feasible solution."

INTERPRETING SOLVER REPORTS AND SENSITIVITY ANALYSIS

Solver generates three reports for linear programs, an *Answer Report*, a *Limits Report*, and a *Sensitivity Report*. For the Colorado Cattle Company problem, these reports are shown in Figure 4.14.

	A	B	C	D	E	F	G
1	Microsoft Excel 8.0a Answer Report						
2	Worksheet: [CCC.xls]CCC						
3	Report Created: 3/13/99 7:39:46 PM						
4							
5							
6	Target Cell (Min)						
7		Cell	Name	Original Value	Final Value		
8		$E$16	Cost Total	$0.00000	$2.59000		
9							
10							
11	Adjustable Cells						
12		Cell	Name	Original Value	Final Value		
13		$B$15	Quantity Grade 1	0.000	8.000		
14		$C$15	Quantity Grade 2	0.000	5.500		
15		$D$15	Quantity Grade 3	0.000	0.500		
16							
17							
18	Constraints						
19		Cell	Name	Cell Value	Formula	Status	Slack
20		$E$17	Calcium Total	10.000	$E$17>=$F$7	Binding	0.000
21		$E$18	Iron Total	12.000	$E$18>=$F$8	Binding	0.000
22		$E$19	Protein Total	15.100	$E$19>=$F$9	Not Binding	0.100
23		$E$20	Fat Total	7.500	$E$20<=$G$10	Binding	0

FIGURE 4.14

Solver reports
for CCC
example

(a) *Answer
Report*

FIGURE 4.14

Solver reports
for CCC
example,
continued

(b) *Limits
Report*

(c) *Sensitivity
Report*

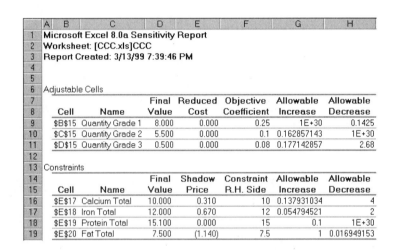

Answer and Limit Reports

The *Answer Report* gives the original and final value of the Target Cell and all Adjustable Cells, a listing of each constraint and its status, and the values of the slack variables. From the *Answer Report*, we see that the solution to this model is to use 8 lbs. of grade 1, 5.5 lbs of grade 2, and .5 lbs. of grade 3, for a total cost of $2.59 per cow per day. We can verify the values of the optimal slack variables by substituting the adjustable cell values into the constraint functions:

$$0.7(8) + 0.8(5.5) + 0(.5) - s_1 = 10 \quad \text{(Calcium)}$$

$$0.9(8) + 0.8(5.5) + 0.8(.5) - s_2 = 12 \quad \text{(Iron)}$$

$$0.8(8) + 1.5(5.5) + 0.9(.5) - s_3 = 15 \quad \text{(Protein)}$$

$$0.5(8) + 0.6(5.5) + 0.4(.5) + s_4 = 7.5 \quad \text{(Fat)}$$

Solving each equation for the slack variables, we obtain $s_1 = 0$, $s_2 = 0$, $s_3 = 0.1$, and $s_4 = 0$. Hence, the minimum requirements on calcium and iron are met exactly, there is 0.1 units of excess protein, and the fat constraint is at its upper limit. Inequality constraints that are satisfied by the optimal solution as equalities are called **binding constraints**. Thus, the calcium, iron, and fat constraints are binding. These are indicated in the *Answer Report* under the *Status* column.

The *Limits Report* gives the lower and upper limits that each adjustable cell can assume when all other adjustable cells are held at their current value, while also ensuring that all constraints are satisfied. For the CCC example, we see that if the optimal values of any two grades of feed are held constant, there is no freedom to change the third grade and still remain feasible. This is why it was very difficult to identify a feasible solution by intuition alone for this problem. The *Limits Report* provides an indication of the degree of "tightness" of a model.

Sensitivity Report

The *Sensitivity Report* provides some very important and useful managerial information. We will discuss the *Constraints* section first. In this section, the **shadow price** for each constraint represents *the change in the objective function as the value of the right-hand side of that constraint is increased by 1 with all other model parameters fixed*. For example, in the CCC problem, the shadow price of the fat constraint is −1.14. Therefore, an increase of 1 unit of fat allowed (to 8.5) will change the objective function by −$1.14 (an improvement, since we are minimizing). Why does this make sense? If you change the right-hand side of the fat constraint to 8.5 and re-solve the model, you will find that the new solution is to use 0 pounds of Grade 1, 12.5 pounds of Grade 2, and 2.5 pounds of Grade 3. Comparing this with the solution in Figure 4.14, we see that the amount of Grade 1 has decreased by 8 pounds, the amount of Grade 2 has increased by 7 pounds, and the amount of Grade 3 has increased by 2 pounds. The net change in the total cost is

$$\text{Net cost change} = \$0.25(-8) + \$0.10(+7) + \$0.08(+2) = -\$1.14$$

Shadow prices can always be found by changing the right-hand side of a constraint by 1, re-solving the model, and comparing the difference in the objective function values—*with one important caveat*. The shadow price remains constant only over a range of values for which *the same constraints remain binding*. When the allowable fat is increased to 8.5, you will find that the new solution still provides 10 units of calcium and 12 units of iron in addition to the 8.5 units of fat; the slacks on these constraints are 0, as before, and therefore are binding. The total amount of protein provided increases to 21 units, well above the minimum requirement, and remains a nonbinding constraint. These ranges can be found in the *Allowable Increase* and *Allowable Decrease* columns in the *Sensitivity Report*. For example, the *Allowable Increase* for the fat constraint is 1.0, and the *Allowable Decrease* is 0.016949153. Thus, if the amount of fat allowed falls between 7.5 − .016949153 = 7.483050847 and 7.5 + 1 = 8.5, then for any change of the right-hand side value within this range, *and with all other elements of the model unchanged*, we can predict the value of the optimal objective function by multiplying the change in the allowable fat content by the shadow price. Outside of this range, we would need to resolve the problem to determine the new value of the objective function. Thus, for instance, an increase of only 0.5 units of fat will cause a change of (0.5)(−1.14) = −0.57 in the objective function. Similarly, a *decrease* of .01 units of fat allowed will change the objective by (−0.01)(−1.14) = +$0.0114 (Watch the negative signs!). By putting a tighter restriction on a constraint, the objective function can never improve.

Applying these ideas to the calcium constraint, we see that if the minimum calcium requirement were to increase to 10.1—a tighter restriction within the *Allowable*

Increase—the objective function would change by $(0.1)(0.31) = \$0.031$, making it more costly to produce the feed. However, if we relax the calcium restriction to 9 units (within the *Allowable Decrease* of 4), the total cost would decrease by \$0.31. In summary, the shadow price simply reflects the simultaneous effects of changes in the decision variables on the objective function, while preserving the integrity of the constraints.

Note that if a constraint is not binding, as we see with the protein constraint, the shadow price will be 0. This will always be true, because if the optimal solution does not use all of an available resource, then increasing the amount of the resource will have no effect, since an excess amount already exists. The same holds true if an optimal solution provides more than a minimum requirement—decreasing the minimum requirement will not change anything since it is already exceeded.

Excel Exercise 4.7

Using the *ccc.xls* spreadsheet model, increase the right-hand side of the iron constraint by 0.01 and re-solve the model. Compute the shadow price by examining the changes in the decision variables and noting their effects on the total cost. Predict the new cost if the iron requirement is 11.5, and verify this by re-solving the model.

In the *Adjustable Cells* section of the *Sensitivity Report*, the **reduced cost** gives the shadow price of the nonnegativity constraint associated with each decision variable (provided that the variable is not constrained by a simple, non-0 lower bound, which we will discuss later). In this case, however, all three variables are positive, so the non-negativity constraints are not binding and the reduced costs are 0. Another interpretation of a non-0 reduced cost (which must be associated with a binding nonnegativity constraint) is the amount by which the objective function coefficient would have to change before it would be profitable to make that variable positive. For example, if we are minimizing and a decision variable is 0 in an optimal solution, then it must be too costly relative to the other variables. The reduced cost will tell how much the cost coefficient would have to decrease before it would be profitable to use this variable.

The *Allowable Increase* and *Allowable Decrease* columns in the *Adjustable Cells* section tell how much the objective function coefficient can change before the current solution (that is, the optimal values of the decision variables) will no longer be optimal (with everything else held fixed). For example, in the CCC problem, the cost of grade 1 may increase without limit ("$1E + 30$" represents infinity) and may decrease by up to 0.1425, and the current solution—that is, the values of the decision variables—will still remain optimal. However, the objective function value *will* change. Thus, if the cost of Grade 1 were reduced to \$0.20 (a decrease of 0.05, which is within the allowable range), the optimal solution would still be to use 8 lbs. of Grade 1, 5.5 lbs of Grade 2, and .5 lbs of Grade 3. However, because the cost of Grade 1 has been changed, the actual total cost would be $.20(8) + .10(5.5) + .08(0.5) = \$2.19$. If the cost of Grade 1 were reduced to, say, \$0.10 (a decrease of 0.15, which is beyond the allowable decrease), then we would have to re-solve the problem to determine a new optimal solution.

MANAGERIAL USES OF SENSITIVITY INFORMATION Why are shadow prices useful to a manager? They provide guidance on how to reallocate resources or change model parameters over which the manager may have control. In LP models, the parameters of some constraints cannot be controlled. For instance, the amount of time available for production or physical limitations on machine capacities would clearly be uncontrollable. Other constraints represent policy decisions, which in essence are arbitrary. The nutritional requirements in the CCC problem are an example of this. These can easily be changed; doing so, however, would affect both the quality of cattle and the cost of doing business. In this example, we saw that raising the allowable fat content would cause a change in the feed mixture that would reduce the total cost of the feed mix. We might also consider reducing the required amount of calcium or iron in order to reduce the total cost, since their shadow prices are positive. To use the shadow prices to predict effects of changes, you must remember to stay within the allowable increase and decrease and keep all other model parameters at their original values. If you want to investigate multiple changes in constraints, you need to re-solve the model.

Shadow prices are often useful to determine the value of additional amounts of a limited resource. For example, if you solve the Santa Cruz MicroProducts example introduced in Chapter 1 and discussed at the beginning of this chapter, you will find that the shadow price for the component B constraint is $3.75. Although it is correct to state that having an additional unit of component B will improve profit by $3.75, does this necessarily mean that the company should spend up to this amount for additional units? This depends on whether the relevant costs have been included in the objective function coefficients. If the costs of the components *have not* been included in the objective function unit profit coefficients, then the company will benefit by paying up to $3.75 for additional components. However, if the component costs *have* been included in the profit calculations, the company should be willing to pay up to an *additional* $3.75, over and above the component costs that have already been included in the unit profit calculations.

The sensitivity ranges for the objective function coefficients provide a manager with some confidence about the stability of the solution in the face of uncertainty. If the allowable ranges are large, then reasonable errors in estimating the coefficients will have no effect on the optimal policy (although they will affect the value of the objective function). Tight ranges suggest that more effort might be spent in ensuring that accurate data or estimates are used in the model.

SOLVER OUTPUT AND LPS WITH LOWER OR UPPER BOUNDS The version of Solver included in Microsoft Office 97 handles simple lower bounds (e.g., $x \geq 10$) and upper bounds (e.g., $x \leq 150$) quite differently than ordinary constraints in the *Sensitivity Report*. To see this, consider a modified version of Kathryn's investment problem (Example 3) to which we have added lower bound constraints (STOCK$i \geq 10$) and upper bound constraints (STOCK$i \leq 50$) for each variable. The *Sensitivity Report* is shown in Figure 4.15.

In Solver, lower bounds are treated in a manner similar to nonnegativity constraints, which do not appear *explicitly* as constraints in the model. We noted that reduced costs are essentially shadow prices of nonnegativity constraints. Thus, a variable at its lower bound is similar to one being 0 in a typical LP model. In the Solver output, only the structural model constraints are listed with shadow prices in the *Constraints* section of the sensitivity report. Any variable that is at its lower bound in the final solution will appear in the *Adjustable Cells* section and have a non-0 reduced cost.

FIGURE 4.15

Sensitivity
report for
Kathryn's
investment
problem

	Cell	Name	Final Value	Reduced Cost	Objective Coefficient	Allowable Increase	Allowable Decrease
Microsoft Excel 8.0a Sensitivity Report							
Worksheet: [Kathryn.xls]Model							
Report Created: 3/13/99 7:45:51 PM							
Adjustable Cells							
$B$14	Number of shares Stock 1	10	0	19.04	0	1E+30
$C$14	Number of shares Stock 2	10	-1	10.2	1	1E+30
$D$14	Number of shares Stock 3	50	0.32	9.28	1E+30	0.32
$E$14	Number of shares Stock 4	13.2	0	28	1	0

Constraints	Cell	Name	Final Value	Shadow Price	Constraint R.H. Side	Allowable Increase	Allowable Decrease
	$F$16	Investment Total	$2,000.00	$0.56	2000	1840	160

In this case, the reduced cost may be interpreted as the shadow price of the bound constraint.

For example, STOCK2 is at its lower bound in the optimal solution, and its reduced cost is -1. This means that if the lower bound is increased from 10 to 11, the expected return will decrease by $1.00.

The same thing holds true for variables at their upper bounds. For example, STOCK3 is at its upper bound of 50. Its reduced cost is 0.32. If the upper bound is increased from 50 to 51, expected profit will increase by $.32.

Excel Exercise 4.8

Solve the market effort allocation model (Example 4 and Excel Exercise 4.2) and determine the shadow prices of the bound constraints.

Scenario Analysis

The model constructed with the initial values of the input parameters is often referred to as a *base case*. From the base case scenario, we may vary model parameters to study changes of the model solution to the input data beyond that provided by the *Sensitivity Report*. Suppose that we want to investigate further the impact of the fat constraint in the Colorado Cattle Company model. The base case fat constraint is

$$0.5GRADE1 + 0.6GRADE2 + 0.4GRADE3 \leq 7.5$$

We know that the solution will remain constant if the allowable fat is between 7.483050847 and 8.5. How does the solution change outside of this range? We can answer this question by varying the right-hand side of the fat constraint from 6 to 9 in increments of 0.1 and solving the model repeatedly.

If we set the fat limitation to 7.4, we find that the problem is infeasible; thus, below the *Allowable Decrease*, it is impossible to meet all the nutritional requirements. If we set the fat right-hand side to 8.6 (just outside the high side of the allowable range for the fat shadow price), we find a new solution: use no Grade 1, 12.5 pounds of Grade 2, and 2.5 pounds of Grade 3 for a cost of $1.45. The *Sensitivity Report* for this problem is shown in Figure 4.16. Notice that the fat constraint is no longer binding (only 8.5 units of fat are used in the optimal solution). The fat constraint now has a shadow price of 0 and an allowable increase of infinity. Hence, for all fat limitations greater than 8.5, we have found the optimal solution. Figure 4.17 shows the minimal cost as a function of allowable fat. As expected, the objective function drops at a rate of $1.14 until allowable fat is 8.5, and then remains constant at $1.45 once the fat constraint is no longer binding. For larger, more complicated models, a graph such as Figure 4.17 might have many more breaks (that is, changes in shadow prices and hence changes in slope of the graph). Figure 4.18 shows the set of solutions in this range. This chart clearly shows that increasing fat content causes a substitution of a combination of Grades 2 and 3 for Grade 1 until Grade 1 is no longer used.

	Microsoft Excel 8.0a Sensitivity Report						
1	Microsoft Excel 8.0a Sensitivity Report						
2	Worksheet: [CCC.xls]CCC						
3	Report Created: 3/13/99 7:48:09 PM						
4							
5							
6	Adjustable Cells						
7			Final	Reduced	Objective	Allowable	Allowable
8	Cell	Name	Value	Cost	Coefficient	Increase	Decrease
9	$B$15	Quantity Grade 1	0.000	0.143	0.25	1E+30	0.1425
10	$C$15	Quantity Grade 2	12.500	0.000	0.1	0.162857143	0.02
11	$D$15	Quantity Grade 3	2.500	0.000	0.08	0.02	0.08
12							
13	Constraints						
14			Final	Shadow	Constraint	Allowable	Allowable
15	Cell	Name	Value	Price	R.H. Side	Increase	Decrease
16	$E$17	Calcium Total	10.000	0.025	10	0.4	8
17	$E$18	Iron Total	12.000	0.100	12	0.2	2
18	$E$19	Protein Total	21.000	0.000	15	6	1E+30
19	$E$20	Fat Total	8.500	0.000	8.6	1E+30	0.1

FIGURE 4.16
Sensitivity report for CCC example with fat limitation = 8.5

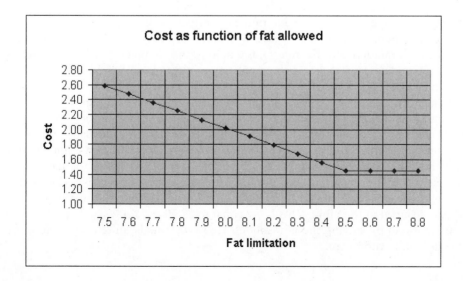

FIGURE 4.17
Graph of total cost as fat limitation is increased

FIGURE 4.18
Changes in
optimal solution
as fat limitation
is increased

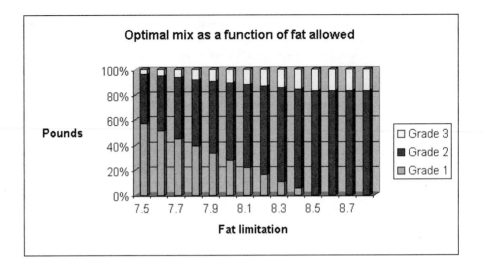

Thus, we may use the allowable shadow price ranges to identify the break points, and then solve new problems just outside of these ranges until either the solution becomes infeasible or the allowable change becomes infinite.

Excel Exercise 4.9

Solve the Santa Cruz MicroProducts model with Solver. Create charts similar to Figures 4.17 and 4.18 that show the change in profit as the numbers of available components A and B vary and how the actual solutions change.

As we noted earlier, if multiple changes are made to the parameters in a model, the model must be re-solved to determine the effect on the objective function value and the solution. For example, Figure 4.19 shows the effect of simultaneously varying the fat limitation and the iron requirement in the Colorado Cattle Company example. The data used to create the graph (by solving $21 \times 15 = 315$ problems) are shown in Table 4.3. From the *Sensitivity Report* shown in Figure 4.14, we know that the shadow price for iron is $.67 (allowable decrease of 2, and note that if we decrease the requirement, the effect will be $-\$.67$) and $-\$1.14$ for fat (allowable increase of 1). Notice from Table 4.3 that these prices do in fact hold if the one parameter is varied and the other is fixed at its original value. For example, if the iron requirement is fixed at 12, the objective function decreases by $1.14(0.1) = .114$ as the fat limit increases to 8.5. However, for multiple changes the effect is not additive. Consider an iron requirement of 11 and fat limit of 8.5. The optimal objective function is in fact $1.35, as shown in the table. This is *not* equal to $2.59 − 1.14 − .67 = \$.78$ and can only be found by solving the new model. However, this is quite easy to do in the spreadsheet environment.

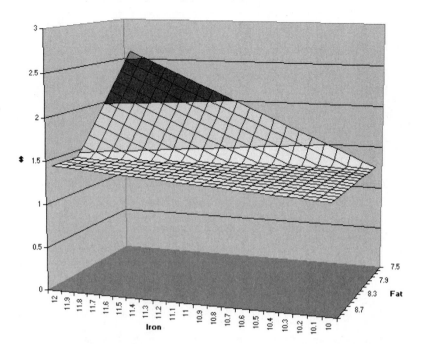

Cost vs Iron and Fat

FIGURE 4.19
3-D graph of
optimal cost as
a function of
fat and iron
constraints

SOLVING MULTIOBJECTIVE MODELS

To assist managers in making decisions with multiple objectives, we need a method of dealing with the tradeoffs among them. We will describe three approaches: *weighted objectives*, *absolute priorities*, and *goal programming*.

Weighted Objective Approach

A simple approach to handle multiple conflicting objectives is to *weight* the different objectives according to their importance, combine them into a single linear objective, and then solve the linear program with this new combined objective. This approach relies on subjective judgement of the decision maker to determine the relative importance of each of the objectives.

We usually assume that the weightings of each objective are nonnegative and that they sum to one. For instance, if objective 1 is four times as important as objective 2, the first objective would receive a weight of $4/5 = 0.8$ while the second objective would be weighted by $1/5 = 0.2$. The combined weighted objective would be 0.8(objective 1) + 0.2(objective 2). It is important to ensure that the parameters in all objectives have roughly the same order of magnitude; that is, their coefficients are of similar size. If this is not true, the relative weights will not cause much differentiation among the objectives. This can be overcome by *scaling* the objectives—that is, dividing by a constant to equalize the magnitude of the parameters.

TABLE 4.3 Optimal cost for various iron and fat amounts

Iron	Fat														
	7.5	7.6	7.7	7.8	7.9	8	8.1	8.2	8.3	8.4	8.5	8.6	8.7	8.8	8.9
10	1.25	1.25	1.25	1.25	1.25	1.25	1.25	1.25	1.25	1.25	1.25	1.25	1.25	1.25	1.25
10.1	1.317	1.26	1.26	1.26	1.26	1.26	1.26	1.26	1.26	1.26	1.26	1.26	1.26	1.26	1.26
10.2	1.384	1.27	1.27	1.27	1.27	1.27	1.27	1.27	1.27	1.27	1.27	1.27	1.27	1.27	1.27
10.3	1.451	1.337	1.28	1.28	1.28	1.28	1.28	1.28	1.28	1.28	1.28	1.28	1.28	1.28	1.28
10.4	1.518	1.404	1.29	1.29	1.29	1.29	1.29	1.29	1.29	1.29	1.29	1.29	1.29	1.29	1.29
10.5	1.585	1.471	1.357	1.3	1.3	1.3	1.3	1.3	1.3	1.3	1.3	1.3	1.3	1.3	1.3
10.6	1.652	1.538	1.424	1.31	1.31	1.31	1.31	1.31	1.31	1.31	1.31	1.31	1.31	1.31	1.31
10.7	1.719	1.605	1.491	1.377	1.32	1.32	1.32	1.32	1.32	1.32	1.32	1.32	1.32	1.32	1.32
10.8	1.786	1.672	1.558	1.444	1.33	1.33	1.33	1.33	1.33	1.33	1.33	1.33	1.33	1.33	1.33
10.9	1.853	1.739	1.625	1.511	1.397	1.34	1.34	1.34	1.34	1.34	1.34	1.34	1.34	1.34	1.34
11	1.92	1.806	1.692	1.578	1.464	1.35	1.35	1.35	1.35	1.35	1.35	1.35	1.35	1.35	1.35
11.1	1.987	1.873	1.759	1.645	1.531	1.417	1.36	1.36	1.36	1.36	1.36	1.36	1.36	1.36	1.36
11.2	2.054	1.94	1.826	1.712	1.598	1.484	1.37	1.37	1.37	1.37	1.37	1.37	1.37	1.37	1.37
11.3	2.121	2.007	1.893	1.779	1.665	1.551	1.437	1.38	1.38	1.38	1.38	1.38	1.38	1.38	1.38
11.4	2.188	2.074	1.96	1.846	1.732	1.618	1.504	1.39	1.39	1.39	1.39	1.39	1.39	1.39	1.39
11.5	2.255	2.141	2.027	1.913	1.799	1.685	1.571	1.457	1.4	1.4	1.4	1.4	1.4	1.4	1.4
11.6	2.322	2.208	2.094	1.98	1.866	1.752	1.638	1.524	1.41	1.41	1.41	1.41	1.41	1.41	1.41
11.7	2.389	2.275	2.161	2.047	1.933	1.819	1.705	1.591	1.477	1.42	1.42	1.42	1.42	1.42	1.42
11.8	2.456	2.342	2.228	2.114	2	1.886	1.772	1.658	1.544	1.43	1.43	1.43	1.43	1.43	1.43
11.9	2.523	2.409	2.295	2.181	2.067	1.953	1.839	1.725	1.611	1.497	1.44	1.44	1.44	1.44	1.44
12	2.59	2.476	2.362	2.248	2.134	2.02	1.906	1.792	1.678	1.564	1.45	1.45	1.45	1.45	1.45

Example 10: An Example of Weighted Objectives

Suppose that Mary Trump, the investor from Example 1, believes that maximizing expected return is four times as important as minimizing risk. To create a single objective, we first need to convert one of them so that both are either maximization or minimization objectives. We will convert the minimization of the risk function to a maximization objective by changing the signs of the coefficients, that is, Minimize $.04x_1$ is the same as Maximize $-.04x_1$. The combined weighted objective function is Maximize $0.8(.06x_1 + .03x_2) + 0.2(-.04x_1) = .04x_1 + .024x_2$. We may now solve Mary's problem with the following linear program:

Maximize $.04x_1 + .024x_2$

subject to

$$x_1 + x_2 \le 1{,}000 \qquad \text{(available funds)}$$

$$200 \le x_1 \le 700 \qquad \text{(investment limits on option 1)}$$

$$200 \le x_2 \le 700 \qquad \text{(investment limits on option 2)}$$

The optimal solution to this problem is to invest $700 in option 1 and $300 in option 2, with an objective function value of 35.2. However, the objective function has no real meaning in the context of this problem. Of real interest is the value of the multiple objectives:

$$\text{Profit} = .06(700) + .03(300) = \$51$$

$$\text{Risk} = .04(700) = \$28$$

Thus, we must evaluate each objective at the optimal point.

The combined weighting approach provides a means for generating a variety of solutions for different importance weightings. For example, suppose that w_1 is the weight for the expected return and w_2 is the weight for risk. Then Mary's objective may be written Maximize $w_1(.06x_1 + .03x_2) + w_2(-.04x_1)$, or Maximize $(.06w_1 - .04w_2)x_1 + .03w_1x_2$. Table 4.4 gives the objective function, solution, its expected return, and risk, as we vary the weights by 0.1.

w_1	w_2	Objective Function	x_1	x_2	Return	Risk
0.0	1.0	$-.04x_1 + .00x_2$	200	700	$33	$ 8
0.1	0.9	$-.03x_1 + .003x_2$	200	700	$33	$ 8
0.2	0.8	$-.02x_1 + .006x_2$	200	700	$33	$ 8
0.3	0.7	$-.01x_1 + .009x_2$	200	700	$33	$ 8
0.4	0.6	$.00x_1 + .012x_2$	200	700	$33	$ 8
0.5	0.5	$.01x_1 + .015x_2$	300	700	$39	$12
0.6	0.4	$.02x_1 + .018x_2$	700	300	$51	$28
0.7	0.3	$.03x_1 + .021x_2$	700	300	$51	$28
0.8	0.2	$.04x_1 + .024x_2$	700	300	$51	$28
0.9	0.1	$.05x_1 + .027x_2$	700	300	$51	$28
1.0	0.0	$.06x_1 + .03x_2$	700	300	$51	$28

TABLE 4.4
Solutions for various weightings of return and risk

Note that the solution for $w_1 = 0$ is what we would have obtained if we ignored the return and simply minimized the risk; the solution for $w_2 = 0$ represents that of maximizing the return while ignoring the risk. Weighted objectives between these two extremes may result in solutions different from those obtained by each of the individual objectives. The solution for equal weightings is an example of this. The solutions are shown graphically in Figure 4.20. The dashed curve is often called the **efficient frontier**. It defines solutions that give maximum expected return for a given level of risk (or alternatively, minimal risk for a given level of expected return).

FIGURE 4.20
Efficient frontier for risk and return

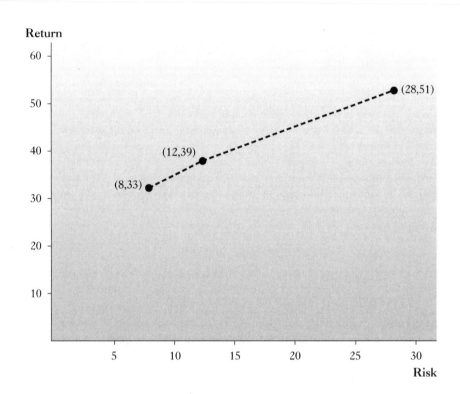

Absolute Priorities Approach

A second approach for dealing with multiple objectives in linear programs is known as the *absolute priorities approach*. In this approach, we rank the objectives from highest to lowest priority and solve a series of linear programs, first using the highest priority objective. In the second LP, we add a constraint by setting the objective function of the highest priority objective equal to the objective function value achieved in the first LP. This ensures that the best value for the highest priority objective will be maintained in subsequent solutions. We then solve the second LP using the second priority objective. We continue in this fashion, adding a constraint that fixes the value of the previous objective function and using the next objective in priority order until all objectives have been evaluated.

Example 11: An Example of Absolute Priorities

In Mary's investment problem, suppose that she determines that maximizing return is a higher-priority objective than minimizing risk:

Priority 1 objective: Max $.06x_1 + .03x_2$

Priority 2 objective: Min $.04x_1$

We solve the first problem using the priority 1 objective:

Maximize $.06x_1 + .03x_2$

subject to

$$x_1 + x_2 \leq 1,000 \qquad \text{(available funds)}$$
$$200 \leq x_1 \leq 700 \qquad \text{(investment limits on option 1)}$$
$$200 \leq x_2 \leq 700 \qquad \text{(investment limits on option 2)}$$

This leads to the solution $x_1 = 700$, $x_2 = 300$ with a maximum return of \$51. Next, we solve for the priority 2 objective, while ensuring that the value of expected return remains \$51:

Minimize $.04x_1$

subject to

$$x_1 + x_2 \leq 1,000 \qquad \text{(available funds)}$$
$$200 \leq x_1 \leq 700 \qquad \text{(investment limits on option 1)}$$
$$200 \leq x_2 \leq 700 \qquad \text{(investment limits on option 2)}$$
$$.06x_1 + .03x_2 = 51 \qquad \text{(priority 1 objective)}$$

This solution is also $x_1 = 700$, $x_2 = 300$, with a risk of \$28. The solution did not change because the priority 1 LP had a unique optimal solution. If alternate optimal solutions existed for the priority 1 problem, then the solution of the priority 2 problem would find the one that also optimized the priority 2 objective.

The advantage of the absolute priorities approach over the weighted objective approach is that only a ranking of the priorities (rather than a specification of weights) is required. However, the absolute priorities approach does not really trade off one objective with another, and as this example shows, once a unique solution is encountered, the lower priority objectives have no effect.

Goal-Programming Approach

In the goal-programming approach, a *target* or *goal* is set for each of the objectives, which are then expressed as *goal constraints* (see the next Management Science Practice box). A **goal constraint** is one that has the form

$$f(x) + s - r = b$$

where $f(x)$ is an objective function, b is the goal, and s and r are nonnegative variables called **deviational variables**. Observe that if both s and r are 0, then the constraint is satisfied as an equality, $f(x) = b$. Hence, minimizing the objective function $s + r$ attempts to force the objective to meet the goal. Of course, this may not happen, and the optimal solution might result in $s + r > 0$, in which case the goal cannot be achieved.

Instead of meeting the goal exactly, our objective might be to ensure that $f(x)$ will not exceed b. This can be accomplished by minimizing r. Note that if the optimal solution is $r = 0$, then for any nonnegative value of s, $f(x) \leq b$. Similarly, if $s = 0$, then for any nonnegative value of r, $f(x) \geq b$, and minimizing s seeks to guarantee that $f(x)$ will be no smaller than b. Thus, the objective function for a goal program is simply a function of the deviational variables of the goal constraints. We may use either the weighted objective approach or the absolute priorities approach to find a solution.

Example 12: Goal Programming Solution

Suppose that Mary sets goals for the expected return to be at least \$45 and the risk no greater than \$12. Her goal constraints would be

$$.06x_1 + .03x_2 + s_1 - r_1 = 45 \qquad \text{(investment goal)}$$
$$.04x_1 + s_2 - r_2 = 12 \qquad \text{(risk goal)}$$

To achieve her objectives, she needs to drive s_1 and r_2 to 0 if possible. If both goals are equal in priority, Mary could solve the LP

Minimize $s_1 + r_2$

subject to

$$x_1 + x_2 \leq 1{,}000 \qquad \text{(available funds)}$$
$$200 \leq x_1 \leq 700 \qquad \text{(investment limits on option 1)}$$
$$200 \leq x_2 \leq 700 \qquad \text{(investment limits on option 2)}$$
$$.06x_1 + .03x_2 + s_1 - r_1 = 45 \qquad \text{(investment goal)}$$
$$.04x_1 + s_2 - r_2 = 12 \qquad \text{(risk goal)}$$

On the other hand, suppose that she considers the return goal three times as important as the risk goal. In this case she would use the objective function

Minimize $.75s_1 + .25r_2$

Finally, we can use the absolute priorities approach to solve this model. If Mary considers the return goal to be higher priority than the risk goal, she could solve two LPs, with the first priority objective function:

Priority 1 Objective: *Minimize s_1*

and the second priority objective function:

Priority 2 Objective: *Minimize r_2*

Excel Exercise 4.10

Solve Mary Trump's problem using the three approaches described in Example 12: equal priority goals, return goal three times as important as the risk goal, and absolute priorities. Compare the results.

Management Science Practice: Goal Programming for Site Location at Truck Transport Corporation

The liquid foods division of Truck Transport Corporation uses five terminals to service its customers. Each terminal must recruit and maintain a force of independent truckers. The independent truckers own their own vehicles, but work for the company delivering goods to its customers. Truck Transport Company, after proceeding through stages one and two of the location analysis, decided on five potential sites for the relocation of its East St. Louis terminal. Because the problem had multiple conflicting objectives, the need for a systematic approach to evaluating each of the five candidates was met with the application of a goal-programming model.

The traditional objective of minimizing the cost of the transportation of goods to the customers had to be considered. However, because Truck Transport used independent truckers, their satisfaction with a new facility also had to be taken into account. The cost of driving a truck from the trucker's home to the terminal is a cost borne by the trucker. If the new facility put an excessive cost on the drivers, they might jump ship and go to another trucking company. The willingness (or lack of willingness) of each trucker currently on the workforce to work with certain managers assigned to the existing or the new facility also needed to be considered. Previous studies had shown that dissatisfaction with managers at different terminals is a major reason for drivers leaving the firm. Likewise, certain customers did not work well with some of the managers. Therefore, Truck Transport needed to develop a model that considered both customer and employee satisfaction, as well as the cost of drivers getting to and from work and the cost of the transport of goods to its customers. The company used data for 12 major customers and 22 drivers in the goal-programming model. The following five goals (in order of priority) were used:

1. Minimize the deviation from the average number of trips required by each of the customers. Minimize the deviation from the requested number of trips for each driver.
2. Minimize the number of undesirable assignments of drivers to terminal managers.
3. Minimize the number of undesirable assignments of customers to terminal managers.
4. Minimize the transportation cost between the drivers' homes and the terminals.
5. Minimize the transportation cost between the terminals and the customers.

The goal-programming model had 170 decision variables, 128 deviational variables, and 65 goal constraints. The model was run with each of the candidate sites added to the existing locations. For each run, the first three priority goals were met; however, significant differences were found for the fourth and fifth goals. The site chosen to replace the current East St. Louis terminal caused the

continued

least amount of increase for both the drivers and the customers. Six months after the move to the new site, the operations manager reported that driver turnover was normal for the period, no significant change in customers occurred, and no complaints concerning excessive transportation costs from home to terminal were registered by the drivers.

Source: Adapted from M. Schniederjans, N. Kwak, and M. Helmer, "An Application of Goal Programming to Resolve a Site Location Problem," *Interfaces*, Vol. 12, No. 3, June 1982, 65–72.

TECHNOLOGY FOR LINEAR OPTIMIZATION

Because mathematical programming has been the foundation of management science, many computer packages for modeling and solving linear and integer (as well as nonlinear) programming problems exist. These range from simple, student-friendly packages to powerful commercial software capable of handling the most complex models. Some of the more popular software programs are described in Table 4.5.

TABLE 4.5
Some popular linear optimization software

Product	Company	Description
AMPL	Duxbury Press Pacific Grove, Cal. **www.duxbury.com**	Developed by AT&T and Bell Laboratories, allows users to formulate models using common algebraic notation
GAMS	GAMS Development Corporation Washington, D.C. **www.gams.com**	Designed for modeling and solving large complex, linear, nonlinear, and mixed integer problems
CPLEX	CPLEX Optimization, Inc. Incline Village, Nev. **www.cplex.com**	A high-performance, robust solver for linear, mixed integer, and certain nonlinear programs
Premium Solver Premium Solver Plus	Frontline Systems, Inc. Incline Village, Nev. **www.frontsys.com**	More powerful extensions to the standard Solver provided in Excel
LINDO	LINDO Systems Chicago, Ill. **www.lindo.com**	Combines large-scale linear and integer optimizers, as well as nonlinear optimizers, and interactive modeling environment
MINOS	Stanford Business Software Mountain View, Cal.	Links with GAMS and AMPL and can accommodate many thousands of constraints and variables
What'sBest!	LINDO Systems Chicago, Ill. **www.lindo.com**	A spreadsheet add-in for linear, integer, and nonlinear models

PROBLEMS

1. Tommy's Tables produces two types of tables, regular and deluxe. These two models use the same legs (each table requires four), but the deluxe is larger because it has a larger surface area. It takes 0.5 hours to finish a regular table and 1 hour to finish a deluxe table. This week Tommy has 45 hours of labor available for finishing and 240 usable table legs on hand in the shop. Each regular table produces a profit of $200, and each deluxe produces a profit of $425.
   a. Develop an LP model that will yield the number of regular and deluxe tables to produce in order to maximize profit.
   b. Implement your model on a spreadsheet and find the best solution you can, using your intuition.
   c. Find an optimal solution using Solver.
2. Kilgore's Deli is a small delicatessen located near a major university. Kilgore's does a large walk-in carry-out lunch business. The deli offers two luncheon chili specials, Wimpy and Dial 911. At the beginning of the day, Kilgore needs to decide how much of each special to make (he always sells out of whatever he makes). The profit on one serving of Wimpy is $.45, on one serving of Dial 911, $.58. Each serving of Wimpy requires .25 pound of beef, .25 cup of onions, and 5 ounces of Kilgore's special sauce. Each serving of Dial 911 requires .25 pound of beef, .4 cup of onions, 2 ounces of Kilgore's special sauce, and 5 ounces of hot sauce. Today, Kilgore has 20 pounds of beef, 15 cups of onions, 88 ounces of Kilgore's special sauce, and 60 ounces of hot sauce on hand.
   a. Develop an LP model that will tell Kilgore how many servings of Wimpy and Dial 911 to make in order to maximize his profit today.
   b. Implement your model on a spreadsheet and find the best solution you can, using your intuition.
   c. Find an optimal solution using Solver.
3. Grad & Sons Insurance carries an investment portfolio of bonds, stocks, and other investment alternatives, such as real estate. At the beginning of next year, $500,000 will be available to invest in four different alternatives. The expected annual rates of return on these four options are 0.06, 0.09, 0.07, and 0.11. The risk factor per dollar invested is a measure of the inherent uncertainty of the expected return, so a low risk factor is desirable. The risk factors per dollar invested for the four investment alternatives are subjective estimates of the possible loss the investor could incur in the worst case analysis (0 means no risk, 1 means the investor could lose everything), which have been provided by a Grad & Sons analyst. They are .02, .05, .04, and .075, respectively. For diversification, no single alternative can be more than $200,000. Management needs a return of at least .08, but would like to minimize the risk taken to achieve that return.
   a. Develop an LP model to determine the amount to invest in each alternative in order to minimize risk.
   b. Implement your model on a spreadsheet model and find a good solution, using your intuition.
   c. Find an optimal solution using Solver.
4. Corry is an undergraduate management major at State U. He has just finished the requirements for the mandatory management science course. The course is composed of a midterm exam, a final exam, individual assignments, and class participation. He has earned an 86 percent on the midterm, a 94 percent on the final, a

93 percent on the individual assignments, and an 85 percent in participation. The benevolent management science instructor is allowing all the students in the class to determine their own weights for each of the four components that will determine their final grade. Of course, there are some restrictions on the weightings used to arrive at an overall course grade:

- The participation grade can be no more than 15 percent.
- The midterm exam score must count at least twice as much as the individual assignment score.
- The final exam score must count at least three times as much as the individual assignment score.
- Each of the four components must count for at least 10 percent of the course grade.
- The weights used must sum to 1 and be nonnegative.

a. Develop a model that will yield the set of valid weights to maximize Corry's score for the course.

b. Implement your model on a spreadsheet model and find a good solution, using your intuition.

c. Find an optimal solution using Solver.

5. The Decatur Nut Company packages and sells three different half-pound canned peanut mixes: Party Nuts, Mixed, and Delightful Mix. These generate a per-can revenue of $2.25, $3.37, and $6.49, respectively. Party Nuts are simply 100 percent peanuts. Mixed consist of 55 percent peanuts, 25 percent cashews, 10 percent Brazil nuts, and 10 percent hazelnuts. Delightful Mix is made up of 40 percent cashews, 20 percent Brazil nuts, and 40 percent hazelnuts. The company has on hand 500 pounds of peanuts, 175 pounds of cashews, 100 pounds of Brazil nuts, and 80 pounds of hazelnuts. The company would like to mix these nuts in a way that will yield maximum revenue.

a. Formulate this problem as a linear program.

b. Implement your model on a spreadsheet and find an optimal solution with Solver.

6. The Blanton Seed Company specializes in food products for birds and other household pets. In developing a new cockatiel mix, company nutritionists have specified that the mixture must contain at least 13 percent protein and 15 percent fat, and no more than 15 percent fiber. The percentages of each of these nutrients in eight types of ingredients that can be used in the mix are given in the next table, along with the wholesale cost per pound.

Ingredient	Protein %	Fat %	Fiber %	Cost/lb.
Sunflower seeds	16.9	26.0	29.0	$0.22
White millet	12.0	4.1	8.3	0.19
Kibble corn	8.5	3.8	2.7	0.10
Oats	15.4	6.3	2.4	0.10
Cracked corn	8.5	3.8	2.7	0.07
Wheat	12.0	1.7	2.3	0.05
Safflower	18.0	17.9	28.8	0.26
Canary grass seed	11.9	4.0	10.9	0.11

a. What is the minimum cost mixture that meets the stated nutritional requirements?

b. Because of weather-related problems, the price of sunflower seeds is expected to rise to $.50 per pound. How will this change the solution?

7. A large California-based consumer goods corporation has asked SALS Marketing, Inc., to develop an advertising campaign for one of its new products. SALS has promised a plan that will yield the highest possible exposure rating, a measurement of the ability to reach the appropriate demographic group, which is correlated with demand produced per dollar spent on advertising. The options for advertisements with their respective costs (per unit of advertising) and per-unit exposure ratings are given in Table 4.6. Of course, certain restrictions exist for the advertising campaign. The corporation has budgeted $800,000 for the campaign, but to restrict overexposure to any particular audience it wants no more than $300,000 put into any one medium. Also, to ensure a broad range of exposure, at least $100,000 must be spent in each medium.

  a. Develop an LP model for this situation.
  b. Implement your model on a spreadsheet and find an optimal solution using Solver.

Media		Cost/Unit	Exposure/Unit
Magazines	Literary	$ 7,500	15,000
	News	$10,000	22,500
	Topical	$15,000	24,000
Newspapers	Major Evening	$ 3,000	75,000
	Major Morning	$ 2,000	37,500
Television	Morning	$20,000	275,000
	Midday	$10,000	180,000
	Evening	$60,000	810,000
Radio	Morning	$15,000	180,000
	Midday	$15,000	17,000
	Evening	$10,000	16,000

TABLE 4.6
SALS Marketing data

8. A company produces four products with current variable costs of $9, $6.50, $5, and $7.50 per pound, respectively. Because the company works on a contractual basis with retailers, it knows the demands for each product for the next three months. These are specified in Table 4.7. Due to a new labor contract, the variable cost of each of the products will increase by 5 percent at the beginning of month 3. There are currently 50 pounds of each product on hand, and company policy dictates that at the end of the coming three-month period, there must also be an inventory of 50 pounds of each product. These four products share a common bottleneck machine that is available for 320 hours per month (two shifts of eight hours each per day, five days per week, four weeks per month). Product 1 needs .05 hour of the bottleneck machine per pound, product 2 requires .05 hour/pound, product 3 requires .2 hr/pound, and product 4 requires .1 hours/pound. The cost of holding inventory per pound per month is 10 percent of the cost of the product.

  a. Develop an LP model to meet demand and minimize the cost of production and inventory.
  b. Implement your model on a spreadsheet and find an optimal solution with Solver.

Product	Month 1	Month 2	Month 3
1	1,000	800	1,000
2	1,000	900	500
3	600	600	500
4	0	200	500

TABLE 4.7
Demand for each product by month

9. The Coger Company is a Midwestern producer of canned soups. The company's food-processing plants are located in Columbus, Ohio; and Indianapolis, Ind. Distribution centers are located in Toledo and Dayton, Ohio; Bowling Green, Ky.; and St. Louis, Mo. The new director of logistics wants to minimize the cost of shipping the goods from the processing plants to the distribution centers. The cost of shipping a case of soup from each processing center to each distribution center is given in Table 4.8. The weekly capacity at the two processing centers as well as the weekly requirements at the distribution centers are also provided in Table 4.8.

   a. Develop an LP model to use to minimize the cost of shipping while meeting capacity and demand requirements.

   b. Implement your model on a spreadsheet and find an optimal solution using Solver.

TABLE 4.8
Coger Company data

	Unit Shipping Costs			
	Toledo	Dayton	Bowling Green	St. Louis
Columbus	$1.75	$1.25	$3.25	$5.60
Indianapolis	$1.90	$1.25	$2.50	$3.00

	Weekly Capacity (cases)		Weekly Demand (cases)	
Columbus	3,000	Toledo	1,200	
Indianapolis	4,000	Dayton	1,800	
		Bowling Green	900	
		St. Louis	2,500	

10. Vohio Oil produces two types of fuels (regular and super) by mixing three ingredients. The major distinguishing feature of the two products is the octane level required. Regular fuel must have a minimum octane level of 90, and super must have a level of at least 100. The octane levels of the three ingredients as well as their availability for the coming two-week period are known. Likewise, the maximum demand for the end products, along with the revenue generated per barrel, are known. These data are summarized in Table 4.9.

    a. Develop an LP model to determine how much of each end product to produce in order to maximize contribution to profit.

    b. Implement your model on a spreadsheet and find the best solution you can, using your intuition.

    c. Find an optimal solution using Solver.

TABLE 4.9
Vohio Oil data

Input	Cost/barrel	Octane	Available barrels
1	$16.50	100	110,000
2	$14.00	87	350,000
3	$17.50	110	300,000

	Revenue/barrel	Max Demand (barrels)
Regular	$18.50	350,000
Super	$20.00	500,000

11. Figure 4.21 shows the Solver *Sensitivity Report* for a product mix example found in the Excel *Solvsamp.xls* workbook included with Excel. The problem involves manufacturing TVs, stereos, and speakers, using limited supplies of common parts. Answer the following questions using only the *Sensitivity Report* information, if possible. If you need to re-solve the model, you can find *Solvsamp.xls* in the Excel Examples folder.

    a. Which constraints are binding?
    b. What is the new solution and total profit if the unit profit for stereos is $60?
    c. How much would the per-unit profit on speakers have to change to make them economical to produce?
    d. What will the total profit be if the number of speaker cones is increased to 850? To 950?
    e. What will happen to the solution if 25 fewer chassis are available? If 75 fewer are available?
    f. Show how the solution changes as the number of electronic components varies.

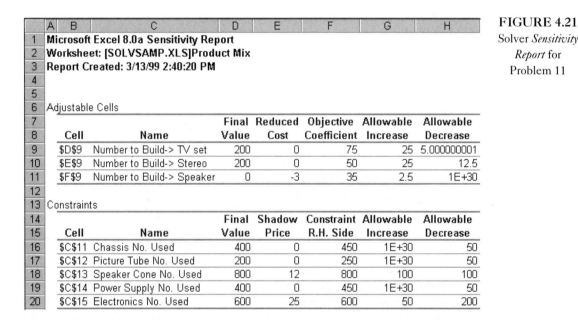

FIGURE 4.21
Solver *Sensitivity Report* for Problem 11

12. The Calhoun Textile Mill is in the process of deciding on a production schedule. It wishes to know how to weave the various fabrics it will produce during the coming quarter. The sales department has confirmed orders for each of the 15 fabrics that are produced by Calhoun. These demands are given in Table 4.10. Also given in this table is the variable cost for each fabric. The mill operates continuously during the quarter: 13 weeks, 7 days a week, and 24 hours a day.

   There are two types of looms: dobbie and regular. Dobbie looms can be used to make all fabrics and are the only looms that can weave certain fabrics, such as plaids. The rate of production for each fabric on both types of looms is also given in the table. Note that if the production rate is 0, the fabric cannot be woven on that type of loom. Also, if a fabric can be woven on either type of loom, the production rates are equal. Calhoun has 15 dobbie looms and 90 regular looms. For this exercise, assume that the time required to change a loom from one fabric to another is negligible.

Fabrics woven at Calhoun proceed to the finishing department in the mill and then are sold. Any fabrics that are not woven in the mill because of limited capacity will be bought on the outside market, finished at the Calhoun Mill, and sold at the selling price. Management would like to know how to allocate the looms to the fabrics and which fabrics to buy on the market.

a. Develop an LP model of Calhoun's problem.
b. Implement your model on a spreadsheet and find the best solution you can, using intuition.
c. Find an optimal solution using Solver.

TABLE 4.10

Calhoun Textile Mill data

Fabric	Demand (yds.)	Dobbie (yds./hr.)	Regular (yds./hr.)	Mill Cost ($/yd.)	Sub. Cost ($/yd.)
1	16,500	4.653	0.000	.6573	.80
2	52,000	4.653	0.000	.5550	.70
3	45,000	4.653	0.000	.6550	.85
4	22,000	4.653	0.000	.5542	.70
5	76,500	5.194	5.194	.6097	.75
6	110,000	3.809	3.809	.6153	.75
7	122,000	4.185	4.185	.6477	.80
8	62,000	5.232	5.232	.4880	.60
9	7,500	5.232	5.232	.5029	.70
10	69,000	5.232	5.232	.4351	.60
11	70,000	3.733	3.733	.6417	.80
12	82,000	4.185	4.185	.5675	.75
13	10,000	4.439	4.439	.4952	.65
14	380,000	5.232	5.232	.3128	.45
15	62,000	4.185	4.185	.5029	.70

13. Very Good (VG) Juice Company produces five different juices: cherry, apple, orange, pineapple, and lemon. VG statisticians report to management with monthly projections of the minimum (based on previous retail orders) and maximum sales for each. In the past, these projections have been accurate, even though the demand is very seasonal. Each month, management must decide how much of each product to produce for the next month. Virtually no finished product inventory is carried, but ample supplies of the raw materials are on hand. What makes the production decision difficult is that the products yield different profits and require different amounts of machine time in the three processes (blending, straining, and bottling) each employs. The relevant data (per 1,000 gallons), as well as the minutes available per process, appear in Table 4.11.

a. Formulate a linear program that will prescribe how much of each product to produce, given next month's projections.
b. Implement your model on a spreadsheet and find an optimal solution with Solver.
c. What effect on profit would an increase of capacity in the straining department have?
d.   Analyze the effect of the straining department capacity on profit by varying the capacity and solving a series of LP models to construct a graph of profit versus capacity.

	Time required in minutes						
	Apple	Cherry	Lemon	Orange	Pineapple	Time Available (minutes)
Blend	23	22	18	19	19	5,000
Strain	22	40	20	34	22	3,000
Bottle	10	10	10	10	10	5,000

	Apple	Cherry	Lemon	Orange	Pineapple
Profit ($/1000 gal.)	$800	$320	$1120	$1440	$800
Max Sales (thousand gals.)	20	30	50	50	20
Min Sales (thousand gals.)	10	15	20	40	10

TABLE 4.11
Very Good Juice data

14. Drew's Sporting Goods, a sporting goods chain, will be sponsoring a gun and knife show in January of next year, and is currently developing a morning radio advertisement plan. Five of the largest radio stations have been flagged as good choices for the campaign. Their various costs per half minute for advertising are shown in Table 4.12. The company has allocated $15,000 for morning radio advertisements. To ensure a broad reach, at least 5 minutes of advertisement should be purchased on each of these five stations and no more than 15 minutes should be bought on any one station. The company has decided that at least as many minutes of advertisements should be purchased on the two country stations (combined) as on the other three stations (combined).

   a. Develop an LP model that will maximize the number of .5 minute advertisements purchased, subject to the stated restrictions.

   b. Implement your model on a spreadsheet and find an optimal solution using Solver.

   c. Which constraints are binding? Interpret the shadow price for each of the binding constraints.

   d. Using the shadow prices, suggest which station is best for lowering the minimum number of .5-minute spots from the current required level of ten?

	Cost/.5 minute
Rock 102.7	$200
Country 100	$150
Country 103.1	$250
All Talk 700	$500
All Sports 55	$100

TABLE 4.12
Drew's Sporting Goods data

15. Welz's Widgets has four production plants, located in Atlanta, Cincinnati, Chicago, and Salt Lake City. It currently has 12 distribution centers, located in Portland, San José, Las Vegas, Tucson, Colorado Springs, Kansas City, St. Paul, Austin, Jackson, Montgomery, Cleveland, and Pittsburgh. The monthly capacity for each plant, monthly demand at each distribution center, and per-unit shipping cost from each plant to each distribution center are given in Table 4.13. Welz is interested in minimizing the cost of transporting its widgets from its plants to its distribution centers.

a. What is the minimum monthly cost of shipping?
b. Which plants will operate at capacity in this solution?
c. Suppose that 500 units of extra capacity are available (and that the cost of this extra capacity is a sunk cost). To which plant should this extra capacity go, and why?
d. Suppose that the cost of shipping from Atlanta to Jackson increases to $0.45 per unit. Will the current solution remain optimal? Why or why not?
e. Of the routes currently chosen for shipment, which seems the most cost sensitive?

TABLE 4.13

Welz's Widgets data: Shipping cost per unit

	Atlanta	Cincinnati	Chicago	Salt Lake City	Demand
Portland	$2.17	$1.97	$1.71	$0.63	5,000
San José	$2.10	$2.02	$1.83	$0.57	15,600
Las Vegas	$1.75	$1.68	$1.53	$0.35	4,250
Tucson	$1.50	$1.53	$1.41	$0.58	3,750
Colorado Springs	$1.20	$1.10	$0.94	$0.39	4,570
Kansas City	$0.68	$0.55	$0.42	$0.93	7,500
St. Paul	$0.92	$0.61	$0.37	$0.99	3,000
Austin	$0.81	$0.97	$0.98	$1.08	8,700
Jackson	$0.35	$0.57	$0.67	$1.34	3,250
Montgomery	$0.14	$0.49	$0.67	$1.55	12,300
Cleveland	$0.56	$0.22	$0.34	$1.57	9,600
Pittsburgh	$0.53	$0.27	$0.41	$1.69	16,700
Supply	15,000	35,000	15,000	35,000	

16. CinTech Co. manufactures seven different products denoted A, B, C, D, E, F, and G. The seven products have three raw materials in common (R1, R2, and R3) and share two processes (P1 and P2). There are 500 pounds of R1, 750 pounds of R2, and 350 pounds of R3 available for next week's production. In addition, there are 60 hours available for P1 and 80 hours available for P2 next week. The per-unit requirements, per-unit contribution to profit, and fixed costs for the seven products are shown in Table 4.14.

Due to prior commitments, a minimum of 300 units of A must be produced. Because of limitations in other processes, a maximum of 700 units of B can be produced per week. For similar reasons, at most 400 units of E can be produced next week.

a. Develop an LP model and find an optimal solution using Solver.
b. Answer each of the following questions independently. (Some of these might require new runs of the model. If so, indicate that, and describe how you would answer the question.)

   i.   The sales manager feels that it would be possible to increase the selling price of product F by $2 per unit. What is your reply?
   ii.   The chief buyer thinks it is unlikely that she can obtain any more of R1 from the current source. However, there is another supplier, but his quoted price is $10 higher. Should she consider buying additional R1 from the other supplier?
   iii. The firm to which CinTech has agreed to supply 300 units of A has requested that it be increased to 350. How should the sales manager respond?

iv.   The production manager thinks he can obtain 20 extra hours of P1 at a total increase in cost of $1,500. Is this a good idea?

v.   The sales manager feels that product B is priced too low. He feels that the price for B should be raised to $14 per unit. However, he estimates that demand at this price will be a maximum of 600. What will be the optimal contribution to profit and fixed costs if this is implemented?

Contribution to Profit	A	B	C	D	E	F	G
($/unit)	$10	$12	$8	$15	$1	$10	$19
R1 (pounds/unit)	.10	3.00	.20	.10	.20	.10	.20
R2 (pounds/unit)	.20	.10	.40	.20	.20	.30	.40
R3 (pounds/unit)	.20	.10	.10	.20	.10	.20	.30
P1 (hours/unit)	.02	.03	.01	.04	.01	.02	.04
P2 (hours/unit)	.04	.00	.02	.02	.06	.03	.05

TABLE 4.14
CinTech data

17. Heil, Inc., produces custom-made stained glass windows. The company employs 35 craftsman, who each work a 40-hour week. Two major customers, Halmart and Window Warehouse, have ordered 150 and 120 windows, respectively, for next week. Each window requires 10 hours of craftsman time. The owner of the company, Tom Heil, realizes that too much overtime would be required to meet these orders, and would like to keep overtime under 200 hours. He has decided on the following goals:

1.   Meet half of each customer's demand.
2.   Keep overtime to 200 hours.
3.   Meet each customer's demand.

Tom has decided that goal 1 is three times as important as goal 3 and that goal 2 is twice as important as goal 3.

a.   Formulate a goal-programming model to meet these goals, using a weighted objective function with the sum of the weights equal to 1.

b.   Use Solver to find an optimal solution. Which goals are or are not met?

18. The conflict between return and risk, as well as the existence of various investor objectives, make goal programming an ideal tool for portfolio construction. Arbitrate pricing theory (APT) provides a means of measuring risk in a form that can be incorporated into a linear model.[4] APT provides estimates of the sensitivity of a particular asset (investment) to specific risk factor. The investor specifies a target level for the weighted risk factor, which leads to a linear constraint that requires that the weighted risk be equal to the desired level. For example, suppose that John has $200,000 he would like to allocate to four assets: life insurance, bond mutual funds, stock mutual funds, and savings. The expected (1 year) returns are 6 percent, 6.5 percent, 11 percent, and 4 percent, respectively. He has established the following lower and upper bounds on the allocations to the alternatives:

Asset	Lower bound	Upper bound
1. Life insurance	$ 5,000	$10,000
2. Bond mutual funds	$60,000	None
3. Stock mutual funds	$30,000	None
4. Savings	None	None

He considers two risk factors important: unexpected inflation and the spread between long- and short-term interest rates. The risk factors per dollar allocated to each of the assets and weighted target for these risk factors are given next:

Factor	Asset				Target
	1	2	3	4	
Inflation	−.5	1.8	2.1	−.3	1.0
Interest Spread	.4	−.5	0.0	−1.1	0.0

John's priorities are (in order of importance) to meet the bound restrictions, achieve his desired risk levels, and maximize return on investments.

a. Develop a goal-programming model to determine the amount to allocate to each asset, using the absolute priorities approach to define the objectives.

b. Implement the model on a spreadsheet and use Solver to find a solution.

19. Anderson Manufacturing produces two products. A unit of product 1 yields a profit of $35, and of product 2, $45. Anderson has promised to provide its customers 100 units of product 1 and 200 units of product 2 next week. However, demand is such that as many as 500 units of each can be sold. Each unit of product 1 requires 3 minutes of machine time and each unit of product 2 requires 4 minutes. Unfortunately, there are only 40 hours (2,400 minutes) of machine time available next week. Anderson's objectives are:

Priority 1: Use 100 percent of available machine capacity.

Priority 2: Make as much product 2 as possible (up to the demand of 500 units).

a. Formulate the constraint set for this scenario.

b. Using Solver, find an optimal solution with Priority 1 as the objective function.

c. Set the first priority equal to the value achieved in part (b) and replace the objective with priority 2. What is the solution to this problem?

d. Suppose that 10 hours of overtime are available for production and that Anderson has the following goals (in order of importance), instead of the priorities presented in the original problem statement:

Goal 1: Minimize the amount of overtime used.

Goal 2: Maximize profit.

Formulate this problem as a goal program and solve it using the absolute priorities approach.

20. Stephanie is considering four investment alternatives:

Alternative	Return
1. Casino bonds	10%
2. Municipal bonds	7%
3. Certificates of deposit	4%
4. Savings bonds	5%

Stephanie needs a return of at least 8 percent and would like to achieve this while satisfying, in order, the following objectives:

Priority 1: Invest as much as possible in certificates of deposit and savings bonds.

Priority 2: Invest as much as possible in savings bonds.

a. Formulate the linear programming constraint set, develop a spreadsheet model, and solve using the absolute priorities approach.

b. Rather than using the priorities specified, Stephanie has decided to use a weighted objective function approach, using the following weights:

Alternative	Objective Weighting
1. Casino bonds	.4
2. Municipal bonds	.3
3. Certificates of deposit	.1
4. Savings bonds	.2

Solve the linear program with the objective of minimizing the dollars invested in each alternative multiplied by their weighting factors and compare your results to part *a*.

21. Milicro Manufacturing[5] has five vendors from which it may purchase its key component part. Historical records provide information with regard to the quality of each vender's product and their on-time delivery record. This information is summarized in Table 4.15. The price quote from each vendor is also given.

	Vendor				
	1	2	3	4	5
Quality (% accepted)	90%	89%	95%	88%	90%
On-time deliveries (%)	80%	70%	90%	85%	87%
Price quote (per unit)	$14.50	$12.00	$15.00	$14.00	$14.50

TABLE 4.15
Milicro Manufacturing data

Milicro needs 5,000 units of this key component next month and is currently trying to decide on a purchase plan. Company policy dictates that no single vendor will receive more than 50 percent of the total purchase. The company has the following goals (in descending order of importance):

Goal 1: An average acceptance rate of 90 percent.
Goal 2: An on-time delivery average of 85 percent.
Goal 3: An average unit cost of $14.25 or lower.

a. Formulate this scenario as a linear goal program.
b. Solve the linear goal program using the absolute priorities approach.
c. Solve the linear goal program using the absolute priorities approach with the following priorities:

Goal 1: An on-time delivery average of 85 percent.
Goal 2: An average acceptance rate of 90 percent.
Goal 3: An average unit cost of $14.25 or lower.

d. Compare and contrast the solutions to parts *b* and *c*.

22. Cincy Sausage[6] produces a variety of meat products, including their specialty, Big Red Sausage. In addition to a secret blend of spices, the Big Red is comprised of beef head, pork jowls, mutton, and water. The cost per pound, amount of fat per pound, and amount of protein per pound of each ingredient is shown in the next table:

	Beef head	Pork jowls	Mutton	Water
Cost/pound	$.15	$.10	$.08	$.00
Fat/pound	.05	.25	.10	.00
Protein/pound	.20	.25	.08	.00

Cincy uses the following goals, in order of importance, when producing its Big Red:

Priority 1: Big Red should consist of at least 18% protein.

Priority 2: Big Red should consist of at most 10% fat.

Priority 3: Roughly equal amounts of beef and pork should be used.

Priority 4: Cost per pound should not exceed $.10.

Cincy is about to produce a 75-pound batch of Big Red.

a. Formulate this production scenario as a linear goal program.

b. Find the solution using the absolute priorities approach.

c. Resolve the problem using a weighted objective function approach with the following weights:

Priority	Weight
1	.4
2	.4
3	.0
4	.2

NOTES

1. This problem is adapted from Philip Kotler, *Marketing Decision Making: A Model Building Approach* (New York: Holt Reinhart and Winston, 1971), 171–173.

2. Some authors call the excess amount associated with a greater-than-or-equal-to constraint a *surplus variable* and use the term *slack variable* for those associated with less-than-or-equal-to constraints. However, because the Excel Solver uses the term "slack" for either case, we do so, too.

3. This section describes the use of the standard Excel Solver. Premium Solver, included on the CD-ROM with this text, is an enhanced version of the standard Excel Solver. Details on the use of the Premium Solver may be found in the appendix to this chapter.

4. Based on M. Schniederjans, T. Zorn, and R. Johnson, "Allocating Total Wealth: A Goal Programming Approach," *Computers and Operations Research*, Vol. 20, No. 7, 1993, 679–685.

5. Based on F. Buffa, and W. Jackson, "A Goal Programming Model for Purchase Planning," *Journal of Purchasing and Materials Management*, Fall 1983, 27–34.

6. Based on Ralph Steuer, "Sausage Blending Using Multiple Objective Programming," *Management Science*, Vol. 30, No. 11, November 1984, 1376–1384.

BIBLIOGRAPHY

Bradley, Stephen P., Arnoldo C. Hax, and Thomas L. Magnanti. 1977. *Applied Mathematical Programming.* Reading, Mass.: Addison-Wesley Publishing Company.

Camm, Jeffrey D., P.M. Dearing, and Suresh K. Tadisina. 1987. "The Calhoun Textile Mill Case: An Exercise on the Significance of Linear Programming Modeling." *IIE Transactions*, Vol. 19, No. 1, March, pp. 23–28.

Dano, Sven. 1974. *Linear Programming in Industry*. Wien: Springer-Verlag.

Dantzig, George B. 1982. "Reminiscences about the Origins of Linear Programming." *Operations Research Letters*, Vol. 1, No. 2, April, pp. 43–48.

———. 1990. "The Diet Problem." *Interfaces*, Vol. 20, No. 4, July–August, pp. 43–47.

Evans, James R., and Jeffrey D. Camm. 1990. "Using Pictorial Representations in Teaching Linear Programming Modeling." *IIE Transactions*, Vol. 22, No. 2, June, pp. 191–195.

Field, Richard C. 1984. "National Forest Planning Is Promoting U.S. Forest Service Acceptance of Operations Research." *Interfaces*, Vol. 14, No. 5, September–October, pp. 67–76.

Gass, Saul I. 1985. *Linear Programming*. 5th edition. New York: McGraw Hill Book Company.

Kotler, Philip. 1971. *Marketing Decision Making: A Model Building Approach*. New York: Holt Reinhart and Winston, pp. 171–173.

Makuch, William M., Jeffrey L. Dodge, Joseph E. Ecker, Donna C. Granfors, and Gerald J. Hahn. 1992. "Managing Consumer Credit Delinquency in the U.S. Economy: A Multi-Billion Dollar Management Science Application." *Interfaces*, Vol. 22, No. 1, January–February, pp. 90–109.

Salvia, A., and W. Ludwig. 1979. "An Application of Goal Programming at Lord Corporation." *Interfaces*, Vol. 9, No. 4, August, pp. 129–133.

Schniederjans, M., N. Kwak, and M. Helmer. 1982. "An Application of Goal Programming to Resolve a Site Location Problem." *Interfaces*, Vol. 12, No. 3, June, pp. 65–72.

APPENDIX: USING PREMIUM SOLVER TO SOLVE LINEAR PROGRAMS

To use the Premium Solver, you must first install the Premium Solver from the CD in the back of this text. The standard Excel solver is still the default solver even after the Premium Solver has been installed. However, the new version of the dialog box has an additional button labeled *Premium*. This is shown in Figure 4A.1. Clicking on this button invokes the Premium Solver dialog box that is shown in Figure 4A.2.

Premium Solver is very similar to the standard Excel Solver with some minor differences in the dialog box and options. Selecting the down arrow just below the *Options* button as shown in Figure 4A.2 shows the available algorithms: *Standard GRG Nonlinear, Standard Simplex LP* and *Standard Evolutionary*. Select *Standard Simplex LP* if you wish to solve a linear program. The other options will be discussed in the next chapter. The objective function, decision variables and constraints are entered into the Premium Solver dialog box in exactly the same manner as described in this chapter for the standard Excel solver.

The Solver Options dialog box (invoked by clicking on the *Options* button) is very similar to that shown in Figure 4.12 for the standard Excel Solver, except that it is tailored only to linear programming. This dialog box is shown in Figure 4A.3. The *Show Iteration Results* option should in general be left unchecked, as it slows the solution process down considerably. The *Use Automatic Scaling* option should be checked if you are having trouble finding a solution and there is a wide range in the magnitude of the data in the problem. *Assume Non-Negative* should be checked if all of the decision

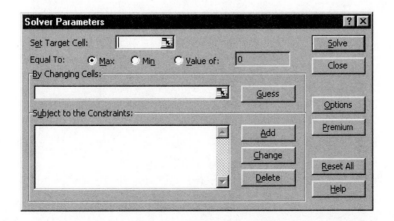

FIGURE 4A.1

Standard Excel Solver Parameters dialog box after Premium Solver installation

FIGURE 4A.2
Premium Solver
Parameters
dialog box

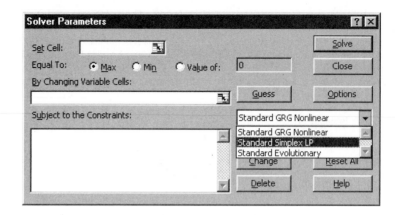

variables must be non-negative. Finally, if the Bypass-Solve Reports option is checked, no reports will be generated and you will not be prompted concerning which reports you desire.

In addition to the standard options for linear programs, the Premium Solver has two new options: *Pivot Tol*, and *Reduced Tol*. These options concern algorithmic parameters that are beyond the scope of this text. In general, these should be kept at their default values.

FIGURE 4A.3
Premium Solver
Options dialog
box for linear
programs

CHAPTER 5

Integer and Nonlinear Optimization Models

O U T L I N E

Applications of Integer and Nonlinear Models
Project Selection at the National Cancer Institute
Crew Scheduling at American Airlines
Mortgage Valuation Models at Prudential Securities
Paper Production

Building Integer-Programming Models
Management Science Practice: Sales Staffing at Qantas

Solving Integer-Programming Models
Using Solver for Integer Programs
Sensitivity Analysis

Building Nonlinear Optimization Models
Management Science Practice: Kanban Sizing at Whirlpool Corporation

Solving Nonlinear Programming Problems
Using Excel Solver
Sensitivity Analysis
The Problems of Nonlinearity
Using Premium Solver's Evolutionary Algorithm for Difficult Problems

Technology for Integer and Nonlinear Optimization Problems

Appendix: Using Premium Solver to Solve Integer and Nonlinear Programs
Solving Linear Integer Programs
Solving Nonlinear Programs

Integer and nonlinear optimization models are extensions of linear models that we studied in the previous chapter. An *integer programming (IP) model* is a linear program in which some or all of the decision variables are restricted to integer values. For many practical applications, fractional values for decision variables are simply unacceptable. For example, in a production planning decision involving low-volume, high-cost items such as ships or airplanes, an optimal value of 10.42 would make little sense, and a difference of one unit would have significant economic consequences. On the other hand, if the problem involved deciding on how many spark plugs to produce next month, we can effectively treat the decision variable as continuous (rounding a value like 3457.63 would have little consequence) and use an LP model.

In other practical problems, objectives and constraints cannot properly be expressed using linear functions. *Nonlinear optimization models* have constraint and/or objective functions that are not linear. For example, purchasing decisions are often subject to economies of scale; the larger the quantity purchased, the smaller the unit price. A function that represents total cost is not linear in the quantity purchased. In industries such as chemical processing, mixing ingredients does not necessarily result in an output equal to the sum of the inputs, as material may be lost in chemical reactions. Modeling such phenomena also requires nonlinear functions.

Integer and nonlinear models allow you to solve a much wider variety of decision problems than you can with linear models. Many important business decisions such as capital budgeting and supply chain design use IP models. Unfortunately, however, both integer and nonlinear models are more difficult to solve and analyze than linear programs. In this chapter we introduce you to a variety of models and applications of integer and nonlinear optimization, and describe how to use Excel Solver to solve them.

APPLICATIONS OF INTEGER AND NONLINEAR MODELS

Project Selection at the National Cancer Institute

The National Cancer Institute (NCI) in Bethesda, Maryland, has used integer programming to help make project-funding decisions. The American Stop Smoking Intervention Study (ASSIST) was designed as a multiple-year, multiple-site demonstration project to reduce smoking. Each project was assigned a value based on the review of the proposals. NCI needed to decide which projects to fund, with the objective being to maximize the total rank function value of the projects selected for funding. Each proposal also had a preference coefficient, obtained from the review process, which was designed to take into account other important criteria besides ranking. Constraints included obtaining a certain total preference score, restricting the amount required to fund the selected proposals to a budgeted amount, and selecting at least one proposal from each of four regions (Gaballa and Pearce, 1979).

Crew Scheduling at American Airlines

American Airlines employs over 8,000 pilots and over 16,000 flight attendants. Total crew cost is the second largest cost incurred by the airline (fuel cost is the largest). Total crew cost includes salaries, fringe benefits, and expenses. Airlines can save quite a bit of money by efficiently scheduling flight crews (pilots and attendants) to flights. Using integer-programming models to model assignment decisions, American can schedule its crews to flights in a way that minimizes the cost of lodging and meals, while also satisfying Federal Aviation Administration rules and pay guarantees. The integer-programming-based scheduling system called TRIP (Trip Evaluation and Improvement Program) took over 15 man-years to develop and refine, but saves an estimated $20 million annually in crew costs. American Airlines Decision Technologies, the management science group at American, is now a separate subsidiary of American's parent corporation and has sold TRIP to ten major airlines and one railroad (Anbil et al., 1991).

Mortgage Valuation Models at Prudential Securities

Prudential Securities, Inc. (PSI), has created a mortgage-backed securities (MBS) department to develop models for managing these complex investments. The department has developed a variety of management science models, including linear, integer, and nonlinear optimization models to help value, trade, and hedge MBSs in inventory and construct portfolios. For example, nonlinear optimization models are used to construct optimal portfolios for clients, matching clients' investment-performance profiles with constraints under a variety of interest rate scenarios. The model inputs required from the client include the portfolio performance target, the securities to consider, diversification restrictions, and view of future interest rates. The resulting portfolio meets a specified performance target while taking into account possible interest rate movement. These models are used hundreds of times per day by PSI personnel (Ben-Dov, Hayre, and Pica, 1992).

Paper Production

In the production of paper, wood chips are cooked in a vat of chemicals, which have a tendency to boil over. Precipitated silica, fine particle silicon dioxide, is used to eliminate this foaming in chemical processes such as this. The proper chemical mix and amount of silicon dioxide to include determines the control over foaming. This, in turn, has obvious production throughput and cost impact. As with many chemical processes, the relationships are nonlinear. A nonlinear performance relationship was estimated to relate foam level to chemical ingredients. The optimization model was then to maximize performance, subject to a mix constraint. In one application, the cost dropped from $.95 per pound to $.86 per pound (Bohl, 1994).

BUILDING INTEGER-PROGRAMMING MODELS

IPs may have one of two types of integer variables: *general integer variables* and *binary variables*. **General integer variables** are decision variables that can take on any *integer* (whole number) value; **binary variables** are restricted to either 0 or 1. Mathematically, we can express a binary variable x as a general integer variable by restricting it to be between 0 and 1 as well as integer:

$$0 \leq x \leq 1 \text{ and integer}$$

However, we usually just write this as

$$x = 0 \text{ or } 1$$

Binary variables are extremely useful because they enable us to model logical yes-or-no decisions in optimization models. For example, binary variables can be used to model decisions such as whether ($x = 1$) or not ($x = 0$) to place a facility at a certain location, whether or not to produce a product this month, and whether or not to invest in a certain mutual fund.

Because an IP is fundamentally a linear program, constraint functions and the objective function must be linear functions, and constraints must be of a $\leq$, $\geq$, or $=$ type. Thus, the model-building approaches we used to formulate LP models in Chapter 4 can be used to build IP models.

Example 1: A General Integer Model

Queen City, Inc., manufactures machine tools. The production planner who oversees the production of two of Queen City's more profitable machines needs to determine how many of each to produce this month. The two machines, TopLathe and BigPress, each require a certain component. Each TopLathe requires 10 of these components, and each BigPress requires 7. Only 49 components are available this month. Production and marketing are not very well coordinated at Queen City, but the planner knows that the salespeople are usually happy if production provides them with 5 machines per month. The profit for a TopLathe is $50,000, and for a BigPress, $34,000. Assuming that labor and other components are available, how many of each product should Queen City produce in order to maximize profit?

What is it that Queen City needs to decide? The planner needs to decide how many TopLathes and how many BigPresses to manufacture this month. We define the following decision variables:

$$x_1 = \text{number of TopLathes to produce}$$
$$x_2 = \text{number of BigPresses to produce}$$

These decision variables must be nonnegative and cannot be fractional, so we must ensure that both variables are integer as well as nonnegative.

What is the objective to be maximized or minimized? The objective is to maximize profit. Since each TopLathe generates a profit of $50,000, and each BigPress generates $34,000, the objective is

$$\textit{Maximize Profit} = 50{,}000x_1 + 34{,}000x_2$$

How is the problem restricted? Queen City has only a limited supply of the component. Each TopLathe requires 10 components and each BigPress requires 7 units; therefore, the total number of components used in production is $10x_1 + 7x_2$. This must not exceed the number available, so we have the constraint

$$10x_1 + 7x_2 \leq 49$$

The second constraint requires us to produce at least five machines to keep the salespeople happy. This leads to the constraint

$$x_1 + x_2 \geq 5$$

The complete model for Queen City's production planning problem is

Maximize Profit = 50,000 x_1 + 34,000 x_2

subject to

$$10x_1 + 7x_2 \leq 49$$

$$x_2 + x_2 \geq 5$$

$$x_1, x_2 \geq 0 \text{ and integer}$$

Figure 5.1 shows a spreadsheet model for this example. The decision variables are in cells B14:C14; the objective function is in cell D16; and the constraint functions, which must meet the requirements in cells D8 and D9 are defined in cells D14 and D15. We will solve this model later in this chapter.

	A	B	C	D
1	Queen City Inc.			
2				
3	*Parameters and uncontrollable inputs*			
4		TopLathe	BigPress	
5	Profit/Unit	$50,000	$34,000	
6	Components/Unit	10	7	
7				
8			Sales Requirement/Month	5
9			Components Available	49
10				
11	*Model*			
12				
13		TopLathe	BigPress	Total
14	Quantity Produced	0	0	0
15	Components Used	0.0	0.0	0
16	Profit	$0.00	$0.00	$0
17				
18	*Key Cell Formulas*			

FIGURE 5.1
Queen City, Inc., spreadsheet model

Excel Exercise 5.1

Open the worksheet *QueenCity.xls*. Use Excel Solver to solve this problem ignoring the integer requirements on the variables. What is the solution to this LP? Round the fractional solution. Is the rounded solution feasible? Find the best integer solution you can, using trial and error.

Binary decision variables allow us to model yes-no decisions within an optimization model. Binary variables are common in many scheduling problems (see the Management Science Practice box on Qantas). A model in which all decision variables are binary is called a *binary integer program*, an example of which is shown next.

Management Science Practice
Sales Staffing at Qantas

Service sector operations such as airlines, hotels, and restaurants must deal constantly with staffing problems in the face of fluctuating demand. If the staff is too small, the firm cannot serve its customers well. This can result in lost sales and the loss of customer good will. A staff which is too large can meet customer demand, but labor costs might be excessive. Qantas Airways uses an integer programming model to determine the least cost staff size in its telephone reservation system to meet projected demand.

The airline industry has been and continues to be an extremely competitive industry. Survival depends on maximizing efficiency in operations and capturing a sufficient share of the customer market. Qantas Airline decided to analyze the size of its reservation staff, since as discussed above, an oversized staff is inefficient, but an undersized staff will result in lost market share. The fluctuation of demand over time makes the search for an optimal staff size a formidable task.

Qantas began its analysis by collecting demand data (number of calls) by month over a two year period. Then, for a three month period, data were collected on a half-hour basis. The data showed that demand varied by time of day and day of the week, but that for a given month, variation over weeks was insignificant. Therefore, a typical or average week could be used for a given month's planning purposes.

The integer-programming model uses demand forecasts to optimize staff size over time. The following assumptions were made:

1. Shifts start only on the hour or half hour.
2. Shifts start during the hours of 7 a.m. to 9:30 a.m., plus one shift that starts at 3 p.m. (7 possible shifts).
3. The length of shifts starting between 8:30 and 9:30 is 8.5 hours with a one-hour lunch; all other shifts are 8 hours with a .5 hour lunch.
4. Lunch breaks are scheduled over a 2-hour interval for the 8-hour shifts and over a 2.5-hour interval for the 8.5-hour shift, commencing on the hour and half-hour. A day runs from 7 a.m. to 11 p.m. (32 half-hour intervals).

Outputs from the model include:

1. Number of staff per shift
2. Start and finish time of each shift
3. Lunch schedule for each shift
4. Total staff needed for the day

Using the output of the daily IP model, a manual approach was developed for devising a minimum workforce schedule that permitted each employee two consecutive days off. This approach is used for each month's typical week.

The use of this IP model and the manual workforce schedule saved over $200,000 over two years in the Sydney office alone. Because of the success of this approach in the reservations sales office, similar approaches were later used in other offices and in other customer contact areas, such as passenger sales and check-in facilities.

Source: Based on A. Gaballa and W. Pearce, "Telephone Sales Manpower Planning at Qantas," *Interfaces*, Vol. 9, No. 3, May 1979, 1–9.

Example 2: Project Selection

Tom Burke, owner of Burke Construction, is trying to determine what projects to undertake during the winter quarter (January, February, and March). Because he has a limited workforce, Tom knows that he will not be able to complete all of his construction projects without subcontracting some of them. He has promised to complete five projects this winter and has estimated the profit to his company for each. In Table 5.1, he has estimated the amount of labor each project will require (in total hours) and what the profit to his company will be if he subcontracts the project to one of the smaller companies in town. Tom has 4,800 labor hours available per quarter (10 workers $\times$ 40 hours/week $\times$ 12 weeks). In order to maximize profit, which jobs should Tom schedule for his company, and which should be subcontracted? We will assume that projects cannot be partially subcontracted; that is, a project will either be completed by Burke Construction or subcontracted in its entirety.

Project	1	2	3	4	5
Profit (Burke)	$30,000	$10,000	$26,000	$18,000	$20,000
Profit (Subcontract)	$ 6,000	$ 2,000	$ 8,000	$ 9,000	$ 4,500
Labor Hours	1,300	950	1,000	1,400	1,600

TABLE 5.1
Data for Burke Construction

What is it that Burke needs to decide? Tom needs to determine which projects to undertake and which to subcontract. This implies that we cannot assign a fraction of a project to either alternative. Therefore, we define the following binary variables:

x_i = 1 if project i is undertaken by Burke, and 0 if it is not $i = 1, 2, 3, 4, 5$

y_i = 1 if project i is subcontracted, and 0 if it is not $i = 1, 2, 3, 4, 5$

What is the objective to be maximized or minimized? The objective is to maximize profit, and is expressed (in thousands of dollars) as

$$\text{Maximize } 30x_1 + 10x_2 + 26x_3 + 18x_4 + 20x_5 + 6y_1 + 2y_2 + 8y_3 + 9y_4 + 4.5y_5$$

Because the variables are binary, a decision to subcontract project 1 for instance ($y_i = 1$), will contribute 6 $\times$ 1 or $6,000 to the objective function.

How is the problem restricted? Because Burke has a limited number of labor hours available, the total number of hours used must not exceed this amount. The number of hours used is computed by multiplying the hours needed by each project by the binary variable that corresponds to performing that project:

$$1,300x_1 + 950x_2 + 1,000x_3 + 1,400x_4 + 1,600x_5 \leq 4,800$$

Again, observe how the binary variables are used to capture the all-or-nothing nature of each term in the constraint.

Finally, each project must either be completed directly by Burke or subcontracted. In terms of our decision variables, this means that exactly one of x_i and y_i must be equal to 1 for each project i. We can express this as

$$x_i + y_i = 1 \quad i = 1, 2, 3, 4, 5$$

The complete model for Burke Construction is

$$\text{Maximize } 30x_1 + 10x_2 + 26x_3 + 18x_4 + 20x_5 + 6y_1 + 2y_2 + 8y_3 + 9y_4 + 4.5y_5$$

subject to

$$1{,}300x_1 + 950x_2 + 1{,}000x_3 + 1{,}400x_4 + 1{,}600x_5 \leq 4{,}800$$
$$x_i + y_i = 1 \qquad i = 1, 2, 3, 4, 5$$
$$x_i, y_i = 0,1$$

Figure 5.2 shows a spreadsheet model for this example. Decision variables are defined in the range B13:F14. Note that cells B17 through F17 are used to enforce the constraints $x_i + y_i = 1$. For example, B17 is the sum of cells B13 and B14. These cells should be set equal to 1 in the constraint section of the Solver dialog box.

FIGURE 5.2 Burke Construction spreadsheet model

	A	B	C	D	E	F	G	H
1	Burke Construction							
2								
3	*Parameters and uncontrollable inputs*							
4								
5	Project	1	2	3	4	5		
6	Profit (Burke)	$30,000	$10,000	$26,000	$18,000	$20,000		
7	Profit (Subcontract)	$6,000	$2,000	$8,000	$9,000	$4,500		Labor Hours Available
8	Labor hours required	1300	950	1000	1400	1600		4,800
9								
10	*Model*							
11								
12	Project	1	2	3	4	5		
13	Burke	0	0	0	0	0		
14	Subcontract	0	0	0	0	0	Total	
15	Profit	$0	$0	$0	$0	$0	$0	Unused Labor Hours
16	Labor hours used	0	0	0	0	0	0	4,800
17	Project chosen?	0	0	0	0	0	0	
18								
19	*Key cell formulas*							

Excel Exercise 5.2

Open the worksheet *Burke.xls*. Find the best solution you can, using trial and error. Use Solver to solve this problem ignoring the integer requirements on the variables (give the variables lower bounds of 0 and upper bounds of 1). What is the solution to this LP? Round the fractional solution. Is it better or worse than the solution you found by trial and error? Is it feasible?

Many models need a combination of both integer and continuous variables. Any model in which at least one decision variable is restricted to be integer (general or binary) and *at least* one variable is allowed to be fractional is called a *mixed integer program (MIP)*. The following is one example.

Example 3: Production Planning with Setups

Chemco, Inc., produces a wide variety of chemical products. One of the firm's most important products is liquid chlorine. Chemco has 100 gallons of liquid chlorine currently in storage, and storage is limited to 1,500 gallons. The weekly production capacity is 1,000 gallons. To meet contractual agreements, Chemco must supply 800, 500, 450, and 900 gallons of liquid chlorine in the next four weeks, respectively. The production manager needs to find the most cost-effective way to meet these demands. The setup and cleaning of the machines costs $500, and it costs $1 per gallon per week to hold inventory. She knows that there is a trade-off between the cost of holding the product in inventory and the cost of setting up and cleaning the processing machines. That is, longer production runs mean fewer setups but higher inventory levels, whereas shorter production runs result in lower inventory levels but more setups. The plant is closed on weekends, so the $500 setup and cleaning cost is incurred for every week there is production.

What is it that Chemco needs to decide? The problem is to determine how much liquid chlorine to produce and when to produce it in order to satisfy demand and minimize the cost of setups and inventory. We can define the following decision variables:

y_j = 1 if we produce in week j, and 0 if we do not $\qquad j = 1, 2, 3, 4$

x_j = number of gallons of chlorine to produce in week j $\qquad j = 1, 2, 3, 4$

I_j = number of gallons to hold in inventory at the end of week j $\qquad j = 1, 2, 3, 4$

Note that we use a binary variable to represent the decision whether or not to produce in a given week; the other variables are continuous.

What is the objective to be maximized or minimized? The objective is to minimize the cost of setups and inventory:

$$Minimize\ 500y_1 + 500\ y_2 + 500y_3 + 500y_4 + 1I_1 + 1I_2 + 1I_3 + 1I_4$$

Setups occur only if we decide to produce; thus, the first four terms are all-or-nothing costs. The remaining terms model the cost of holding inventory.

How is the problem restricted? We would like to find how the inventory is related to demand and the amount produced. As we have seen in Chapter 4 in the Suzie's Sweatshirt production planning problem (Example 6), this can be accomplished through the use of inventory balance equations. Recall that these balance equations state that for every period,

Beginning Inventory + amount produced = ending inventory + demand

These balance equations are

$$100 + x_1 = I_1 + 800$$
$$I_1 + x_2 = I_2 + 500$$
$$I_2 + x_3 = I_3 + 450$$
$$I_3 + x_4 = I_4 + 900$$

We must also ensure that the model forces y_i = 1 whenever the production in week i (x_i) is positive. This is done with the constraint

$$x_j \le 1{,}000y_j$$

for each period j. Note that if $y_j = 0$ (no setup occurs), then $x_j \leq 0$, and because of nonnegativity restrictions, x_j must equal zero. On the other hand, if $y_j = 1$, then x_j is *allowed* to be less than or equal to 1,000 (the production capacity). This constraint does not allow production unless the fixed (setup) cost is incurred. Note that although it is feasible for $y_j = 1$ and $x_j = 0$, this will *never* be optimal if y_j has a positive cost. Hence, in any optimal solution, $y_j = 1$ only when $x_j > 0$.

Finally, inventory must not exceed 1,500 gallons in each period:

$$I_j \leq 1,500 \qquad j = 1, 2, 3, 4$$

The complete model for Chemco can be written as

Minimize $500y_1 + 500y_2 + 500y_3 + 500y_4 + 1I_1 + 1I_2 + 1I_3 + 1I_4$

subject to

$$100 + x_1 = I_1 + 800$$
$$I_1 + x_2 = I_2 + 500$$
$$I_2 + x_3 = I_3 + 450$$
$$I_3 + x_4 = I_4 + 900$$
$$x_j \leq 1{,}000y_j \qquad j = 1,2,3,4$$
$$I_j \leq 1{,}500 \qquad j = 1,2,3,4$$
$$x_j \geq 0 \qquad j = 1,2,3,4$$
$$y_j = 0 \text{ or } 1 \qquad j = 1,2,3,4$$

Figure 5.3 shows a spreadsheet model for Chemco. Note that the inventory variables I_j are not explicitly defined as decision variables in this implementation because they are uniquely determined by the values of production decisions using the inventory balance equations (in cells B19:E19). A solution is feasible if these inventory cells are nonnegative.

FIGURE 5.3

Chemco, Inc., spreadsheet model

	A	B	C	D	E	F
1	Chemco Inc.					
2						
3	*Parameters and uncontrollable inputs*					
4						
5	Week	1	2	3	4	
6	Demand	800	500	450	900	
7						
8	Setup Cost	$500				
9	Holding Cost/Unit/Week	$1.00				
10	Weekly Production Capacity	1000				
11	Current Inventory	100				
12	Inventory Limit	1500				
13						
14	*Model*					
15						
16	Week	1	2	3	4	
17	Setup? (1 = yes, 0 =no)	0	0	0	0	
18	Production	0	0	0	0	
19	Ending Inventory	-700	-1200	-1650	-2550	Total
20	Inventory Cost	($700.00)	($1,200.00)	($1,650.00)	($2,550.00)	($6,100.00)
21	Setup Cost	$0	$0	$0	$0	$0.00
22	Total Cost	($700.00)	($1,200.00)	($1,650.00)	($2,550.00)	($6,100.00)
23						
24	*Key cell formulas*					

Excel Exercise 5.3

Open the worksheet *Chemco.xls*. Familiarize yourself with the structure of this spreadsheet. Find a low-cost feasible solution, using trial and error.

Example 3 illustrates how we might link a binary variable and a continuous or general integer variable. The following example shows a slightly different application of binary variables.

Example 4: A Location Problem[1]

In January 1979, the banking laws in the state of Ohio were changed to allow greater freedom in the placement of branches throughout the state. The previous law only allowed banks to establish branches in the county of the bank's principal place of business (PPB). The new law allowed a bank to establish branches (upon approval of the superintendent of banks) in any one of the state's 88 counties contiguous to one in which it has a PPB. Suppose that a large western institution, Sun Bank, is considering entering the Ohio market and would like to know where to place 5 PPBs so as to maximize the population reachable with new branches.

What is it that needs to be decided? The problem is to decide where to put the PPBs so as to maximize the population reached. Hence, we have to keep track of two things: whether or not a county is chosen for a PPB, and if its population can be counted as *reached* (i.e., it is a PPB location or adjacent to one). Let us define the following:

$$x_j = 1 \text{ if county } j \text{ is chosen for a PPB, 0 if not} \quad j = 1, 2,....88$$
$$y_i = 1 \text{ if county } i \text{ is reached, 0 if not} \quad i = 1, 2,....88$$

What is the objective to be maximized or minimized? We would like to maximize the population reached, which is modeled as:

$$Maximize \sum_{i=1}^{88} pop_i y_i$$

where pop_i is the population of county i.

How is the problem constrained? First, this problem is similar to several others we have seen, in that we must link the variables so that they take on the proper values as we have defined them. In particular, we must relate the x_j and y_i variables. Figure 5.4 shows a county map of Ohio with the counties numbered and shaded corresponding to population. This map was constructed using the map tool discussed in Chapter 2. Consider county 1. If we do not chose county 1 or any of its adjacent counties for a PPB, then we cannot count county 1's population. This means that if x_1, x_2, x_{18} and x_{19} are all 0, then we must force $y_1 = 0$. This is accomplished by the following constraint:

$$y_1 \leq x_1 + x_2 + x_{18} + x_{19}$$

FIGURE 5.4
Ohio county
map

If one or more of these x's are 1, then the constraint allows y_1 to be equal to 1 (or 0). Similarly, the constraint for county 2 is:

$$y_2 \leq x_1 + x_2 + x_3 + x_{17} + x_{18}$$

In general, we can develop a constraint for each county, in the following generic fashion:

$$y_i \leq \sum_{j=1}^{88} a_{ij} x_j \qquad i = 1, 2, \ldots 88$$

where a_{ij} parameters are such that $a_{ij} = 1$ if county i shares a border with county j, and 0 if not.

The second type of constraint simply limits the number of PPBs to a fixed number (here that number is fixed to 5):

$$\sum_{j=1}^{88} x_j \leq 5$$

The complete integer programming model is

$$Maximize \sum_{i=1}^{88} \text{pop}_i y_i$$

subject to

$$y_i \leq \sum_{j=1}^{88} a_{ij} x_j \qquad i = 1, 2, \ldots 88$$

$$\sum_{j=1}^{88} x_j \leq 5$$

$$x_j = 0 \text{ or } 1 \qquad j = 1, 2, \ldots 88$$

$$y_i = 0 \text{ or } 1 \qquad i = 1, 2, \ldots 88$$

Observe that we may replace the binary restriction on y_i by the simple constraints $0 \le y_i \le 1$, because if county i or any of its adjacent counties are chosen for a PPB, then y_i will assume its upper bound of 1, since it has a positive coefficient in a maximization objective function. This changes the model from a binary integer program with 176 variables to a mixed integer program with 88 binary and 88 real variables, which requires less computational effort to solve. Experience in recognizing such things is an important aspect of the art of modeling.

A spreadsheet model for this example is provided in the Excel workbook *Sun.xls*. The model is very large, so we are able to show only portions of it in Figure 5.5. A matrix of zeros and ones with 88 rows and 88 columns (one row and one column for each county in Ohio) starts in cell B6. These are the a_{ij} values in the model. The population values of the counties appear in cells B94 through CK94. Cells B100 through CK100 contain decision variables for deciding to place a PPB in a county (the x_j variables in the model), and cells B101 through CK101 correspond to the y_i variables—whether or not a county has been covered. The number of covers for a given county are calculated using Excel's *Sumproduct* function starting in cell B107. For example, the formula in B107 is $= \text{SUMPRODUCT}(\text{B6:CK6},\$\text{B}\$100:\$\text{CK}\$100)$. These correspond to the right-hand side of the first set of constraints in the model. The number of PPBs used is in cell E107 [$= \text{SUM}(\text{B100:CK100})$] and the number allowed, in F107. Finally, the population reached, the objective function of the model, is computed in cell E110, using the formula $= \text{SUMPRODUCT}(\text{B94:CK94},\text{B101:CK101})$.

	A	B	C	D	E	F
1	**Sun Bank Location Problem**					
2						
3	*Data:*					
4						
5	Counties:	Williams	Defiance	Paulding	Van Wert	Mercer
6	Williams	1	1	0	0	0
7	Defiance	1	1	1	0	0
8	Paulding	0	1	1	1	0
9	Van Wert	0	0	1	1	1
10	Mercer	0	0	0	1	1

(a)

	A	B	C	D	E	F
94	Population:	36,369	39,987	21,302	30,458	38,334
95						
96						
97	*Model:*					
98						
99	Counties:	Williams	Defiance	Paulding	Van Wert	Mercer
100	PPB ?	0	0	0	0	0
101	Covered ?	0	0	0	0	0
102						
103						
104						
105						
106	Counties:	# Coverings			# PPBs Used	Allowed
107	Williams	0			0	5
108	Defiance	0				
109	Paulding	0			Population Covered	
110	Van Wert	0			0	
111	Mercer	0				

(b)

FIGURE 5.5
Portions of Sun Bank spreadsheet model

As you can see from these examples, binary variables provide a great deal of modeling flexibility to expand the types of optimization problems that we can formulate.

SOLVING INTEGER-PROGRAMMING MODELS

At first thought, you might think that solving an IP model might be as simple as solving the corresponding LP model and rounding the solution. If you did Excel Exercise 5.1 for the Queen City, Inc., example, you should have found that the solution to the LP is $x_1^* = 4.66667$, $x_2^* = .33333$, and profit = 244,666.67. The problem we get when we ignore, or *relax*, the integer restrictions of an IP is called the *linear programming relaxation* of the integer program. Note that rounding the solution to the LP relaxation in this example [to point (5,0)] is not feasible. In fact, even truncating the solution to the LP relaxation (that is, dropping the decimal fraction) to the point (4,0) is also infeasible. Even if rounding or truncating the solution to the LP relaxation results in a feasible integer solution, it may be far from the optimal solution to the original integer problem. The closest integer point to the LP solution is $x_1 = 4$ and $x_2 = 1$, with an objective value of $234,000. The optimal solution, however, is $x_1 = 0$ and $x_2 = 7$, with an objective function value of $238,000, far from the LP relaxation.

When you solve an integer program on the computer, you will notice immediately that the solution takes more time than the solution of a linear program. This is because the program actually solves a series of linear programs. The most common approach is called *branch and bound*. This technique begins by solving the LP relaxation. If the optimal decision variables to the LP relaxation turn out to be integer, then that solution *must* also be optimal to the integer program. However, if the solution to the LP relaxation has fractional values, then the solution to the LP relaxation is not feasible for the integer program. In this case, branch and bound uses a "divide and conquer" strategy. The set of solutions to the LP relaxation is broken into subsets (by a process known as *branching*) in such a way that only noninteger solutions are discarded. These subsets then become the basis for the series of LPs solved. The subsets are further divided until, eventually, solutions to the LPs will be integer. The term "bound" stems from the fact that by solving an LP relaxation, we can determine the best possible objective function value for any integer solution in a subset. If we have already found a better integer solution, then we may discard that subset from further consideration. When all subsets can be discarded, we know we have found the optimal IP solution. This makes the technique far more efficient than trying to enumerate all possible solutions.

Using Solver for Integer Programs[2]

Solver uses the branch and bound approach. To use Solver for integer programs, start by setting up the model as a linear program as described in Chapter 4. Changing cells are defined as integer in the *Add Constraint* dialog box by selecting the *int* option for general integer variables, or *bin* for binary variables from the drop-down box. As with LPs, you should also select the *Options* button from the *Solver Parameters* dialog box and check *Assume Linear Model* and *Assume Non-Negative*. For IPs, it is also important to check the setting of the *Tolerance* option. For example, with a *Tolerance* setting of 10 percent, Solver only guarantees that the solution it reports will be within 10 percent of the true optimal solution. By requesting a solution within some tolerance (for example, 5 percent or 10 percent) of optimality rather than requiring the true optimal solution, the solution time may be drastically cut. *However, if the true optimal solution is desired*, Tolerance *must be set to 0%*. As with linear programming models, the solution process is started by selecting the *Solve* button in the *Solver Parameters* dialog box.

Example 5: Solving an Integer-Programming Model with Solver

We will solve the Queen City production planning problem from Example 1 (see Figure 5.1). Figure 5.6 shows the *Solver Parameters* dialog box. Note that the changing cells in the range $B$14:$C$14 are defined as integer; this is done in the *Add Constraint* dialog box, as shown in Figure 5.7. We set the *Tolerance* to 0 in the *Solver Options* box to ensure that we find the optimal solution, and checked the *Assume Linear Model* and *Assume Nonnegative* options.

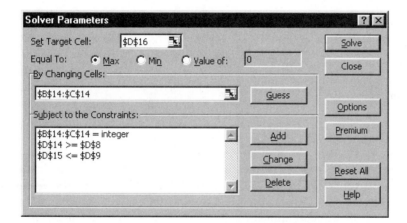

FIGURE 5.6
Solver parameters dialog box for the Queen City example

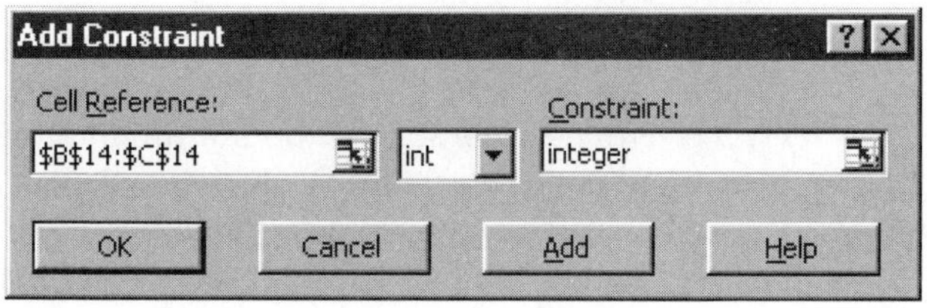

FIGURE 5.7
Defining integer variables for the Queen City example

When solving models with integer variables, only the *Answer Report* (Figure 5.8) is generated. The *Sensitivity* and *Limits* reports discussed in Chapter 4 for linear programs are not available, because the solution approach used for integer programs does not allow for easy calculation of shadow prices and reduced costs. The *Answer Report* shows that the component constraint is binding and that the maximum profit achievable is $238,000, by making 0 TopLathes and 7 BigPresses.

FIGURE 5.8

Solver answer report for the Queen City example

	A B	C	D	E	F	G
1	Microsoft Excel 8.0e Answer Report					
2	Worksheet: [QueenCity.xls]Queen City Model					
3	Report Created: 7/31/99 9:24:29 AM					
4						
5						
6	Target Cell (Max)					
7	Cell	Name	Original Value	Final Value		
8	$D$16	Profit Total	$0	$238,000		
9						
10						
11	Adjustable Cells					
12	Cell	Name	Original Value	Final Value		
13	$B$14	Quantity Produced TopLathe	0	0		
14	$C$14	Quantity Produced BigPress	0	7		
15						
16						
17	Constraints					
18	Cell	Name	Cell Value	Formula	Status	Slack
19	$D$14	Quantity Produced Total	7	$D$14>=$D$8	Not Binding	2
20	$D$15	Components Used Total	49	$D$15<=$D$9	Binding	0

Excel Exercise 5.4

Open the worksheet *Burke.xls*. Use Solver to find an optimal solution for the binary integer program in Example 2 by enforcing the binary restrictions on the decision variables. Compare the optimal solution to those solutions you found in Excel Exercise 5.2.

Sensitivity Analysis

Since shadow prices and reduced costs are not available for integer and mixed integer programs, the only way to obtain information about the stability of the solution is to change the model input parameters and re-solve the model. For example, if we want to know the effects of additional component availability in the Queen City example, we will have to change the entry in cell D9 in Figure 5.1 and re-solve the model. Figure 5.9 shows the impact of increased component availability on profit (under the assumption that we can sell everything produced).

To understand these results, observe that when the component availability is increased to 50, the minimum sales requirement $x_1 + x_2 \geq 5$ becomes redundant, since we can produce at least 5 of either machine if we do not produce any of the other. Thus, the problem reduces to a single constraint IP, which is called an *integer knapsack problem*. We saw an example of a *continuous* knapsack problem in Example 3 of Chapter 4. In that case, we could identify the optimal solution by simply taking ratios of the objective function coefficients to the constraint coefficients and choosing the best (a "greedy" approach). Intuitively, you would suspect that TopLathes are preferred over BigPresses because the profit per component is higher ($50,000 ÷ 10 = $5,000 for TopLathes, versus $34,000 ÷ 7 = $4,857.14 for BigPresses). The greedy approach would suggest producing as many TopLathes as possible, and then any BigPresses if sufficient components are available. Unfortunately, this does not always work for IPs. For example, when the component availability is 50, the constraint becomes $10x_1 + 7x_2 \leq 50$, the optimal solution is $x_1 = 5$, $x_2 = 0$, a case where the greedy approach works. However, when it is increased to 51, the optimal solution is $x_1 = 3$, $x_2 = 3$; here the greedy approach (which would identify $x_1 = 5$, $x_2 = 0$ as optimal) yields an infe-

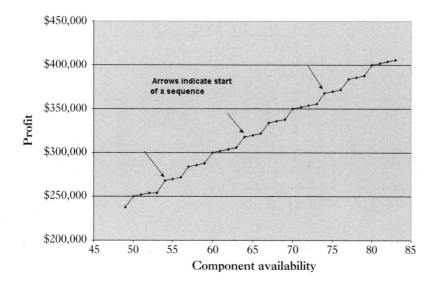

FIGURE 5.9
Profit as a
function of
component
availability

rior solution. For component availabilities of 52 and 53, the solution $x_1 = 1$, $x_2 = 6$ is optimal, again showing that intuition is wrong.

After 54 or more components are available, however, a clear pattern emerges in the solutions (the beginning of each pattern is labeled with an arrow). As the pattern repeats itself in Figure 5.9, it does so as more one more unit of TopLathe than in the last pattern can be produced. Table 5.2 shows some of these solutions. By investigating these solutions we can develop a decision rule that defines the optimal solution for any value of 54 or more components. If the last digit of the component availability is

4, make 2 BigPresses and the rest TopLathes until all components are used

5, make 5 BigPresses and the rest TopLathes until all components are used

6, make 8 BigPresses and the rest TopLathes until all components are used

7, make 1 BigPress and the rest TopLathes until all components are used

8, make 4 BigPresses and the rest TopLathes until all components are used

9, make 7 BigPresses and the rest TopLathes until all components are used

0, make all TopLathes until all components are used

1, make 3 BigPresses and the rest TopLathes until all components are used

2, make 6 BigPresses and the rest TopLathes until all components are used

3, make 9 BigPresses and no TopLathes

In general, we see that the optimal solution is always one for which we make as many TopLathes as possible, while leaving a remainder of components that are divisible by 7 (so that we get an integer number of BigPresses). Can you predict what the solution will be if Queen City has 84 components available?

As we have just seen, we can learn a lot about a process and optimal solutions by solving the problem for a variety of input parameter values. Investigating various solutions helps our understanding of those solutions and can build our intuition where it was previously lacking.

For a larger, more realistic example of the importance of sensitivity analysis, let us return now to the Sun Bank location problem discussed in Example 4.

	Components	Profit	TopLathes	BigPresses
TABLE 5.2 Solutions to Queen City for various component availability	54	$268,000	4	2
	55	$270,000	2	5
	56	$272,000	0	8
	57	$284,000	5	1
	58	$286,000	3	4
	59	$288,000	1	7
	60	$300,000	6	0
	61	$302,000	4	3
	62	$304,000	2	6
	63	$306,000	0	9
	64	$318,000	5	2
	65	$320,000	3	5
	66	$322,000	1	8
	67	$334,000	6	1
	68	$336,000	4	4
	69	$338,000	2	7
	70	$350,000	7	0
	71	$352,000	5	3
	72	$354,000	3	6
	73	$356,000	1	9
	74	$368,000	6	2
	75	$370,000	4	5
	76	$372,000	2	8
	77	$384,000	7	1
	78	$386,000	5	4
	79	$388,000	3	7
	80	$400,000	8	0
	81	$402,000	6	3
	82	$404,000	4	6
	83	$406,000	2	9

Example 6: Solving the Sun Bank Location Problem

In the problem described in Example 4, Sun Bank wishes to locate five principal places of business (PPBs) in the state of Ohio to maximize the population reached with its branches. Because state law says that bank branches may be placed in a county containing a PPB or in any county adjacent to a county with a PPB, the problem involves finding the five counties in Ohio with maximum total population in these counties and their adjacent counties.

Although the model is quite large, it is easily formulated in the Solver dialog box, as shown in Figure 5.10. Note that we have set the PPB variables to be binary and have allowed the coverage variables to be continuous with upper bounds of 1 (as discussed in Example 4). The second constraint set corresponds to the coverage constraints in the model and the last constraint set limits the number of PPBs to 5.

The results of this optimization model are linked to a map of the state of Ohio in Excel (this map is not available with Excel and was purchased by a third party). The map displays the location of counties chosen for PPBs by Solver (data for the map start in row 199 and are linked to the results of the optimization through row

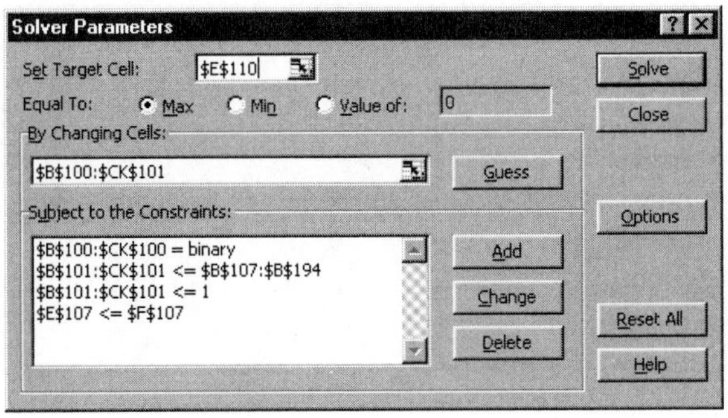

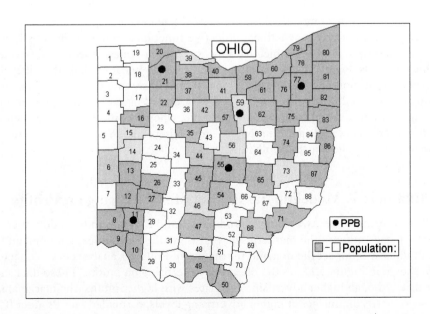

FIGURE 5.10
Solver Parameters dialog box for Sun Bank example

100). As we discussed in Chapter 3, being able to view the results quickly on the map is helpful in building intuition and confidence in the model results. For example, the solution for 5 PPBs is shown in Figure 5.11. This solution, which consists of counties 11, 21, 55, 59 and 77, covers a population of 8,190,731. The map helps us to understand the solution in that it is clear that the chosen counties are in the heavily populated sections of the state and they each have a relatively large number of adjacent counties. The choice of county 59 illustrates that a less heavily populated county may be chosen for a PPB if it is adjacent to heavily populated counties, something that a manager might not realize.

We might wish to conduct some sensitivity analysis on the solution, for example, to examine how the population reached increases as we increase the number of PPBs allowed. We can obtain this information by changing the data in cell F107, and rerunning the optimization model. The results of varying F107 from 1 to 16 appear in Figure 5.12. What do we learn from this graph? First, since there is no increase in the population reached when the number of PPBs is increased from 15 to 16, we conclude that we can reach the total population with only 15 PPBs. Second, the incremental increase in population reached diminishes as we increase the number of

FIGURE 5.11
Solution to the Sun Bank problem for 5 PPBs.

FIGURE 5.12

Population reached as a function of number of PPBs allowed

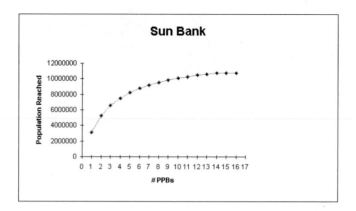

PPBs, particularly after about 8. Further analysis is needed to determine if the cost of PPBs beyond 8 is worth the investment.

Excel Exercise 5.5

Open the workbook *Sun.xls*. Enter 15 in cell F107 and solve the model. Which counties have been chosen? (You will not be able to display your new results on the map, however, since the map file is not included with Excel.) Can you find— by inspection of the map—any other solutions of 15 PPBs that cover all counties?

BUILDING NONLINEAR OPTIMIZATION MODELS

Many types of nonlinear models exist. Some have a nonlinear objective function and linear constraints, others have a linear objective function and nonlinear constraints, and still others have nonlinear objective functions and nonlinear constraints. Because nonlinear models do not have a common structure as do linear models, the difficulty of developing accurate and appropriate models, and of solving them, both increase. Although many nonlinear optimization models, are used in science and engineering, many applications exist in business. For example, when empirical data are available, statistical regression can be a useful tool for identifying an appropriate function to use in a model, as the following example illustrates.

Example 7: Modeling the Impact of Advertising on Profits

Phillips, Inc., produces two distinct products, A and B. The products do not compete with each other in the marketplace; that is, neither cost, price, nor demand for one product will impact the demand for the other. Phillips's analysts have collected data, shown in Figure 5.13, on the effects of advertising on profits. These data suggest that, although higher advertising correlates with higher profits, the marginal increase in profits diminishes at higher advertising levels, particularly for Product B.

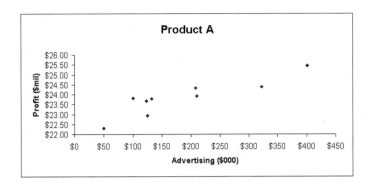

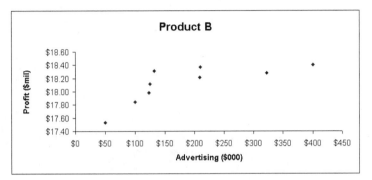

FIGURE 5.13
Data for
Phillips
example

Regression analysis can help us determine a nonlinear function that relates profit and advertising. In Excel, we may use the *Add Trendline* option to fit certain forms of nonlinear functions to a scatterplot by placing the cursor over the data series and right-clicking the mouse. This brings up the dialog box shown in Figure 5.14. Of the options available, the logarithmic option appears to be the most appropriate, based on the form of the data series. Before clicking on OK to create the trend, select the *Options* tab and check *Display equation on chart* and *Display R-squared value on the chart*. The results are shown in Figure 5.15. In the functional equation, $\text{Ln}(x)$ is the natural logarithm of x (the advertising variable).

The R^2 value—the percentage of variation in the data explained by the fitted model—suggests that these models provide a good fit over the range of the data.

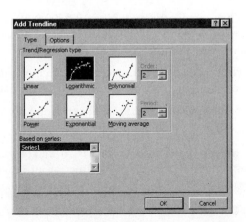

FIGURE 5.14
Add Trendline
dialog box

FIGURE 5.15
Logarithmic
trendlines for
Phillips data

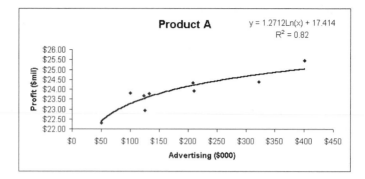

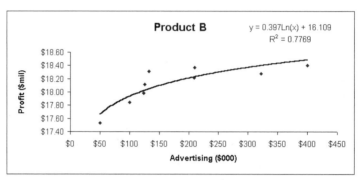

Note that these might not be the best theoretical models, since Ln(0) is undefined, implying that an allocation of $0 advertising pushes profits toward negative infinity! Nonetheless, if we are operating roughly within the range of the data, these functions reliably predict the effects on profits of increasing advertising expenditures.

An optimization model that has any nonlinear function in it is called a *nonlinear programming model* (NLP model). The approach we use for modeling nonlinear programming problems is the same as that we have used for linear and linear integer programming models. We first define the decision variables, then the objective function, and finally the constraints.

Example 8: Optimal Advertising Budget Allocation Model

Phillips (Example 7) has an advertising budget of $500,000. At least $50,000 must be allocated to each product. How should this budget be allocated between products A and B to maximize total profit?

What is Phillips trying to decide? The decisions are how much of the $500,000 budget to allocate to each product. We define two decision variables:

x_A = amount ($000) to allocate for advertising for product A

x_B = amount ($000) to allocate for advertising for product B

What is Phillips's objective? The objective is to maximize total profit. Since we modeled the profit for each product as a function of its advertising level, in Example 7, the objective function for total profit is simply their sum:

Maximize $17.414 + 1.2712 \, \mathrm{Ln}(x_A) + 16.109 + .397\mathrm{Ln}(x_B)$

This may be simplified to

$$Maximize\ 33.523 + 1.2712\ Ln(x_A) + .397Ln(x_B)$$

How is this problem constrained? First, the total budget is limited to $500,000; therefore:

$$x_A + x_B \leq 500$$

Second, at least $50,000 must be allocated to each product:

$$x_A \geq 50$$
$$x_B \geq 50$$

Finally, advertising must be nonnegative:

$$x_A, x_B \geq 0$$

Note, however, that the nonnegativity constraints are redundant because of the lower bounds of 50 on each variable. The complete allocation model follows:

   $Maximize\ 33.523 + 1.2712\ Ln(x_A) + .397Ln(x_B)$

   subject to

$$x_A + x_B \leq 500$$
$$x_A \geq 50$$
$$x_B \geq 50$$

This is a nonlinear programming problem with linear constraints.

Nonlinear models play an important role in finance because continuous discounting and covariance among variables lead to nonlinear relationships. One important nonlinear model in finance that we first introduced in Chapter 2 is the Markowitz portfolio model.

Example 9: The Markowitz Portfolio Model

The Markowitz portfolio model[3] seeks to minimize the risk of a portfolio of stocks subject to a constraint on the portfolio's expected return. For example, suppose an investor is considering 3 stocks. The expected return for Stock 1 is 15 percent, for Stock 2, 20 percent, and for Stock 3, 10 percent, and we would like an expected return of at least 14 percent. The portfolio's risk is measured by its variance. Research has found the variance-covariance matrix of the individual stocks to be:

	Stock 1	Stock 2	Stock 3
Stock 1	.0225	.0135	−.0015
Stock 2		.0324	.0050
Stock 3			.0025

The investor has a fixed amount to invest and would like to know what percentage of the investment budget to allocated to each stock.

What are we are trying to decide? Because we do not know the actual investment budget, it would not make sense to define the decision variables in terms of dollars.

However, we could define the variables as the *proportion* of the budget to allocate to each stock. (You might be familiar with the term "asset allocation model" that many financial investment companies suggest to their clients; for example, "maintain 60 percent equities, 30 percent bonds, and 10 percent cash." This makes the decision independent of the actual investment.) Therefore, define x_j = fraction of the portfolio to invest in stock j.

What is our objective? We would like to minimize the risk of the portfolio as measured by its variance. Because stocks' prices are correlated with one another, the variance of the portfolio must reflect not only variances of the stocks in the portfolio, but also the covariance between stocks. The variance of a portfolio is the weighted sum of the variances and covariances:

$$\text{Variance of Portfolio} = \sum_{i=1}^{k} s_i^2 x_i^2 + \sum_{i=1}^{k}\sum_{j>i} 2s_{ij}x_i x_j$$

where s_i^2 = the sample variance in the return of stock i, and s_{ij} = the sample covariance between stocks i and j. Using these data, the objective function for this problem is

Minimize $.0225x_1^2 + .0324x_2^2 + .0025x_3^2 + 2(.0135)x_1 x_2 + 2(-.0015)x_1 x_3 + 2(.005)x_2 x_3$

How is the problem constrained? First, we must invest 100 percent of our budget. Since the variables are defined as fractions, we have:

$$x_1 + x_2 + x_3 = 1$$

Second, the portfolio must have an expected return of at least 14 percent. The expected return on a portfolio is simply the weighted sum of the expected returns of the stocks in the portfolio.

$$.15x_1 + .20x_2 + .10x_3 \geq .14$$

Finally, we cannot invest negative amounts:

$$x_1, x_2, x_3 \geq 0$$

The complete model is

Minimize $.0225x_1^2 + .0324x_2^2 + .0025x_3^2 + .027x_1 x_2 - .003x_1 x_3 + .01x_2 x_3$

subject to

$$x_1 + x_2 + x_3 = 1$$
$$.15x_1 + .20x_2 + .10x_3 \geq .14$$
$$x_1, x_2, x_3 \geq 0$$

Like the previous example, this model has a nonlinear objective function and linear constraints. Figure 5.16 shows a spreadsheet model for this example. The decision variables (percentage of portfolio for each stock) are in cells B15:B17. The expected return and variance of the portfolio are in cells B19 and B20. The spreadsheet contains a pie chart showing the percentage allocation of stocks in the portfolio.

An alternative approach would be to maximize the return subject to a constraint on risk (see the discussion on multiobjective models in Chapter 4.) For example, suppose the investor wants to maximize expected return subject to a risk (variance) no greater than 0.5 percent. This form of the model would be

Maximize $.15x_1 + .20x_2 + .10x_3$

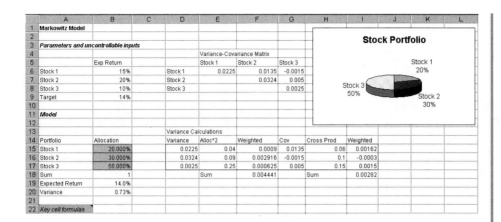

FIGURE 5.16
Markowitz
portfolio
spreadsheet
model

subject to

$$x_1 + x_2 + x_3 = 1$$

$$.0225x_1^2 + .0324x_2^2 + .0025x_3^2 + .027x_1x_2 - .003x_1x_3 + .01x_2x_3 \le .005$$

$$x_1, x_2, x_3 \ge 0$$

In this case, we would have a linear objective function and a mixture of linear and nonlinear constraints.

Excel Exercise 5.6

Open the worksheet *Markowitz.xls*. Use trial and error to find a low-risk portfolio with an expected return of at least 14 percent. How does the nonlinear nature of this problem change your approach, as compared with linear models?

The next example involves locating a facility within a geographic area to optimize its relationship with respect to other locations. This type of model can be applied to many different types of situations—for example, locating a warehouse to optimize the flow of goods from plants and customer markets, or finding the best location for a radiology department in a hospital.

Example 10: A Facility Location Model

Jack's Job Shop is studying where to locate its tool bin facility on the shop floor. The locations of five production cells—fabrication, paint, subassembly 1, subassembly 2, and assembly, expressed as *x* and *y* coordinates on a shop floor grid—are fixed, and the daily demand for tools (measured in number of trips to the tool bin) at each production cell is also known. One approach to this problem is to find the location that minimizes the sum of the distances from each production department to the tool bin location. However, this ignores the fact that some stations make more trips for tools than others. By minimizing the sum of the *weighted distances* from each station to the tool bin, where the weights are the demand from each department, we essentially find the "center of gravity."

What is it we are trying to decide? We would like to find a location for the tool bin. Define the following decision variables:

$$X = \text{horizontal location } (x \text{ coordinate}) \text{ of the tool bin}$$

$$Y = \text{vertical location } (y \text{ coordinate}) \text{ of the tool bin}$$

What is our objective? We would like to minimize the sum of the demand-weighted distances from the manufacturing stations to the tool bin. We may measure the straight-line distance from a station to the tool bin located at (X,Y) by using Euclidean (straight-line) distance. For example, the distance from fabrication to the tool bin is $\sqrt{(X-1)^2+(Y-4)^2}$. We weight these distances by demand to get the following objective function:

$$\textit{Minimize}\ \Big\{12\sqrt{(X-1)^2+(Y-4)^2}\ +24\sqrt{(X-1)^2+(Y-2)^2}\ + 13\sqrt{(X-2.5)^2+(Y-2)^2}\ +\ 7\sqrt{(X-3)^2+(Y-5)^2}\ +17\sqrt{(X-4)^2+(Y-4)^2}\Big\}$$

How is the problem constrained? This problem has no explicit constraints. You might at first think that the location variables should be nonnegative, but this is not necessary. Intuitively, it seems that the solution to this problem must be somewhere in the interior of the set of points under consideration. Thus, we have an *unconstrained optimization problem.*

Figure 5.17 shows a spreadsheet model for this problem. The location of the five production stations and their relative average daily demand for tools appear in the bubble chart. The size of the bubbles corresponds to the level of daily demand. An arbitrary demand value of 7 is given to the tool bin so that it would be plotted as a bubble on the graph (the unlabeled bubble). This value does not enter into the calculations, however. Distances between the bin location and each production station are calculated in cells G13:G17, and the weighted distances are in cells H13:H17. The decision variable cells are B18 and C18.

FIGURE 5.17
Jack's Job Shop spreadsheet model

	A	B	C	D	E	F	G	H	I	J
1	**Jack's Job Shop**									
2										
3	***Parameters and uncotrollable inputs***									
4	Cell	x-coordinate	y-coordinate	Demand						
5	Fabrication	1	4	12						
6	Paint	1	2	24						
7	Subassembly 1	2.5	2	13						
8	Subassembly 2	3	5	7						
9	Assembly	4	4	17						
10										
11	***Model***									
12		x-coordinate	y-coordinate	Demand			Distance	Weighted Distance		
13	Fabrication	1	4	12		Fabrication	1.581	18.974		
14	Paint	1	2	24		Paint	2.121	50.912		
15	Subassembly 1	2.5	2	13		Subassembly 1	1.500	19.500		
16	Subassembly 2	3	5	7		Subassembly 2	1.581	11.068		
17	Assembly	4	4	17		Assembly	1.581	26.879		
18	Bin Location	2.5	3.5	7		Sum	8.365	127.333		

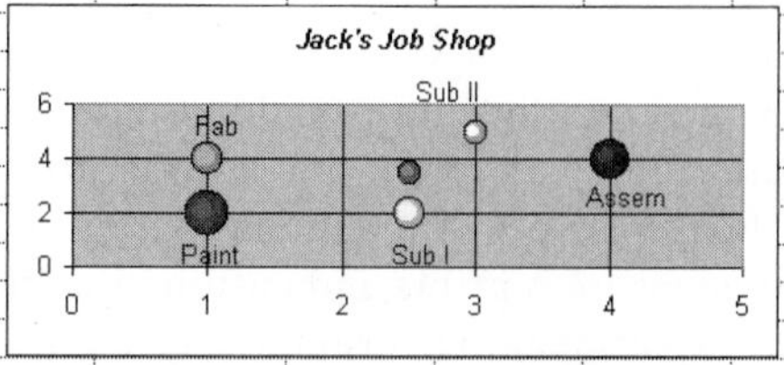

Excel Exercise 5.7

Open the worksheet *Jack's Job Shop.xls*. Use trial and error guided by the bubble graph to determine a good solution (a low value of the sum of the demand-weighted distances given in cell H18).

Some models might include both nonlinearities and integer variables. The Management Science Practice box for Whirlpool illustrates one practical example.

Management Science Practice:
Kanban Sizing at Whirlpool Corporation

Kanban (the Japanese word for "card") is an information system used in just-in-time (JIT) production systems for scheduling delivery or production of items only when they are required, thus "pulling" production through the system to create a smooth, rapid flow of products from purchasing through shipping. The type and number of units required by a process are written on kanbans and used to initiate withdrawal and production of items through the production process. Parts are transported between production operations, using standard size containers. An important design issue of kanban systems is determining the minimum "trigger" values required to meet production targets for parts. Trigger values represent the number of kanbans that must accumulate before a machine cell is authorized to begin production, and thus control the frequency of machine setups. While the ideal trigger value is 1, many processes do not allow for this, because of the time required for machine setup. To minimize work in process, it is necessary to minimize the trigger values for all parts.

Whirlpool Corporation developed a nonlinear, integer optimization model to estimate the minimum trigger values while maintaining cell utilization below a specified threshold. The primary driver for the model is a formula for machine cell utilization:

$$\text{TotU} = \frac{\sum_{i=1}^{n}\left(\dfrac{D_i*(1 + \text{Scrap})*\text{Setup}_i}{\text{CSize}_i*Trigger_i} + \dfrac{D_i*(1 + \text{Scrap})}{\text{PRate}_i}\right)}{\text{ACap}}$$

where

  Acap = available daily capacity for the machine cell under consideration (hours)
  CSize_i = Container size for part i assigned to the cell (parts)
  D_i = Daily demand for part i assigned to the cell (parts)
  PRate_i = Expected production rate for part i assigned to the cell (parts/hour)
  Scrap = scrap allowance (shrinkage factor) common to all assigned parts (percentage)
  Setup_i = Expected time for one setup operation for part i assigned to the cell (hours)
  TotU = total utilization of the cell required to process all assigned parts (percentage)

And the decision variables are:

  $Trigger_i$ = Trigger value for part i assigned to the cell

continued

Each term in the numerator represents the estimated average daily capacity required to produce part i in the cell (hours).

The number of kanbans is derived from the trigger values. These trigger values represent the absolute minimum number of kanbans (and therefore the minimum amount of work in process) for the given parts. A safety stock level is specified to ensure that subsequent production operations are not starved for parts due to inherent variation in the process. This is computed by the formula:

$$\text{Kanban}_i = \textit{Trigger}_i + (\text{SStock}_i * \textit{Trigger}_i)$$

where

Kanban_i = total kanbans for part i assigned to the cell
SStock_i = safety stock specified for part i assigned to the cell (percentage)

The optimization model is

$$\textit{Minimize} \sum_{i=1}^{n} \textit{Trigger}_i$$

subject to

$$\text{TotU} = \frac{\sum_{i=1}^{n} \left(\dfrac{D_i^*(1 + \text{Scrap})^*\text{Setup}_i}{\text{CSize}_i^* \textit{Trigger}_i} + \dfrac{D_i^*(1 + \text{Scrap})}{\text{PRate}_i} \right)}{\text{ACap}} \leq \text{MaxU}$$

$$\textit{Trigger}_i \geq 1 \text{ and integer for all } i$$

In this model, *MaxU* is the specified threshold for cell utilization. Because the decision variables (trigger values) appear in the denominator of the constraint, it is easy to see that the model is not linear.

Use of this optimization model has resulted in dramatically lower work-in-process for Whirlpool. Setup times in many manufacturing cells have been reduced from over two hours to less than 30 minutes, and batch sizes of some parts have been reduced from a ten-day supply to a half-day supply.

Source: Reprinted with permission of APICS—The Educational Society for Resource Management, Alexandria, VA. John D. Hall, Royce O. Bowden, Richard S. Grant, and William H. Hadley, "An Optimizer for the Kanban Sizing Problem: A Spreadsheet Application for Whirlpool Corporation," *Production and Inventory Management Journal*, Vol. 39, No. 1, First Quarter 1998, 17–22.

SOLVING NONLINEAR PROGRAMMING PROBLEMS

Because of the complexities that might exist in a nonlinear programming problem, we may not always be able to find a true optimal solution, except under certain theoretical conditions that are beyond the scope of this book. We will discuss some of these issues and provide practical advice for helping to ensure that you get the best solution possible.

Using Excel Solver[4]

Using Excel Solver for NLPs is similar to solving linear and integer models. We specify the objective function, and constraints are specified in the Solver dialog box in the same fashion as the linear case. However, a different solution algorithm (the "generalized reduced gradient algorithm") is used. Therefore, *do not* check the *Assume Linear Model* option. Solver provides special options for nonlinear problems, which control how it estimates the effects of changing the decision variables. In general, the defaults work quite well and should not be changed unless you are having trouble finding an answer in a reasonable amount of time. We will use Example 9, the Markowitz portfolio model, to illustrate the use of Solver for NLPs.

The Solver dialog box for this model (see Figure 5.16), shown in Figure 5.18, shows the model structure in the spreadsheet. You should click the *Options* button, ensure that the *Assume Linear Model* option is *not* checked, but check *Assume Nonnegative,* since negative percentages are unacceptable. As with linear models, three reports are generated when you terminate Solver. The *Answer Report* in Figure 5.19 reports the solution—a portfolio containing 24.074 percent of stock 1, 27.963 percent of stock 2 and 47.963 percent of stock 3 with a variance of .72 percent. As shown in the constraint section, both constraints are binding.

The Sensitivity Report (Figure 5.20) looks quite different from the linear case, although it provides similar information. The interpretation of *reduced gradient* for each variable is the same as reduced cost in a linear model; that is, it measures how the objective function will be worsened if a variable not in the solution is forced into the solution. Since all three stocks are in the portfolio, the reduced gradients are all 0. In the constraint section, the *Lagrange multiplier* is analogous to the shadow price in a linear model. The *Lagrange multiplier* indicates *approximately* how much the objective function changes if the right-hand side of the associated constraint is increased by 1. For example, the variance of the portfolio is expected to increase by 24.9 percent as

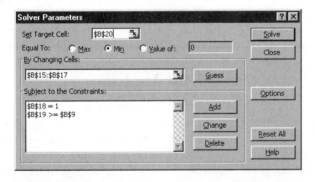

FIGURE 5.18 Solver parameters dialog box for Markowitz Portfolio example

Microsoft Excel 8.0 Answer Report
Worksheet: [Portfolio3.xls]Sheet1
Report Created: 1/30/99 3:10:07 PM

Target Cell (Min)

Cell	Name	Original Value	Final Value
$B$20	Variance Allocation	1.16%	0.72%

Adjustable Cells

Cell	Name	Original Value	Final Value
$B$15	Stock 1 Allocation	30.000%	24.074%
$B$16	Stock 2 Allocation	40.000%	27.963%
$B$17	Stock 3 Allocation	30.000%	47.963%

Constraints

Cell	Name	Cell Value	Formula	Status	Slack
$B$18	Sum Allocation	1	$B$18=1	Binding	0
$B$19	Expected Return Allocation	14.0%	$B$19>=$B$9	Binding	0.0%

FIGURE 5.19 Answer report for Markowitz Portfolio example

FIGURE 5.20

Sensitivity
report for
Markowitz
Portfolio
example

	A	B	C	D	E
1	Microsoft Excel 8.0 Sensitivity Report				
2	Worksheet: [Portfolio3.xls]Sheet1				
3	Report Created: 1/30/99 3:10:08 PM				
4					
5					
6	Adjustable Cells				
7				Final	Reduced
8		Cell	Name	Value	Gradient
9		$B$15	Stock 1 Allocation	24.074%	0.000%
10		$B$16	Stock 2 Allocation	27.963%	0.000%
11		$B$17	Stock 3 Allocation	47.963%	0.000%
12					
13	Constraints				
14				Final	Lagrange
15		Cell	Name	Value	Multiplier
16		$B$18	Sum Allocation	1	-0.02047185
17		$B$19	Expected Return Allocation	14.0%	24.9%
18					

the required expected return increases by 1. Your intuition might suggest that this seems rather high, since the optimal objective function value is only 0.72 percent. However, since we have used percentages in the model, the return constraint is

$$.15x_1 + .20x_2 + .10x_3 \geq .14$$

An increase of 1 in the required return changes it from 14 percent to 114 percent—not a small increase. On the other hand, if we increase the requirement by .01 to .15, we would expect an increase in the variance of .01(24.9 percent) = .249 percent, a more reasonable value. The key word in interpreting *Lagrange multipliers* is *approximate*. While shadow prices provide *actual* changes in the objective function over relevant ranges, *Lagrange multipliers* do not because of the nonlinearities in the model. Therefore, no ranges are specified for the validity of the *Lagrange multipliers* in the *Sensitivity Report*. For NLPs, we recommend that sensitivity analyses be conducted by actually changing problem parameters and rerunning Solver to determine their impact.

Sensitivity Analysis

We will illustrate sensitivity analysis for NLPs using the Markowitz portfolio model. An investor might be interested in the trade-off between the variance of the portfolio and its required expected return. Since the expected return on stocks under consideration ranges from 10 percent to 20 percent, requiring an expected return greater than 20 percent results in an infeasible problem. Figure 5.21 shows the minimum variance for required expected returns from 10 percent to 20 percent found by re-

FIGURE 5.21

Risk-return
trade-off graph

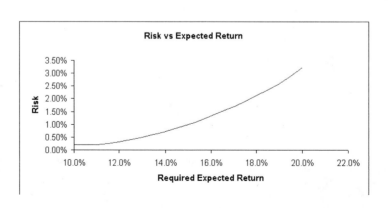

solving the model and changing the right-hand side target of the return constraint (cell B9 in Figure 5.16).

Figure 5.22 helps explain this behavior by showing the portfolio mix over the same range of problems. For low values of return, the portfolio is made up of stocks 1 and 3. They have lower variance than stock 2, and they have negative covariance, which further reduces the variance of the portfolio. Note that for required return of 10 percent, the return constraint is not binding, as the portfolio containing stocks 1 and 3 gives a return greater than 10 percent and has smaller variance than a portfolio containing only stock 3. As higher expected returns are required, the higher expected return of stock 2 is needed. The increase in variance seen in Figure 5.21 directly parallels the use of the riskier stock 2.

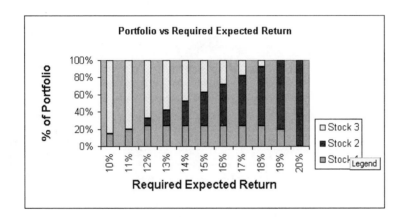

FIGURE 5.22
Portfolio composition for various risk and returns

The Problems of Nonlinearity

As we noted earlier, Solver (or any other optimization tool) cannot guarantee finding an optimal solution for every nonlinear optimization problem. Whether it does converge to an optimal solution depends on the nature of the nonlinear function. Examine the functions shown in Figure 5.23 over the feasible region $0 \le x \le 1$ and $0 \le y \le 1$. The first is the linear function $z = 3x + y$. The second is a "nice" nonlinear function $z = 3 - (x - .5)^2 - (y - .5)^2$, and the third is a rather nasty nonlinear function $z = x \sin(5\pi x) + y \sin(5\pi y)$, where $\pi \approx 3.1416$ and "sin" is the sine trigonometric function.

We see that both the linear and the "nice" nonlinear functions have a single maximum point; the third function, however, has many peaks, but only one is the best (*a global optimal solution*). Stopping rules for many optimization tools are based on the simple idea that if they find a point where all its close neighbors are no better (called a *local optimal solution*), they identify that point as an optimal solution (think of the analogy of standing at the top of a small hill in a mountain range in a dense fog!). For linear and "nice" nonlinear cases, this is enough to guarantee optimality, but it clearly does not hold for the difficult nonlinear case.

Most optimization algorithms such as the one used in Solver can only find local optimal solutions. In the difficult nonlinear case, which local optimal solution is found depends on the starting point of the algorithm. Table 5.3 shows the solutions found by Solver from various starting points (Solver starts from whatever point you enter into the decision variable cells in your model). The solution found from the starting point $x = 1$, $y = 1$ is the optimal solution; the others are simply locally optimal.

FIGURE 5.23
Examples of
linear and non-
linear functions

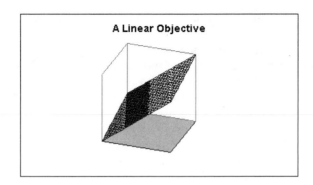

A Linear Objective

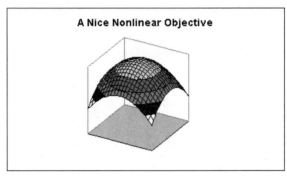

A Nice Nonlinear Objective

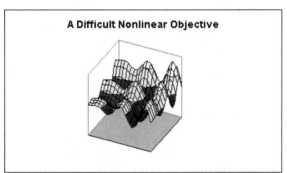

A Difficult Nonlinear Objective

TABLE 5.3
Solutions
found by
Solver for hard
nonlinear
problem

Starting Point		Solution		
X	Y	X	Y	Objective
0.0000	0.0000	0.1292	0.1292	0.2317
1.0000	0.0000	0.9045	0.0000	0.9022
0.0000	1.0000	0.0000	0.9045	0.9022
0.5000	0.5000	0.5079	0.5079	1.0080
1.0000	1.0000	0.9045	0.9045	1.8045

Highly nonlinear constraints can also cause trouble for algorithms. Just as Solver stopped at some local solutions because they appeared to be optimal, it can likewise become trapped at nonoptimal points because the feasible region is nonlinear in shape, and there appears to be no way to improve the objective function value.

Unless you are fairly skilled mathematically, there is no easy way to know if your nonlinear function is well behaved. Practically speaking, if you are unsure about the nature of the functions involved, we advise you to run Solver from a variety of starting points to test the stability of the solution, or use an algorithm specifically designed to overcome local optimality.

Several new approaches for solving difficult nonlinear optimization problems, called *metaheuristics*, are designed to overcome the problem of local optimal solutions. These approaches include genetic algorithms, neural networks, tabu search, and other exotic-sounding heuristic methods. Evolutionary Solver, which is part of the Premium Solver software packaged with this text, uses some of these approaches.

Using Premium Solver's Evolutionary Algorithm for Difficult Problems[5]

The Premium Solver evolutionary algorithm is based on a heuristic that remembers the best solutions it finds, then modifies and combines these solutions to try to find better solutions. We will illustrate the use of evolutionary solver with the "nasty" nonlinear function shown in Figure 5.23. It is not unusual to see such functions in applications such as engineering design. The spreadsheet model, shown in Figure 5.24, simply has the decision variables X and Y in cells B3 and B4, respectively, and the objective function, =B3*SIN(3.1416*5*B3) + B4*SIN(3.1416*5*B4), in cell B2. The optimization model is shown in the solver dialog box (although not shown, the *Assume Nonnegative* option has been selected) in Figure 5.25. To use the evolutionary algorithm on this problem we click on the *Premium* button, which invokes the Premium Solver dialog box shown in Figure 5.26. From this dialog box, we select *Standard Evolutionary*, followed by a click of the *Solve* button.

The Answer report for this example problem is shown in Figure 5.27. The best solution found is $X = 0.9045$, $Y = 0.9045$ and objective function value 1.80448 was the best found. This is in fact the global solution to the problem. Figure 5.28 shows the population report. This report gives statistics on the search process used by the Standard Evolutionary Solver. Here we see that the range of searched values for X was from .7046 to .90457 and for Y from .64 to .9045. The average and standard deviations for all search

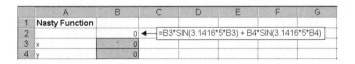

FIGURE 5.24
Spreadsheet for optimizing the nasty nonlinear function

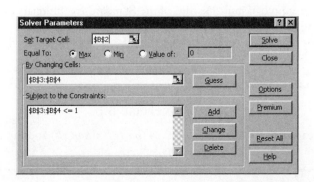

FIGURE 5.25
Solver parameters dialog box for the nasty nonlinear function

FIGURE 5.26 Premium Solver dialog box

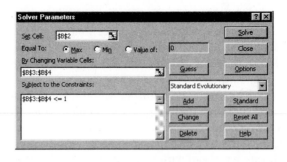

FIGURE 5.27 Standard Evolutionary Solver answer report

Microsoft Excel 8.0e Answer Report
Worksheet: [nasty function.xls]Nasty function
Report Created: 6/14/99 4:14:09 PM

Target Cell (Max)

Cell	Name	Original Value	Final Value
$B$2		0	1.804478497

Adjustable Cells

Cell	Name	Original Value	Final Value
$B$3	x	0	0.904471415
$B$4	y	0	0.904471376

Constraints

Cell	Name	Cell Value	Formula	Status	Slack
$B$3	x	0.904471415	$B$3<=1	Not Binding	0.095528585
$B$4	y	0.904471376	$B$4<=1	Not Binding	0.095528624

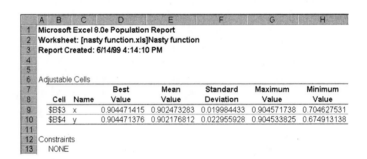

FIGURE 5.28 Standard Evolutionary Solver population report

Microsoft Excel 8.0e Population Report
Worksheet: [nasty function.xls]Nasty function
Report Created: 6/14/99 4:14:10 PM

Adjustable Cells

Cell	Name	Best Value	Mean Value	Standard Deviation	Maximum Value	Minimum Value
$B$3	x	0.904471415	0.902473283	0.019984433	0.904571738	0.704627531
$B$4	y	0.904471376	0.902176812	0.022955928	0.904533825	0.674913138

Constraints
NONE

values for each decision variable are also given. Note that no constraints information is given because Solver handles the simple upper and lower bound constraints implicitly. The population report is sometimes useful to determine if more runs of the model would be prudent. For example, if multiple runs of the Standard Evolutionary Solver are made and the resulting best solutions are the same or very close, with small standard deviations, this may be reason for confidence that the solution is close to the global solution.

Excel Exercise 5.8

Use the spreadsheet model in Figure 5.24 (*nasty function.xls*) and run it with the Premium Solver Standard Evolutionary solver.

TECHNOLOGY FOR INTEGER AND NONLINEAR OPTIMIZATION

Like Excel Solver, most computer packages that solve linear programs also solve integer programs. Nonlinear programming, however, requires significantly different solution algorithms and often special interface languages. Although such differences are transparent in spreadsheet models, special purpose commercial programs might require you to write the actual model terms and possibly the derivatives (from a calculus perspective) of the model functions. For an LP or IP, on the other hand, you need only specify the objective function and constraint coefficients and the right-hand side. This is one of the reasons why nonlinear optimization has never been as popular as linear or even integer programming.[6] However, optimization modeling languages, which are linked to nonlinear solvers, are becoming increasingly more sophisticated and are simplifying the solution process. Table 5.4 lists some of the nonlinear optimization technology that is often used in business and industry.

Product	Company	Description	
AMPL	Duxbury Press Pacific Grove, Calif. **www.duxbury.com**	Developed by AT&T and Bell Laboratories, allows users to formulate models using common algebraic notation	**TABLE 5.4** Some popular nonlinear optimization software
GAMS	GAMS Development Corporation Washington, DC **www.gams.com**	Designed for modeling and solving large complex, linear, nonlinear, and mixed integer problems	
GINO	LINDO Systems Chicago, Ill. **www.lindo.com**	A robust package for nonlinear optimization that includes severa built-in mathematical, financial, and probability functions	
GRG2	Optimal Methods Inc. Kirtland, Ohio	A widely used NLP code that interfaces with GAMS, Excel (used in Solver), GINO, and others	
LINGO	LINDO Systems Chicago, Ill. **www.lindo.com**	Combines large-scale linear, nonlinear, and integer optimizers and interactive modeling environment	
MINOS	Stanford Business Software Mountain View, Calif.	Links with GAMS and AMPL and can accommodate many thousands of constraints and variables	
Problem Solver and Premium Solver Plus	Frontline Systems Inc. Incline Village, Nev. **www.frontsys.com**	More powerful extensions of the Standard Excel Solver	
What'sBest!	LINDO Systems Chicago, Ill. **www.lindo.com**	A spreadsheet add-in for linear, nonlinear, and integer models	

PROBLEMS

1. Camco, Inc., is planning the purchase of one of its component parts needed for its finished product. The anticipated demands for the component for the next twelve periods are shown in the following table. The cost to order the component (labor, shipping, and paperwork) is $150. The cost to hold these items in inventory is $1 per component per period. The price of the component is expected to remain stable at $12 per unit, and no quantity discounts are available. Develop a model to minimize the cost of satisfying the company demand for this component.

Period	1	2	3	4	5	6	7	8	9	10	11	12
Demand	20	20	30	40	140	360	500	540	460	80	0	20

2. STAR Co. provides paper to smaller companies whose volumes are not large enough to warrant dealing directly with the paper mill. STAR receives 100-feet-wide paper rolls from the mill and cuts the rolls into smaller rolls 12, 15, and 30 feet wide. The demands for these widths vary from week to week. The following cutting patterns have been established:

Pattern	Number of rolls		
	12'	15'	30'
1	0	6	0
2	0	0	3
3	8	0	0
4	2	1	2
5	7	1	0

Demands this week are 5,670 12' rolls, 1,680 15' rolls, and 3,350 30' rolls. Develop a model that will determine how many 100' rolls to cut into each of the five patterns in order to meet demand and minimize trim loss (leftover paper from a pattern).

3. An investment firm is considering six options for the investment of $2 million that will become available at the beginning of next month. The expected return over the next year, along with the cost of each investment, is shown in the following table, along with conditions placed on each investment due to company policy (considerations of risk, company goals, politics, etc.). Any cash not invested will earn a no-risk fixed rate and be rolled into the funds that will become available next month. Formulate a model to determine which investments to select in order to maximize expected return.

Investment	Cost	Expected Return	Condition
I	$200,000	$25,000	Only if III
II	$1,000,000	$100,000	Not if I
III	$750,000	$150,000	None
IV	$1,000,000	$270,000	Only if I and III
V	$500,000	$300,000	Not if III
VI	$500,000	$100,000	None

4. The producers of cassette tapes (for those who still buy them!) are faced with the problem of placing songs on both sides of the tape so that the total time on each side is as close as possible to being equal in length (they sometimes fail in this endeavor, as evidenced by those blank portions we must endure on certain tapes). Formulate this problem as a linear integer program for the following example:

Song	Run Time (Minutes: Seconds)
1	2:56
2	3:37
3	3:44
4	3:50
5	4:00
6	4:05
7	4:06
8	4:08
9	4:16
10	4:20

5. Hospital administrators must schedule nurses so that the hospital's patients are provided with adequate care. At the same time, in the face of tighter competition in the health care industry, careful attention must be paid to keeping costs down. From historical records, administrators can project the *minimum* number of nurses to have on hand for the various times of day and days of the week. The nurse scheduling problem seeks to find the minimum total number of nurses required to provide adequate care. Nurses start work at the beginning of one of the four-hour shifts given next, and work for eight hours. Formulate and solve the nurse scheduling problem as an integer program for one day for the data given here.

Shift	Time	Minimum Number of Nurses Needed
1	midnight–4 a.m.	5
2	4 a.m.–8 a.m.	12
3	8 a.m.–noon	14
4	noon–4 p.m.	8
5	4 p.m.–8 p.m.	14
6	8 p.m.–midnight	10

6. Consider an investor who has $100,000 to invest in 10 mutual fund alternatives with the following restrictions. For diversification, no more than $25,000 can be invested in any one fund. If a fund is chosen for investment, then at least $10,000 will be invested in it. No more than two of the funds can be pure growth funds, and at least one pure bond fund must be selected. The total amount invested in pure bond funds must be at least as much as the amount invested in pure growth funds. Using the following expected returns, formulate and solve a model that will determine the investment strategy that will maximize expected return.

Fund	Type	Expected Return
1	growth	13.4%
2	growth	15.3%
3	growth	15.1%
4	growth	14.9%
5	growth & income	15.0%
6	growth & income	12.9%
7	growth & income	14.1%
8	stock & bond	13.8%
9	bond	10.4%
10	bond	11.8%

Note: Some of the remaining problems ask you to solve integer programs using Excel Solver. Be sure to set the *Solver Tolerance* option to zero!

7. Develop an Excel Spreadsheet model for the Camco planning problem (Problem 1).
   a. Solve the problem using Excel Solver.
   b. Investigate the sensitivity of the solution to changes in the holding cost per unit per time period. In particular, how does holding cost impact total cost and the number of orders placed?

8. Develop an Excel spreadsheet model for the STAR Co. cutting pattern problem (Problem 2).
   a. Solve the problem using Excel Solver.
   b. Change the objective of your model to be to minimize the number of rolls used to meet demand. Solve this new problem using Excel Solver and compare the the two different objectives and their solutions.

9. Develop an Excel spreadsheet model for the investment problem in Problem 3. Solve this problem using Excel Solver.

10. Tindall Bookstores[7] is a major national retail chain with stores principally located in shopping malls. For many years, the company has published a Christmas catalog, which was sent to current customers on file. This strategy generated additional mail order business and also attracted customers to the stores. However, the cost effectiveness of this strategy was never determined. In 1998 John Harris, vice-president of marketing, conducted a major study on the effectiveness of direct mail delivery of Tindall's Christmas catalog. The results were favorable: Patrons who were catalog recipients spent more, on average, than did comparable nonrecipients. These revenue gains more than compensated for the costs of production, handling, and mailing, which had been substantially reduced by cooperative allowances from suppliers.

With the continuing interest in direct mail as a vehicle for delivering holiday catalogs, Harris continued to investigate how new customers could most effectively be reached. One of these ideas involved purchasing mailing lists of magazine subscribers through a list broker. In order to determine which magazines might be most appropriate, a mail questionnaire was administered to a sample of 6,625 current customers to ascertain which magazines they regularly read. Seventy-four magazines were selected for the survey. The assumption behind this strategy is that subscribers of magazines read by a high proportion of current customers would be viable targets for future purchases at Tindall stores. The question is which magazine lists should be purchased to maximize the reach of potential customers (number of unique customers ithe lists) in the presence of a limited budget for the purchase of lists.

Data from the customer survey have begun to trickle in. Which of the 74 magazines a customer subscribes to is provided on the returned questionnaire. Harris has asked us to develop a prototype model that can be used later to decide which lists to purchase. So far only 53 surveys have been returned. To keep the prototype model manageable, Harris has instructed us to go ahead with the model development using the data from the 53 returned surveys and using only the first ten magazines in the questionnaire. The costs of the first ten lists are given next, and your budget is $3,000. These data can be found in file *Tindall.xls*. Develop a Solver model that will allow us to maximize reach for a budget of $3,000. Solve the model using Solver.

Data for Tindall Bookstores Survey

List	1	2	3	4	5	6	7	8	9	10
Cost(000)	$1	$1	$1	$1.5	$1.5	$1.5	$1	$1.2	$.5	$1.1

Customer	Magazines	Customer	Magazines
1	10	28	4, 7
2	1, 4	29	6
3	1	30	3, 4, 5, 10
4	5, 6	31	4
5	5	32	8
6	10	33	1, 3, 10
7	2, 9	34	4, 5
8	5, 8	35	1, 5, 6
9	1, 5, 10	36	1, 3
10	4, 6, 8, 10	37	3, 5, 8
11	6	38	3
12	3	39	2, 7
13	5	40	2, 7
14	2, 6	41	7
15	8	42	4, 5, 6
16	6	43	NONE
17	4, 5	44	5, 10
18	7	45	1, 2
19	5, 6	46	7
20	2, 8	47	1, 5, 10
21	7, 9	48	3
22	6	49	1, 3, 4
23	3, 6, 10	50	NONE
24	NONE	51	2, 6
25	5, 8	52	NONE
26	3, 10	53	2, 5, 8, 9, 10
27	2, 8		

Current Survey
Results

11. Conduct a budget sensitivity analysis on the Tindall magazine list selection problem. Solve the problem for a variety of budgets and graph percentage of total reach (number reached/53) versus budget amount. In your opinion, when is an increment in budget no longer warranted (based on this limited data)?

12. Refamiliarize yourself with the Hanover plant location problem described in Chapter 2. The data for this problem are in the file *Hanover.xls*. These include forecasted demand, variable costs, duty costs, freight & warehousing costs, fixed cost by capacity level, and price. Recall that Hanover has two existing plants, in Austin, Texas, and Paris, France. Hanover is considering seven other locations for a new plant: Australia, India, Malaysia, South Africa, and Spain, and U.S. locations at Charleston, S.C., and Mobile, Ala. Different capacities are available at the two current locations and the proposed sites.

   a. Develop a mixed integer programming model to evaluate the cost of a given site. The model should include the current sites, with capacity options and a single new site with its capacity options.

   b. Implement your model from part *a* in Excel. Solve the model seven times (once for each proposed location), using Excel Solver. What are the top three locations in terms of cost minimization?

13. In Chapter 1 we used a case study in inventory management to introduce the Economic Order Quantity (EOQ) model. The assumptions for that model are that:

(1) Only a single item is considered.
(2) The entire quantity ordered arrives at one time.
(3) The demand for the item is constant over time.
(4) No shortages are allowed.

Suppose we relax the first assumption and allow for multiple items that are independent except for a budget restriction. The following model describes this situation:

Let D_j = annual demand for item j
  C_j = unit cost of item j
  S_j = cost per order placed for item j
  i = inventory carrying charge per unit
  B = the maximum amount of investment in goods
  N = number of items

The decision variables are Q_j, the amount of item j to order. The model is

$$Minimize \sum_{j=1}^{N} [C_j D_j + \frac{S_j D_j}{Q_j} + iC_j \frac{Q_j}{2}]$$

subject to

$$\sum_{j=1}^{N} C_j Q_j \leq B$$

$$Q_j \geq 0 \quad j = 1,2,...N$$

Set up a spreadsheet model and solve this problem using Solver for the following data:

	Item 1	Item 2	Item 3
Annual Demand	2,000	2,000	1,000
Item Cost	$100	$50	$80
Order Cost	$150	$135	$125

$B = \$20,000$
$i = 0.20$

14. Open *Jack's Job Shop.xls*. Solve the problem as described in Example 10, using Excel Solver. Investigate how sensitive the solution is to demand from the paint cell by resolving the model for demands of 12 and 36. Use a bubble chart to display the results and reinforce your intuition.

15. Consider the following data from Markowitz:

		Growth	
Year	ATT	GMC	USX
1	.300	.225	.149
2	.103	.290	.260
3	.216	.216	.419
4	−.046	−.272	−.078
5	−.071	.144	.169
6	.056	.107	−.035
7	.038	.321	.133
8	.089	.305	.732
9	.090	.195	.021
10	.083	.390	.131
11	.035	−.072	.006
12	.176	.715	.908

These data are stored in *Markowitz data.xls*.

a. Use the *Covariance* option in Excel (under *Tools, Data Analysis*) to construct the covariance matrix for these data.

b. Construct the Markowitz portfolio model for these data, using a minimum expected return of 0.15. Solve the problem using Excel Solver.

c. Solve the model for various expected returns to develop the frontier of risk versus return.

16. Bees Books is thinking of placing an advertisement in two magazines. Five magazines are under consideration. Through a sample survey, data on 30 individuals' propensity to purchase each of the magazines have been obtained and can be found in the worksheet *Bees.xls*. We refer to these data as p_{ij}, the probability that person i purchases magazine j. Bees would like to minimize the expected number of people *not* reached. What is the expected number of people not reached given the data we have available? Consider person 1 from the survey. Let $x_j = 1$, if magazine j is selected, and 0, if not. Under the assumption of independence, the associated expectation that this person will not be reached is

$$1(\text{probability of not being reached}) + 0(\text{probability of being reached}) =$$

$$\text{probability of not being reached} = (1 - p_{11}x_1)(1 - p_{12}x_2)(1 - p_{13}x_3)(1 - p_{14}x_4)(1 - p_{15}x_5)$$

Hence the overall problem can be formulated as follows for the thirty respondents:

$$\begin{aligned} \textit{Minimize } & (1 - p_{11}x_{11})(1 - p_{12}x_2)(1 - p_{13}x_3)(1 - p_{14}x_4)(1 - p_{15}x_5) + \\ & (1 - p_{21}x_1)(1 - p_{22}x_2)(1 - p_{23}x_3)(1 - p_{24}x_4)(1 - p_{25}x_5) + \\ & (1 - p_{31}x1)(1 - p_{32}x_2)(1 - p_{33}x_3)(1 - p_{34}x_4)(1 - p_{35}x_5) + \dots \\ \dots + & (1 - p_{301}x_1)(1 - p_{302}x_2)(1 - p_{303}x_3)(1 - p_{304}x_4)(1 - p_{305}x_5) \end{aligned}$$

subject to

$$x_1 + x_2 + x_3 + x_4 + x_5 = 2$$
$$x_j = \{0,1\} \qquad j = 1, 2, 3, 4, 5$$

Note that this model is an integer nonlinear program. Develop this model in *Bees.xls* and solve, using Excel Solver.

17. For the constrained EOQ model in problem 13, perform a sensitivity analysis of this model with respect to the parameter B. Graph total cost versus B. What value of B would you recommend?

18. Refer to the Jack's Job Shop problem. In problem 14 we asked you to use Excel Solver to find the optimal solution. Now consider a different objective:

$$\begin{aligned} \textit{Min } \{ & 12(|x - 1| + |y - 4|) + 24(|x - 1| + |y - 2|) + 24(|x - 2.5| + |y - 2|) \\ & + 7(|x - 3| + |y - 5|) + 17(|x - 4| + |y - 4|) \} \end{aligned}$$

a. Explain the differences between this objective and the one used in Example 10.

b. Solve the problem with this new objective function, using Premium Solver Evolutionary algorithm.

c. Compare the solutions to the two different objectives.

19. Consider the marketing model of problem 16. This problem has many local optimal solutions. Use Premium Solver Evolutionary algorithm to find a solution. Is the solution better than the one you found in problem 16?

20. Many labor-intensive production operations experience a learning curve effect. The learning curve specifies that the cost to produce a unit is a function of the unit number; that is, as production volume increases, the cost to produce each

unit drops. One form of the learning curve is as follows: $C_i = ai^b$ where C_i is the cost of unit i, a is the cost of the first unit cost and b is the learning parameter. The total cost of producing a batch of size x can then be approximated by $(ax^{1+b}) \div (1 + b)$. Now consider a production setting where there is learning. We have the following single-product production-planning data: demands for the next five periods are 100, 150, 300, 200, and 400. Holding cost per unit per period is $.30 and production cost follows a learning curve.

a. Open the worksheet *learning curve data.xls*. Use the *Add Trendline* option on the data in the graph to fit a learning curve to the data (see Example 7). Hint: Use the *power form* option and be sure to choose the *display equation on the graph* option. What are the fitted values of a and b?

b. Set up a spreadsheet for the following production-planning problem. Let $x_t =$ amount to produce in period t, $I_t =$ amount to hold in inventory at the end of period t, and $d_t =$ demand in period t. If there is complete loss of learning between time periods, the production problem can be modeled as:

$$\text{Minimize } \sum_{t=1}^{5} \frac{ax_t^{1+b}}{1 + b} + \sum_{t=1}^{5} .3I_t$$

subject to

$$I_{t-1} + x_t - I_t = d_t \qquad t = 1, 2, 3, 4, 5$$
$$I_t, x_t \geq 0 \qquad t = 1, 2, 3, 4, 5$$

Attempt to solve this nonlinear model using Solver.

c. Attempt to solve this model using Premium Solver Evolutionary algorithm.

d. Solve this problem ignoring learning, that is, solve this problem using the linear objective:

$$\text{Min } \sum_{t=1}^{5} ax_i + \sum_{t=1}^{5} .3I_t$$

e. Compare the solutions you obtained in c and d. Explain any differences.

21. Course and examination scheduling can present a formidable challenge to university administrators. Courses and examinations must be scheduled so as to avoid conflicts between required courses. Suppose we have six courses that must be scheduled in three time slots (two courses per time slot). Based on the number of students in each major, we have estimates of the number of students in common between pairs of courses. How should the courses be scheduled so as to minimize conflicts?

Consider the following nonlinear model for this problem (where c_{ij} is the number of students in both classes i and j):

Let $x_{it} = 1$ if class i is assigned to time slot t, 0 if not $\qquad i = 1, 2 \dots 6, t = 1, 2, 3$

$$\text{Minimize } \sum_{t=1}^{3} \sum_{i=1}^{6} \sum_{j>i} c_{ij} x_{it} x_{jt}$$

subject to

$$\sum_{i=1}^{6} x_{it} = 2 \qquad t = 1, 2, 3$$

$$\sum_{t=1}^{3} x_{it} = 1 \qquad i = 1, 2, 3, 4, 5, 6$$

$$x_{it} = 0 \text{ or } 1 \qquad i = 1, 2, 3, 4, 5, 6 \text{ and } t = 1, 2, 3$$

The objective function minimizes the sum of the conflicts (note that when x_{it} and x_{jt} are both 1, we count the conflicts). The first set of constraints make sure that each time slot has two courses. The second set of constraints ensure that each course is scheduled to exactly one time period. Data for this problem is contained in the file *course scheduling.xls*.

a. Construct a spreadsheet model using the data and the above optimization model.
b. Solve the model using the Standard GRG Nonlinear algorithm in Premium Solver with integer restrictions.
c. Solve the model using the Standard Evolutionary algorithm in Premium Solver with integer restrictions. Compare your results to those found in part *b*.

NOTES

1. Based on D. Sweeney, R. Mairose, and R. K. Martin, "Strategic Planning in Bank Location," American Institute for Decision Sciences Proceedings, November 1979.
2. Information on using Premium Solver for integer programs may be found in the appendix to this chapter.
3. H. M. Markowitz, *Portfolio Selection: Efficient Diversification of Investments* (New York: John Wiley and Sons, 1959).
4. Information on using Premium Solver for nonlinear programs may be found in the appendix to this chapter.
5. This section assumes that you have loaded the Premium Solver add-in included with this text. More details on the Standard Evolutionary algorithm may be found in the appendix to this chapter.
6. Stephen G. Nash, "Software Survey: NLP," *OR/MS Today*, April 1995, 60–63.
7. This problem is based on the case by the same name developed by James R. Evans of the University of Cincinnati and was sponsored by the Direct Marketing Policy Center.

BIBLIOGRAPHY

Anbil, R., E. Gelman, B. Patty, and R. Tanga. 1991. "Recent Advances in Crew-Pairing Optimization at American Airlines." *Interfaces*, Vol. 21, No. 1, January–February, pp. 62–74.

Ben-Dov, Yosi, Lakhbir Hayre, and Vincent Pica. 1992. "Mortgage Valuation Models at Prudential Securities." *Interfaces*, Vol. 22, No. 1, January–February, pp. 55–71.

Bohl, Alan, H. 1994. "Computer-Aided Formulation of Silica Defoamers for the Paper Industry." *Interfaces*, Vol. 24, No. 5, September–October, pp. 41–48.

Camm, J., R. Raturi, and S. Tsubakitani. 1990. "Cutting Big M Down to Size." *Interfaces*, Vol. 20, No. 5, September–October, pp. 61–66.

Carraway, R., J. Cummins, and J. Freeland. 1990. "Solving Spreadsheet-Based Integer Programming Models: An Example from International Telecommunications." *Decision Sciences*, Vol. 21, No. 4, pp. 808–824.

Dano, S. 1974. *Linear Programming in Industry*, 4th edition. Wien: Springer-Verlag.

Farley, A. 1991. "Planning the Cutting of Photographic Color Paper Rolls for Kodak (Australasia) Pty. Ltd." *Interfaces*, Vol. 21, No. 1, January–February, pp. 92–106.

Gaballa, A., and W. Pearce. 1979. "Telephone Sales Manpower Planning at Qantas." *Interfaces*, Vol. 9, No. 3, May, pp. 1–9.

Hall, N., J. Hershey, L. Kessler, and R. Stotts. 1992. " A Model for Making Project Funding Decisions at the National Cancer Institute." *Operations Research*, Vol. 40, No. 6, November–December, pp. 1040–1052.

Markowitz, H. M. 1959. *Portfolio Selection, Efficient Diversification of Investments*. New York: John Wiley and Sons.

Nemhauser, G. L., and L. A. Wolsey. 1988. *Integer and Combinatorial Optimization*. New York: John Wiley and Sons.

Salkin, H. M., and K. Mathur. 1989. *Foundations of Integer Programming*. New York: Elsevier Science Publishing, North-Holland.

APPENDIX: USING PREMIUM SOLVER TO SOLVE INTEGER AND NONLINEAR PROGRAMS

To use the Premium Solver, you must install the Premium Solver from the CD in the back of this text. In this appendix we discuss the use of Premium Solver for linear integer and nonlinear programs.

Solving Linear Integer Programs

Solving linear integer programs using Premium Solver is very similar to using the standard Excel Solver. From the standard Solver dialog box, select *Premium* and then *Standard Simplex LP* (see Figures 4A.1 and 4A.2 in the Appendix to Chapter 4). Adding general integer or binary restrictions is accomplished through the *Add constraint* using *int* and *bin*. The Premium Solver has several options not available in the standard Excel Solver. These are found by selecting the *Integer Options* button in the *Solver Options* dialog box. This invokes the *Integer Options* dialog box, shown in Figure 5A.1. The options *Max Subproblems* and *Max Integer Sols* place limits on the number of linear programming subproblems and candidate integer solutions, respectively, before the Solver pauses to ask if you would like to continue. Solver may have to solve many linear programs to find the optimal integer solution, and it may find many integer solutions along the way. These parameters offer the option of stopping the solution process at these limits so as to avoid long solution runs. In general, the default values are fine.

The *Tolerance* parameter is as defined for the standard Excel Solver. The solution reported by Solver is guaranteed to be within *Tolerance* percent of the optimal solution. Hence, the *Tolerance* parameter needs to be set to zero to guarantee finding the optimal solution. If long solution times are a problem and a solution within, say, 5 percent of optimality is acceptable, set *Tolerance* to .05.

FIGURE 5A.1

Premium Solver integer options dialog box

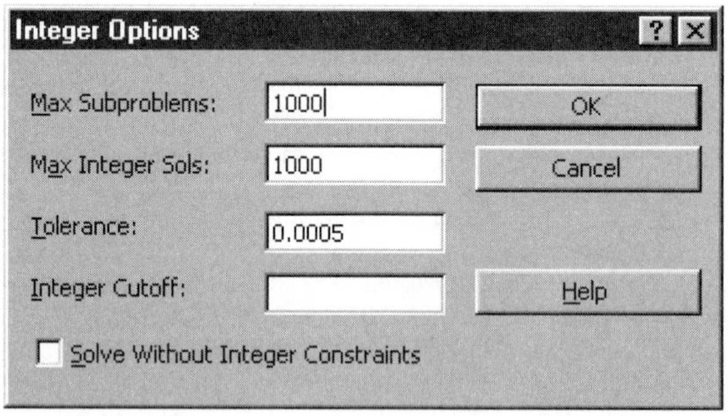

Another option is the option to *Solve Without Integer Constraints*. This option allows you to solve to linear programming relaxation of the problem without going into the Solver dialog box and changing the model. This might be useful in generating a feasible integer solution manually. For example, for some models, you may solve the linear programming relaxation and round or truncate the decision variables to get a feasible integer solution. The *Integer Cutoff* parameter allows you to specify the objective function value of a known integer solution. This may speed up the solution process.

Solving Nonlinear Programs

Solving nonlinear programs using Premium Solver is very similar to using the standard Excel Solver with some minor differences in the dialog box and options. The objective function, decision variables, and constraints are entered into the Premium Solver dialog box in exactly the same manner as described in this chapter for the standard Excel Solver.

From the standard Solver dialog box, select *Premium* and then the algorithm you wish to use. Premium Solver provides two algorithms for solving nonlinear problems, *Standard GRG Nonlinear* and *Standard Evolutionary* (see Figures 4A.1 and 4A.2 in the Appendix to Chapter 4).

USING THE STANDARD GRG NONLINEAR The Standard GRG Nonlinear algorithm may be used to solve nonlinear problems. The Solver Options dialog box for GRG is shown in Figure 5A.2. Most of these parameters play the same role as in the linear case. However, the *Convergence* option is the value used to stop the algorithm. In particular, when the absolute value of the relative change in the objective function is less than the value set in *Convergence*, the algorithm halts. This value can be increased if you are having trouble converging to a solution. The *Estimates*, *Derivative* and *Search* parameters deal with approaches used by the GRG algorithm and in general can be set to their default values.

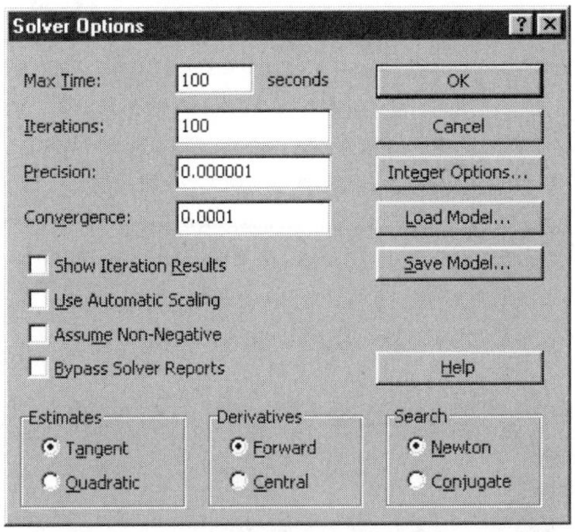

FIGURE 5A.2
Premium Solver options dialog box for the Standard GRG Nonlinear algorithm

USING THE STANDARD EVOLUTIONARY SOLVER The Standard Evolutionary Solver is quite different than the Standard Simplex LP and the Standard GRG Nonlinear algorithms, both of which are based on the mathematical structure of the model. The Standard Evolutionary Algorithm is a heuristic—a "genetic" algorithm, specifically designed to overcome the problems of local optimal solutions as discussed in this chapter. It is based on the following genetic principles:

1. A population of candidate solutions is always maintained.
2. Random "mutations" (changes in subsets of solution values) are made to current population members to create new solutions.
3. "Reproduction" of new solutions is also accomplished through "crossover" (the combining of two existing candidates in some fashion to create a new candidate).

4. The "survival of the fittest" principle is enforced (the "most fit" elements in the population survive and go to the next generation, whereas the "least fit" elements are eliminated). Fitness is generally measured by the objective function of the problem, with some penalty for any infeasibility.

It is through the random moves that the algorithm hopes to break out of local optimal solutions and sample other regions for better solutions. Because of these random moves, it is quite possible that multiple runs of the same model will yield different final solutions. Also, unlike the Simplex LP and GRG Nonlinear algorithms, the algorithm cannot guarantee that it has even a local optimal solution (it will simply report the best solution from the current population). Nonetheless, this type of genetic algorithm can be very useful on hard problems that cannot be solved by the classical linear of nonlinear algorithms.

The *Solver Options* dialog box for the Standard Evolutionary algorithm is shown in Figure 5A.3. The *Max Time, Iterations* and *Precision* parameters play the same roles as in the linear and GRG case. The *Convergence* parameter has a slightly different meaning. In particular, when all members of population have fitness levels within the convergence tolerance, the algorithm stops. The *Population Size* is the number of candidate solutions maintained at any point in time. It is recommended that this be 70 to 100 members. The *Mutation Rate* is the probability that some member of the population will be mutated. If you believe that the algorithm is returning suboptimal solutions, increasing the *Population Size* and the *Mutation Rate* might help broaden the search to locate improving solutions.

The other options in Figure 5A.3 are self-explanatory, with the possible exception of the *Require Bounds on the Variables*. Bounds on variables are important for the performance of evolutionary algorithms (for example, the initial population list is generated by randomly sampling values between each variable's bounds). The default is to require bounds. If bounds are not required, the algorithm will still run (however, the initial range will be essentially infinite for each variable).

FIGURE 5A.3

Premium Solver options dialog box for the Standard Evolutionary Algorithm

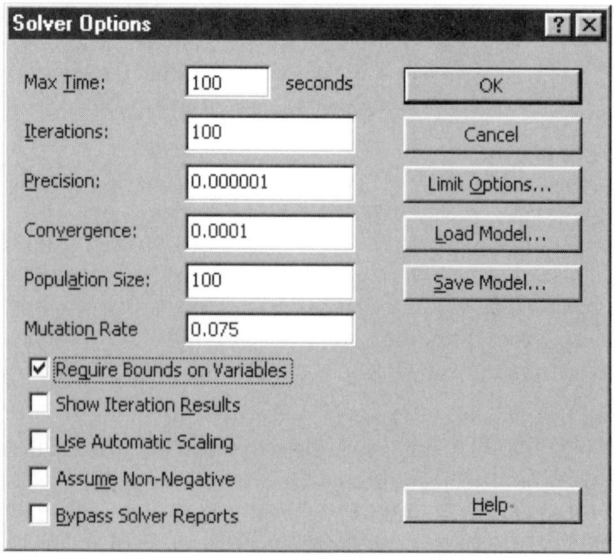

The *Limit Options* button invokes the dialog box shown in Figure 5A.4. For the Standard Evolutionary Algorithm, the *Tolerance* and *Max Time w/o Improvement* work as follows. If the relative improvement in the best solution is less than the *Tolerance* value for the number of seconds in the *Max Time w/o Improvement* edit box, the algorithm terminates.

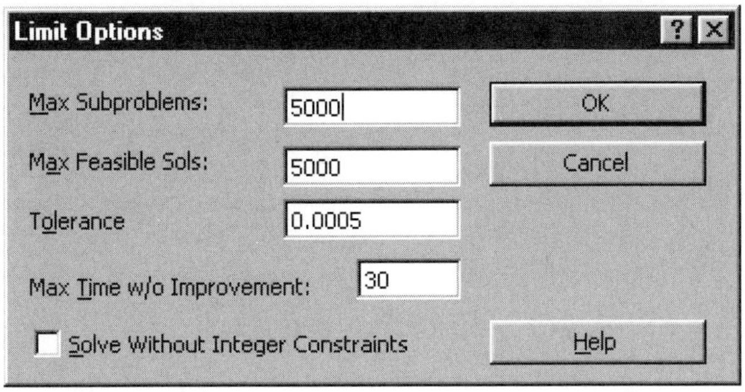

FIGURE 5A.4 Premium Solver limit options dialog box for the Standard Evolutionary Algorithm

CHAPTER 6

Decision and Risk Analysis

OUTLINE

Applications of Decision and Risk Analysis
Environmental Impact Assessment
Assessment of Catastrophic Risk
Sports Strategies
Risk Assessment for the Space Shuttle

Structuring Decision Problems
Generating Alternatives
Defining Outcomes
Decision Criteria
Decision Trees
Management Science Practice: Collegiate Athletic Drug Testing
Decision Strategies

Understanding Risk in Making Decisions
Average Payoff Strategy
Aggressive Strategy
Conservative Strategy
Opportunity Loss Strategy
Quantifying Risk—Insights from Finance
An Application of Decision and Risk Analysis: Evaluating Put and Call Options

Expected Value Decision Making
Management Science Practice: The Overbooking Problem at American Airlines
An Application of Expected Value Analysis: The "Newsvendor" Problem
Expected Value of Perfect Information

Optimal Expected Value Decision Strategies
Sensitivity Analysis of Decision Strategies

Technology for Decision Analysis

Risk Trade-offs and Multiobjective Decisions

Utility and Decision Making
Exponential Utility Functions

Problems

E verybody makes decisions, both personal and professional. Individuals make decisions about purchasing automobiles, choosing colleges, selecting mortgage instruments, and investing their money. Physicians make decisions regarding surgery and treatment strategies. Managers face decisions involving new product introductions, plant locations, choice of suppliers, new equipment selection, and software upgrades, to name just a few. For example, a company like Hanover, Inc., might need to decide on increasing capacity to meet forecasted increases in demand. In such a highly volatile industry, demand may change drastically as global economic conditions change. Clearly, the decision is significant because deciding wrongly may lead to significant unused capacity with associated financial losses, or missed sales opportunities, and future market share losses if demand cannot be met.

Making good decisions involves *defining alternatives, assessing the possible impacts of each alternative, determining the preferences or values by which the decision will be based, and evaluating the alternatives.* Models provide information to assist in the decision process. For example, by constructing data tables, tornado charts, and spider diagrams we can better understand the risks associated with uncertain information. Examining the sensitivity analysis information for a linear program provided by Solver may help us to identify other good solutions that incorporate various intangible factors. However, models do not make decisions—people do. The ability to make good decisions is the mark of a successful (and promotable) manager. In today's complex business world, intuition and gut feel alone are not sufficient; models of the decision process can help to provide structure and better insight into decisions.

Decision analysis is the formal study of how people make decisions, particularly when faced with uncertain information, as well as a collection of techniques to support the analysis of decision problems. The types of decision situations for which decision analysis techniques apply have the following characteristics:[1]

1. *They must be important.* Decision analysis techniques would not be appropriate for minor decisions, where the consequences of a mistake are so small that it is not worth our time to study the situation carefully. The consequences of many decisions, such as building a major facility, are not felt immediately but may cover a long time period.
2. *They are probably unique.* Decisions that recur can be programmed and then delegated. But the ones that are unusual and perhaps occur only one time cannot be handled this way.
3. *They allow some time for study.* For example, decision analysis techniques would not be useful in making a decision in the emergency room or when a jet fighter flames out during takeoff.
4. *They are complex.* Practical decision problems involve multiple objectives, requiring the evaluation of trade-offs among the objectives. For example, in evaluating routes for proposed pipelines, a decision maker would want to minimize environmental impact, minimize health and safety hazards, maximize economic benefit, and maximize social impact. Decisions involve many intangibles, such as the goodwill of a client, employee morale, and governmental regulations, and may involve several stakeholders. For instance, to build a plant in a new area, corporate management may require approval from stockholders, regulatory agencies, community zoning boards, and perhaps even the courts. Finally, most decisions are

closely allied to other decisions. Choices today affect both the alternatives available in the future and the desirability of those alternatives. Thus, a sequence of decisions must often be made.

5. *They involve uncertainty and risk.* **Uncertainty** refers to not knowing what will happen in the future. An advertising campaign may fail, a reservoir may break, or a new product may be a complete failure. Uncertainty is further complicated when little or no data are available, or necessary data are very expensive or time-consuming to obtain. Faced with such uncertainties, different people view the same set of information in different ways. **Risk** is the uncertainty associated with an undesirable outcome, such as financial loss. To appreciate the importance of risk, consider the fact that it takes hundreds of millions of dollars and about ten years for a pharmaceutical company to bring a drug to market. Once there, 7 of 10 products fail to return the company's cost of capital. Decisions involving capital investment and continuation of research over the long development cycle do not lend themselves to traditional financial analysis.[2]

In this chapter we will learn how to structure decision problems, quantify the expected results of decisions, assess risk, and select good decision strategies.

APPLICATIONS OF DECISION AND RISK ANALYSIS

Environmental Impact Assessment

Electric utilities face decisions that can have important impacts on the environment. The impacts stem from the by-products of combustion and other chemicals, equipment, and processes that utilities use to produce electricity. Electric utilities have used decision analysis to make better environmental decisions. For example, utilities use large boilers to boil water and make steam to generate electricity. The cleaning process results in a waste solution that may be hazardous. Whether or not the waste stream will be hazardous is uncertain, as are the costs and effects of the various management strategies. Several courses of action—choice of cleaning agent, whether or not to include a pre-rinse stage, treatment and disposal method, and cleaning frequency—are available. Using techniques of decision analysis, the consulting firm Decision Focus Incorporated developed a strategy that would save a utility $119,000 for one boiler over a 20-year horizon (Balson et al., 1992).

Assessment of Catastrophic Risk

The executive vice president of a major bank asked its management science group to develop an approach to identify potentially catastrophic events and select the best alternatives to deal with the risks. In a pilot study, the group studied threats of fire and power failure to several critical services at a data-processing facility at the operation's headquarters. The group determined that the potential loss resulting from these threats could exceed $100 million. A lengthy disruption would also have an unfavorable impact on customer relations and future profits. Using techniques of decision analysis, the group evaluated the benefits of alternatives such as fire control modifications and an emergency generator as well as their costs, which ranged upward to $20 million. The study indicated that a small generator for the funds-transfer division was cost justified, but that a large generator for all operations was not (Engemann and Miller, 1992).

Sports Strategies

Coaches of sports teams face uncertain decisions all the time. In professional football, for instance, an important decision is whether to attempt a field goal with short yardage required for first down, or to try for the first down. Such decisions are usually made by gut feelings rather than quantitative analysis. Virgil Carter, a former NFL quarterback, and Robert Machol applied statistical analysis to evaluate football strategies. For example, they showed that the expected value of having the ball with first down and 10 yards to go varies from about −1.64 points at one's own goal line and improves at the rate of about 1 point per yard up to about 10 yards from the opponent's goal line, with a larger increase per yard over the last 10 yards. A further analysis of field goal attempts showed that inside the 30-yard line, the run is preferred to the field goal attempt if there are 1 or 2 yards to go, and possibly with 3. Inside the 10-yard line, the run is preferred to the field goal attempt with up to 5 yards to go. These results were contrary to practice, but many coaches continued to employ the field goal far more than the analysis indicated (Carter and Machol, 1971, 1978).

Risk Assessment for the Space Shuttle

The tiles of the space shuttle orbiter are critical to its safety at reentry, and their maintenance between flights is time-consuming. NASA consultants developed a model to identify the most risk-critical tiles and set priorities in the management of the heat shield. The model included data on the probability of debonding due either to debris hits or to a poor bond, the probability of losing adjacent tiles once the first one was lost, the probability of burn-through given tile loss, and the probability of failure of a critical subsystem under the skin of the orbiter if a burn-through occurs. The model found that 15 percent of the tiles account for about 85 percent of the risk, and that some of the most critical tiles are not in the hottest areas of the orbiter's surface. This helps to set priorities for maintenance, which are estimated to reduce by about 70 percent the probability of a shuttle accident attributable to tile failure (Pate-Cornell and Fischbeck, 1994).

STRUCTURING DECISION PROBLEMS

Structuring decision problems involves defining alternative decisions that can be made, uncertain outcomes that may result, and criteria by which to evaluate the value of the various combinations of decisions and outcomes.

Generating Alternatives

Decision alternatives represent the choices that a decision maker can make. They might be a simple set of decisions, such as locating a factory from five potential sites or choosing one of three corporate health plan options. Other situations require a more complex sequence of decisions. For example, in new product introduction, a marketing manager might have to decide whether to test market a new product and then, based on the test market results, whether to conduct further tests, begin a full-scale marketing effort, or drop the product from further consideration. In any case, the manager must list the options that are available. Generating viable alternatives might involve some prescreening (perhaps using optimization models). For instance, a company might develop, solve, and perform sensitivity analysis on a mathematical programming model to generate potential plant location sites based on total distribution costs. However, making the final decision would involve many qualitative factors, such as labor supply, tax incentives, environmental regulations, future uncertainties, and so on.

Managers must ensure that they have considered all possible options so that the "best" one will be included in the list. A good deal of creativity may be required to define unusual options that otherwise might not be considered. Managers must put aside the tendency to jump right into the process of finding a solution to consider creative alternatives.

Defining Outcomes

The second task in structuring decision problems is defining the outcomes, or **events,** that may occur once a decision is made and over which the decision maker does not have control. These outcomes provide the basis for evaluating risks associated with decisions. Outcomes may be quantitative or qualitative. For instance, in selecting the size of a new factory, a company needs to consider the future demand for the product. The demand might be expressed quantitatively in sales units or dollars. If you are planning in January a spring break vacation to Florida, you might define uncertain weather-related outcomes qualitatively: sunny and warm, sunny and cold, rainy and warm, or rainy and cold.

Decision Criteria

Decision makers must have well-defined criteria on which to evaluate potential options. Decision criteria might be to maximize discounted net profits or social benefits, or to minimize costs or some measure of loss. A different criterion might involve environmental impact.

To completely structure a decision problem, we must be able to express the value of every combination of decisions and events. The value of making a decision D and having event S occur is called the *payoff* and is expressed as $V(D,S)$. Payoffs are summarized in a *payoff table*, a matrix whose rows correspond to decisions and whose columns correspond to events.

Example 1: Forest Fire Management

Fire is an important tool in contemporary forest management.[3] Prescribed fires are often ignited by forest management personnel under controlled conditions to achieve certain objectives, such as reducing fire hazards, enhancing wildlife habitat, facilitating site preparation for planting seedlings, and controlling diseases and insects. Although it is a highly effective management technique, uncertainties inherent in its use make planning and execution of successful burns challenging. A simple decision that fire managers face is choosing between committing resources to a burn (decision D_1), with an unavoidable cost if the burn is subsequently cancelled, or postponing the burn (decision D_2), with a smaller cost associated with the delay in meeting the resource management objectives.

Two uncertainties affect this decision. First is the actual weather conditions on the day of the burn. If the weather is good, the burn can be carried out; if not, it must be canceled. The second uncertainty is the results—will all objectives be met at no additional cost, or will it be only partially successful? These uncertainties lead to three potential events:

S_1: Good weather and successful outcome

S_2: Good weather and marginal outcome

S_3: Poor weather (the outcome is irrelevant since the burn would be canceled)

The payoffs are a function of the costs of preparing and executing the burn and the value of the resources saved from the burn. The cost of preparing for a burn is $1,200, and the cost of executing it is $2,000. If the burn is wholly successful, the value of the resources saved is $6,000; if the burn is only marginally successful, the value of resources saved is $3,000. If the burn is postponed, costs are estimated to be $300. The payoff table for this problem is

	S_1	S_2	S_3
D_1	($2,800)	$200	$1,200
D_2	$300	$300	$300

Note that the cost associated with D_1 and S_1 is negative, since the value of the resources saved exceeds the cost of the burn (a net gain).

Decision Trees

A **decision tree** is a graphical representation of a decision problem. A decision tree consists of the following:

- *Decision nodes*, represented by squares. These represent points at which decisions must be made.
- *Decision branches*, which stem from decision nodes. Each branch corresponds to a decision alternative.
- *Event nodes*, represented by circles. These represent points at which uncertain outcomes will occur.
- *Event branches*, which stem from event nodes. Each branch corresponds to an event.
- *Terminal nodes*, which represent the result (payoff) of a combination of decisions and events.

Decision trees that contain a single set of decision nodes followed by a set of event nodes are called **single stage decision trees**; those that contain sequences of decision and event nodes are called **multistage decision trees**. Figure 6.1 shows a single stage decision tree for the situation in Example 1. We "read" the tree from left to right. At node 1, a decision node, we have a choice of two decisions. If the decision is to commit resources, we observe which of the three events may occur. If the decision is to postpone the burn, the weather-related events are irrelevant. Thus, each path from node 1 to the end of the tree represents a possible outcome—a sequence of decisions and events. At the end of each path we list the payoff associated with each event.

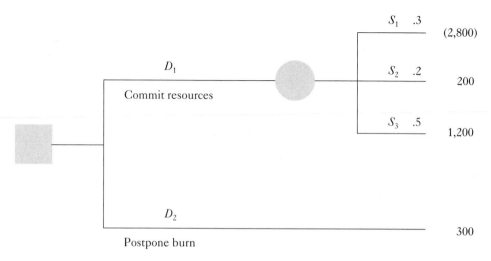

FIGURE 6.1
A decision tree
for the fire
management
example

Multistage decision trees provide a more visual and logical description of complex decision problems (see the Management Science Practice box on collegiate drug testing). One example of this is a situation in which a decision maker considers acquiring additional information on which to base a key decision. The decision maker must first decide whether it is worth the cost and effort to obtain the information, and if so, how to use the information to improve the decision. The next example shows a typical multistage decision tree model where consumer research is used to obtain information on which to base a decision.

Management Science Practice:
Collegiate Athletic Drug Testing

The athletic board of Santa Clara University had to decide whether to recommend implementing a drug-testing program for intercollegiate athletes. One of the board members, who was a management science professor, developed a simple decision model to address the question whether or not to test a single individual for the presence of drugs. The model focused on the key issue of the reliability of the testing procedures, consequences of testing errors, and the benefits of identifying a drug user compared with the costs of false accusations and nonidentification of users.

Figure 6.2 shows the decision tree developed for testing an individual for drug use. The two main alternatives are "test" and "don't test." The model evaluates the expected cost of testing for drug use compared with that of not testing. If testing is chosen, the test is given and the result, positive or negative, is observed. If the result is positive, action is taken. Since not all those who test positively are actually users, there is some chance of a false accusation, which costs an amount C_1. If the result is negative, then some drug users are not iden-

continued

tified, which costs C_2. Nonusers who test negatively might be expected to experience some cost C_3, perhaps based on invasion of privacy. Following the lower path of the tree, if the alternative "don't test" were selected, the expected cost is just the cost of an unidentified user, C_2, multiplied by the prior probability that an individual is a drug user.

FIGURE 6.2 Decision tree for drug use testing

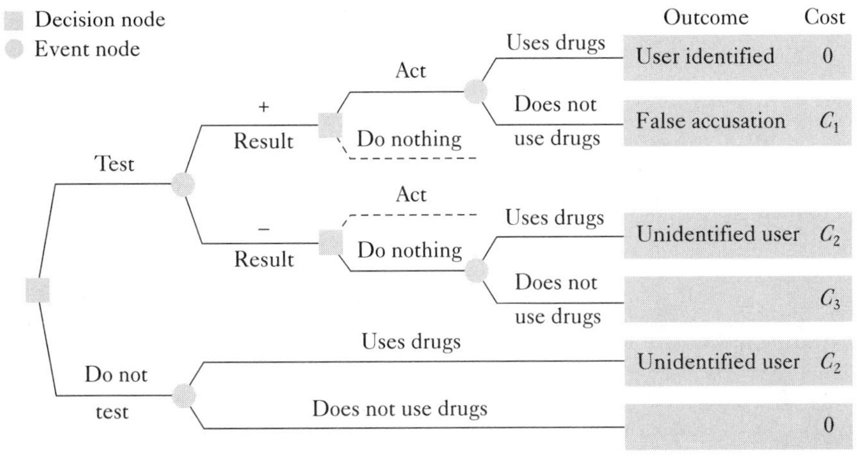

The model's results surprised many board members. For instance, the model showed that if a test that is a 95 percent reliable is applied to a population of 5 percent drug users, only 50 percent of all those that tested positively will actually be drug users. Most board members had read about the reliability of drug tests in various publications and agreed that 95 percent reliability was a representative value. As a result, the board concluded that a false accusation was more serious than not identifying drug users and rejected the proposal. The university administration later accepted this recommendation.

Source: Adapted from Charles D. Feinstein, "Deciding Whether to Test Student Athletes for Drug Use," *Interfaces*, Vol. 20, No. 3, May–June 1990, pp. 80–87.

Example 2: New Product Introduction

A national chain of quick-service restaurants has developed a new specialty sandwich. Initially, it faces two possible decisions: introduce the sandwich nationally, or evaluate it in a regional test market. If it introduces the sandwich nationally, the chain might find either a high or low response to the idea. If it starts with a regional marketing strategy, it might find a low response or a high response at the regional level. This may or may not reflect the national market potential. In any case, the chain needs to decide whether to remain regional, market nationally, or drop the product. Figure 6.3 shows a decision tree for this example.

Decision trees provide useful models for gaining insight into decision problems. They force decision makers to think carefully about the decisions they must make

FIGURE 6.3
New product
introduction
decision tree

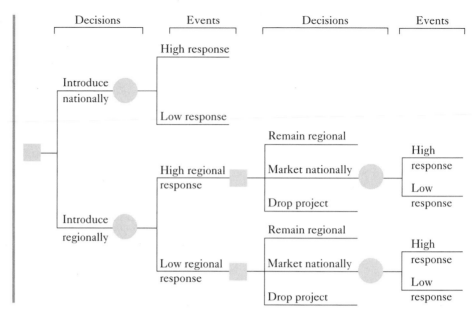

and the consequences of each decision. Although we will discuss some quantitative approaches for analyzing the information in a decision tree model, the real value of decision trees lies in structuring important information in a logical fashion.

Decision Strategies

A *decision strategy* refers to the sequence of decisions that a decision maker would make based upon the events that occur. For instance, in Example 2, one strategy is simply to introduce the product nationally and then observe whether the response would be high or low. Another strategy would be to introduce it regionally first, observe the regional response, and then decide to remain regional, market nationally, or drop the product. Selecting a decision strategy depends on several factors, the most important being the decision maker's attitudes toward risk, an issue we address next.

UNDERSTANDING RISK IN MAKING DECISIONS

Making decisions with uncertain future consequences is often quite frustrating and thus a source of anxiety for individuals and managers alike. We run the risk that any decision we choose may result in undesirable consequences once we see what the future holds in store. (Just think of making vacation plans before you know what the weather will be!)

Risk probably has the most meaning to individuals in the context of investments. For example, if you sell a bond before its maturity date, you will probably receive an amount different from your purchase price. Bond prices are sensitive to interest rates; as rates fall, prices rise and vice versa. Uncertainty exists about future interest rates, and therefore investors run the risk of a loss in principal should rates rise significantly. Understanding our own attitudes and preferences regarding risk helps us to make more informed decisions; this is the focus of this section.

Example 3: A Personal Investment Decision

Rick Martin has $10,000 to invest. As he is expecting to buy a new car in a year, he can tie the money up for only 12 months. Rick is considering three options: a bank CD paying 4 percent, a bond mutual fund, and a stock fund. Both the bond and stock funds are sensitive to changing interest rates. If rates remain the same over the coming year, the share price of the bond fund is expected to remain the same, and Rick expects to earn $840. He expects the stock fund to return about $600 in dividends and capital gains. However, if interest rates rise, he anticipates losing about $500 from the bond fund after taking into account the drop in share price, and likewise expects to lose $900 from the stock fund. If interest rates fall, however, his yield from the bond fund would be $1,000, and the stock fund would net $1,700. As economic forecasters and investment advisers are giving conflicting advice, Rick is unsure as to the best decision. Table 6.1 summarizes the payoff table for his decision problem. Before continuing, determine which decision *you* would make and why. Keep your answer in mind as you read further.

Decision/ Event	Rates Rise	Rates Stable	Rates Fall
Bank CD	$400	$400	$400
Bond fund	($500)	$840	$1,000
Stock fund	($900)	$600	$1,700

TABLE 6.1 Payoff table for Rick's investment problem

The best decision for Rick's problem clearly depends on what event occurs. If rates remain stable, then obviously the bond fund would be the right decision. However, if he selects the bond fund and rates rise, he faces a loss of $500; if rates fall, he will make a bit more, but he could have done much better with the stock fund. Clearly, there is no "optimal" solution—because the future is uncertain. Not all decision makers view risk in the same fashion. However, several different strategies exist that reflect different risk attitudes.

Average Payoff Strategy

Since the events are unpredictable, we might simply assume that each one is as likely to occur as the others. This approach was proposed by the French mathematician Laplace, who stated the *principle of insufficient reason:* if there is no reason for one event to be more likely than another, treat them as equally likely. Under this assumption, we may evaluate each decision by simply averaging the payoffs. We then select the decision with the best average payoff. For Rick's investment problem we have the following:

Decision	Average Payoff
D_1: Bank CD	$400.00
D_2: Bonds	$446.67
D_3: Stocks	$466.67

Using this approach, the best decision would be to invest in stocks.

Aggressive Strategy

The aggressive decision maker might seek the option that holds the promise of maximizing his or her potential return. This type of decision maker would ask the question, "What is the best that could result from each decision?" For Rick's investment problem, this is summarized as follows:

Decision	Maximum Potential Return
Bank CD	$400
Bond fund	$1,000
Stock fund	$1,700

The aggressive decision maker would select the decision that maximizes the best potential return, or in this case, select the stock fund. Aggressive decision makers are often called **speculators,** particularly in financial arenas, because they increase their exposure to risk in the hope of increasing their return.

Conservative Strategy

A conservative decision maker, on the other hand, might take a more pessimistic attitude and ask, "What is the worst thing that might result from my decision?" For Rick's investment problem, the minimum potential returns for each option are as follows:

Decision	Minimum Potential Return
Bank CD	$400
Bond fund	($500)
Stock fund	($900)

A conservative decision maker would choose the decision that maximizes the smallest potential return. In this case, Rick would choose the CD alternative. Thus, no matter what event occurs, Rick is guaranteed a return of $400. Conservative decision makers are often called **risk averse** (or **hedgers** in finance) and are willing to forgo potential returns in order to reduce their exposure to risk.

Opportunity Loss Strategy

A fourth approach that underlies decision selections for many individuals is to consider the *opportunity loss* associated with a decision. Opportunity loss represents the regret that people often feel after making a nonoptimal decision ("I should have bought that stock years ago . . ."). In Rick's investment decision, suppose he chooses the bond fund. If rates remain stable, then he could not have done any better by selecting a different decision; in this case, his opportunity loss is zero. However, if rates rise, his best decision would have been to choose the CD. In this case he would have gained an additional $900 (the difference between $400 and −$500); this is the opportunity loss for selecting the bond fund instead of the CD. If rates fall, he should have chosen the stock fund. In this case, he lost the opportunity to earn an additional $1,700 − $1,000 = $700.

In general, the opportunity loss associated with any decision and event is the difference between the *best* decision for that particular event and the payoff for the decision that was made. *Opportunity losses can only be nonnegative values!* Thus, you need to be careful when computing these, especially if some payoffs are negative. For Rick's investment problem, the opportunity losses associated with each decision and event are shown in Table 6.2.

TABLE 6.2 Opportunity loss table for Rick's investment

	Rates Rise	Rates Stable	Rates Fall
Bank CD	$400 − $400 = 0	$840 − $400 = $440	$1,700 − $400 = $1,300
Bond fund	$400 − (−$500) = $900	$840 − $840 = 0	$1,700 − $1,000 = $700
Stock fund	$400 − (−$900) = $1,300	$840 − $600 = $240	$1,700 − $1,700 = 0

Once opportunity losses are computed, the decision strategy is similar to the conservative strategy discussed earlier. The decision maker would select the decision that minimizes the largest opportunity loss. For Rick's investment problem, this is summarized next:

Decision	Largest Opportunity Loss
Bank CD	$1,300
Bond fund	$900
Stock fund	$1,300

Using this strategy, Rick would choose the bond fund. No matter what event occurs, this decision ensures that Rick will never be further than $900 away from the best return he might have realized.

Different criteria, different decisions. Think of the decisions you have made; what type of decision maker are you?

Excel Exercise 6.1

Create an Excel template for Rick's investment problem that implements each of these decision strategies.

Quantifying Risk—Insights from Finance

A serious drawback of the average payoff strategy is that it neglects to consider the actual outcomes that can occur. For a one-time decision, the average outcome will *never* occur (with the sole exception of equal payoffs). For instance, an investment in the stock fund will never produce a return of $466.67; the return will either be −$900, $600, or $1,700. Most individuals will weigh the potential losses against the potential gains. Thus, the bond fund might be more attractive than a CD if Rick can afford to take the chance of losing $500, but views the potential of making about twice as much as the CD a good gamble. On the other hand, although the upside potential of the stock fund is much higher, the risk of losing $900 may not be acceptable.

In financial investment analysis, a key measure of risk is the standard deviation. For example, *Fortune* magazine's 1999 Investor's Guide (December 21, 1998) evaluated mutual fund risk using the standard deviation, because it measures the tendency of a fund's monthly returns to vary from their long-term average. (As *Fortune* stated: "Standard deviation tells you what to expect in the way of dips and rolls. It tells you how scared you'll be.") For example, a mutual fund's return might have averaged 11 percent with a standard deviation of 10 percent. Thus, about two-thirds of the time the annualized monthly return was between 1 percent and 21 percent. By contrast, another fund's average return might be 14 percent, but have a standard deviation of 20 percent. Its returns would have fallen in a range of −6 percent to 34 percent, and therefore it is more risky.

However, a large standard deviation alone does not necessarily imply high risk. For instance, suppose that you are offered a chance to win a $40,000 car in a charity raffle for $100, and only 1,000 tickets are sold. With a probability of winning of 0.001, we compute the expected value of this distribution of outcomes to be −$59.9 and the standard deviation to be 1267.44. Although the standard deviation appears to be high, most individuals would not view this to be very risky because the loss of $100 would not be viewed as catastrophic (ignoring the charitable issues!).

The statistical measure of **coefficient of variance,** which is the ratio of the standard deviation to the mean, provides a relative measure of risk to return. The smaller the coefficient of variation, the smaller the relative risk is for the return provided. The reciprocal of the coefficient of variation, called **return-to-risk,** is often used because it is easier to interpret. A related measure in finance is the **Sharpe ratio,** which is the ratio of a fund's excess returns (annualized total returns minus Treasury bill returns) to its standard deviation. *Fortune* noted that, for example, although both the American Century Equity-Income fund and Fidelity Equity-Income fund had three-year returns of about 21 percent per year, the Sharpe ratio for the American Century fund was 1.43 compared with 0.98 for Fidelity. If several investment opportunities have the same mean but different variances, a rational (risk averse) investor will select the one that has the smallest variance.[4] This approach to formalizing risk is the basis for modern portfolio theory, which seeks to construct minimum-variance portfolios. As *Fortune* noted, "It's not that risk is always bad . . . It's just that when you take chances with your money, you want to be paid for it."

In Rick's situation, the standard deviation and return-to-risk associated with each decision are

Decision	Average Payoff	Standard Deviation	Return to Risk
D_1: Bank CD	$400.00	0	—
D_2: Bonds	$446.67	672.57	.66
D_3: Stocks	$466.67	1,065.62	.43

Although D_3 has the highest average payoff, it is also the riskiest. D_2 has a larger return-to-risk value, and D_1 is completely risk-free. Thus, if Rick is willing to take some risk, the bond fund might be more attractive than the CD.

An Application of Decision and Risk Analysis: Evaluating Put and Call Options

One approach to protecting stock investments in volatile markets is to purchase options.[5] A **put option** is like an insurance policy; for a fixed fee, it entitles you to sell a security at a given price, called the **strike price,** on or before a specified date. (This is called an *American option;* a European option can be exercised only on the expiration date.) For example, suppose a stock is valued at $100 per share and you buy a put option with a strike price of $92. If the stock falls, you bear the first $8 of losses, but if it falls below $92, you cannot lose any more money. However, if the option expires and the stock is worth at least $92, the option is worthless and you lose the cost of the option. The cash flow from a put option is

Max(strike price − stock price on exercise date, 0) − option cost

A **call option** is the opposite of a put, and gives the holder the right to buy a stock at the strike price. If the stock rises, you will make money, but if it falls, you lose only the price of the option. The cash flow from a call option is

Max(stock price on exercise date − strike price, 0) − option cost

The following example shows how decision analysis may be applied to evaluating these strategies.

Example 4: Decision Analysis of Put and Call Options

Suppose that a NASDAQ stock currently sells for $3. A put option is available at 52 cents/share at a strike price of $2.50, and a call option is also available at a price of 1.2 cents/share with the strike price of $3. An investor is considering five decision alternatives:

1. Buy 1,000 shares of the stock.
2. Buy a put option on 1,000 shares.
3. Buy a call option on 1,000 shares.
4. Buy 1,000 shares of the stock along with a put option.
5. Buy 1,000 shares of the stock along with a call option.

Figure 6.4 shows the payoff table and the analysis using the decision criteria we have discussed. Only the call and stock and call alternatives have positive return to risk ratios, and the stock and call decision is much riskier. However, it is important to realize that these are based on the average payoffs, assuming that the price of the stock may move in either direction with equal chances. If there is reason to believe that the stock is more likely to move in a certain direction, the best decision becomes more clear. Figure 6.5 shows a radar chart of the payoffs for each decision. It is easy to see that the stock and put option is *dominated* by the call option in the sense that no matter what event occurs, the call option is always better. (In the chart, the lines connecting the returns for the stock and put option fall inside the call option profile.) Therefore, in this case, it would not make sense to consider the

FIGURE 6.4
Put and call options evaluation

	A	B	C	D	E	F	G
1	Options Evaluation						
2				Future stock value			
3	Decision	$ 1.00	$ 2.00	$ 3.00	$ 4.00	$ 5.00	
4	Buy stock	$ (2,000.00)	$ (1,000.00)	$ -	$ 1,000.00	$ 2,000.00	
5	Put option	$ 980.00	$ (20.00)	$ (520.00)	$ (520.00)	$ (520.00)	
6	Call option	$ (12.00)	$ (12.00)	$ (12.00)	$ 988.00	$ 1,988.00	
7	Stock and put option	$ (1,020.00)	$ (1,020.00)	$ (520.00)	$ 480.00	$ 1,480.00	
8	Stock and call	$ (2,012.00)	$ (1,012.00)	$ (12.00)	$ 1,988.00	$ 3,988.00	
9							
10		Average	Aggressive	Conservative	Opportunity	Standard	Average
11	Decision	Payoff	Strategy	Strategy	Loss	Deviation	Return-to-risk
12	Buy stock	$ -	$ 2,000.00	$ (2,000.00)	$ 2,980.00	$ 1,414.21	0.000
13	Put option	$ (120.00)	$ 980.00	$ (520.00)	$ 4,508.00	$ 583.10	-0.206
14	Call option	$ 588.00	$ 1,988.00	$ (12.00)	$ 2,000.00	$ 800.00	0.735
15	Stock and put option	$ (120.00)	$ 1,480.00	$ (1,020.00)	$ 2,508.00	$ 969.54	-0.124
16	Stock and call	$ 588.00	$ 3,988.00	$ (2,012.00)	$ 2,992.00	$ 2,154.07	0.273
17							
18				Opportunity Loss Table			
19	Decision	$ 1.00	$ 2.00	$ 3.00	$ 4.00	$ 5.00	Maximum
20	Buy stock	$ 2,980.00	$ 988.00	$ -	$ 988.00	$ 1,988.00	$ 2,980.00
21	Put option	$ -	$ 8.00	$ 520.00	$ 2,508.00	$ 4,508.00	$ 4,508.00
22	Call option	$ 992.00	$ -	$ 12.00	$ 1,000.00	$ 2,000.00	$ 2,000.00
23	Stock and put option	$ 2,000.00	$ 1,008.00	$ 520.00	$ 1,508.00	$ 2,508.00	$ 2,508.00
24	Stock and call	$ 2,992.00	$ 1,000.00	$ 12.00	$ -	$ -	$ 2,992.00

FIGURE 6.5
Radar chart for
options strate-
gies (each axis
represents the
stock price
event)

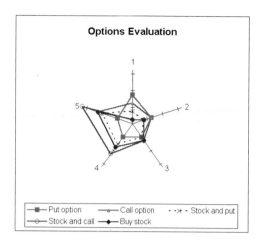

stock and put decision. Each of the other four alternatives can be the best, depending on the future stock price.

EXPECTED VALUE DECISION MAKING

When faced with a one-time decision, attitudes toward risk play a critical role, and the
strategies described in the previous section provide a basis for making a decision.
However, if an individual or business faces the same decision problem repeatedly,
then over the long run, the decision can be made based on expected value. The
expected value approach is to select the decision alternative with the best expected
payoff. Suppose that $P(S_j)$ = the probability that event S_j occurs, $V(D_i, S_j)$ is the payoff for D_i under S_j, and n = the number of events. The expected value approach is to
calculate the expected payoff for each decision alternative D_i:

$$E(D_i) = \sum_{j=1}^{n} P(S_j)V(D_i,S_j)$$

The decision alternative with the highest expected payoff is selected as the optimal
decision.

Expected value decisions require good estimates of event probabilities. If historical data on past occurrences of events are available, as is probably the case with an airline, then we can estimate probabilities objectively. In other business decisions, such
historical data may not be available. Nevertheless, good managers will be able to estimate the likelihood of events from their experience (that's usually why they're being
paid the big bucks).

Most airlines, for example, offer discount fares for advanced purchase. Assume that
only two fares are available: full and discount. The airline must decide whether or not
to accept the next request for a discount seat. If it accepts the discount request, the revenue it earns is the discount fare. If it rejects the discount request, two outcomes are
possible. First, the seat may remain empty and the airline will not realize additional revenue. Alternatively, the remaining seat may be filled by a full fare passenger, either
because full-fare passenger demand is sufficient to fill the seats or because discount-
fare passengers choose to pay full fare when told the discount fare is not available.

This decision situation is illustrated in Figure 6.6. The decision depends on the
probability p of getting a full-fare request when a discount request is rejected. The

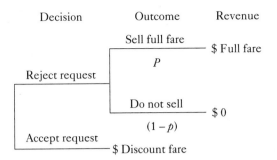

FIGURE 6.6
Airline
discount-fare
request
decision

expected value of rejecting the discount seat request is p times the full fare value. Thus, if a full-fare ticket is $560, the discount fare is $400, and $p = 0.75$, the expected value of rejecting the discount request is $0.25(0) + 0.75 ($560) = 420. Since this is higher than the discount fare, the discount request should be rejected. Since an airline makes hundreds or thousands of such decisions each day, the expected value criterion is appropriate. (See the *Management Science Practice Box* for American Airlines for further information on its use of models for yield management.)

Management Science Practice:
The Overbooking Problem at American Airlines

Airlines allow customers to cancel unpaid reservations with no penalty. Even after purchasing a ticket, many passengers may cancel or miss their flights and receive at least partial refunds. Those who do not formally cancel a reservation and do not show up are considered no-shows. On average, about half of all reservations made for a flight are cancelled or become no-shows. American Airlines estimates that about 15 percent of seats on sold-out flights would be unused if reservation sales were limited to aircraft capacity.

By properly setting reservation levels higher than seating capacity, American is able to compensate for passenger cancellations and no-shows. However, poor overbooking decisions can be costly. If reservation levels are set too low (more passengers than expected cancel or do not show up), then flights depart with empty seats that could have been filled. With overbooking, the airline takes the risk that more passengers may show up for a flight than there are seats on the aircraft. The airline must compensate such passengers for their inconvenience and must accommodate them on other flights.

The cost of oversales consists of compensation for the inconvenience, in the form of vouchers that can be redeemed on a future flight; hotel and meal accommodations, if necessary; and a seat on a later flight, either on American or some other airline. If the passenger is put on another airline, American must pay the other airline for transportation. The oversale cost is not constant. The more oversales that occur on a flight, the higher the voucher offer will be and the more likely that unaccommodated passengers will have to be transported on other airlines. Therefore, the total oversale cost grows at an increasing rate.

continued

American Airlines Decision Technologies (AADT), the management science consulting branch of the company, developed an optimization model that maximizes net revenue associated with overbooking decisions. In choosing the best overbooking level for a flight, the model balances the additional revenue that can be gained by selling a reservation against the cost of the additional oversale risk. This is shown in Figure 6.7. When there is little or no overbooking, total revenue equals net revenue. The optimal overbooking level occurs at the maximum of the net revenue curve, where net revenue is the difference between total passenger revenue and oversale cost.

FIGURE 6.7 Overbooking model

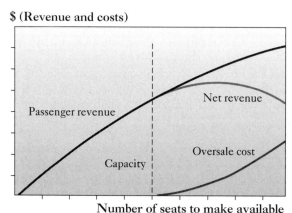

The number of oversales allowed by using this model, however, may sometimes degrade passenger service to an unacceptable level. To prevent this problem, AADT added a constraint to limit the expected number of oversales on each flight. The complete model accounts for

- The additional revenue generated by adding more reservations
- The probability distribution of passenger cancellations and no-shows
- The expected number of oversales
- The maximum number of oversales allowed
- The likelihood that a passenger who cannot secure a reservation on one flight will choose another American Airlines flight (called the recapture probability)

Because these factors vary with the amount of time before departure, overbooking levels are recalculated several times before departure. Four forecasts are needed for the model:

1. The probability that a passenger will cancel
2. The probability that a passenger with an active reservation will not show up
3. The recapture probability
4. The oversale cost

AADT uses forecasting models to estimate the probabilities.

Source: Adaptation and figure from Smith et al., "Yield Management at American Airlines," *Interfaces*, Vol. 22, No. 1, January–Feburary 1992, pp. 8–31.

An Application of Expected Value Analysis: The "Newsvendor" Problem

The *newsvendor problem* is the following: a street newsvendor sells daily newspapers and must decide how many to purchase. Purchasing too few results in lost opportunity to increase profits, but purchasing too many will result in a loss, since the excess must be discarded at the end of the day. This problem applies to a variety of practical situations in which a one-time purchase decision must be made in the face of uncertain demand. For example, department store buyers must purchase seasonal clothing well in advance of the buying season, and candy shops must decide how many special holiday gift boxes to assemble. In these cases, any leftovers will probably not be discarded, but will be sold at a discount, perhaps even at a loss.

We will first develop a general model for this problem and then illustrate it with an example. Let us assume that each item costs $\$c$ to purchase and is sold for $\$r$. At the end of the period, any unsold items can be disposed of at $\$s$ each (the salvage value). Clearly, it makes sense to assume that $r > c > s$. Let d be the number of units demanded during the period and x be the number purchased.

Notice that we cannot sell more than the minimum of the actual demand and the amount purchased. If demand is known, then the optimal decision is obvious: Choose $x = d$. However, if d is not known in advance, we run the risk of overpurchasing or underpurchasing. If $x < d$, then we lose the opportunity of realizing additional profit (since we assume that $r > c$), and if $x > d$, we incur a loss (since $c > s$). Because d is uncertain, the value of profit may assume a range of values for a fixed purchase quantity Q. We consider two cases:

Case 1: *Quantity purchased (x) greater than or equal to quantity demanded (d).* If x is greater than or equal to d, then $(x - d)$ units will be left over and disposed of at a loss. Specifically, the revenue received is $rd + s(x - d)$. Since the purchase cost is cx, the net profit will be

$$\text{Net profit} = rd + s(x - d) - cx$$

Case 2: *Quantity purchased (x) less than quantity demanded (d).* If x is less than d, then x units will be sold and we will have lost the opportunity to sell an additional $(d - x)$ units at a profit. The net profit would be

$$\text{Net profit} = rx - cx = (r - c)x$$

The problem is to choose the best value of x. To simplify our presentation, we will assume that demand is discrete, that is, it follows a discrete probability distribution, with $p(d)$ being the probability that demand is d. We develop an expression for the expected profit when x units are purchased. The expected profit is computed by multiplying the net profit obtained for a specific demand d by the probability that the demand is d and then summing these terms over all possible values of d. Using the formulas for the two cases just described, we have

$$\text{Expected profit} = \sum_{d=0}^{x} [rd + s(x - d) - cx]p(d) + \sum_{d=x+1}^{\infty} [(r - c)x]p(d)$$

The first term corresponds to case 1, when $d \le x$; and the second term to case 2, when $d > x$.

Example 5: A Newsvendor Problem

A newsvendor purchases papers for $0.20 each and sells them for $0.35. At the end of the day, any unsold papers can be disposed of at $0.05 each. If the newsvendor purchases at least enough papers to meet the demand, the net profit will be $0.35d + 0.05(x - d) - 0.20x$. If x is less than d, then the newsvendor will sell only x papers and will have lost the opportunity to sell an additional $(d - x)$ papers at a profit. The net profit would be $0.35x - 0.20x = 0.15x$.

Let us assume that the demand will vary between 40 and 49, and that the probability of any number of papers within this range is equal to 0.1. Suppose that 43 papers are purchased. If $d = 40$, the net profit is

$$0.35d + 0.05(x - d) - .20x = 0.35(40) + 0.05(43 - 40) - 0.20(43) = \$5.55$$

Likewise, the net profit for a demand of 41 is

$$0.35d + 0.05(x - d) - .20x = 0.35(41) + 0.05(43 - 41) - 0.20(43) = \$5.85$$

If the demand is greater than 43, the net profit is

$$0.15x = 0.15(43) = \$6.45$$

because the newsvendor can only sell 43 papers.

Figure 6.8 shows a spreadsheet that computes the payoffs associated with each combination of order quantity and demand using a data table, as well as the expected profits, standard deviation,[6] and return-to-risk. We see that the optimal purchase quantity is either 44 or 45, having an expected profit of $6.30. However, as the order quantity increases, the decisions become more risky. Using the spreadsheet, it would be easy to investigate changes in the probability distribution or other input data.

FIGURE 6.8
Spreadsheet analysis for newsvendor example

	A	B	C	D	E	F	G	H	I	J	K	L	M	N	O
1	Newsvendor Problem							Order Quantity							Best
2			Probability	$ 5.55	40	41	42	43	44	45	46	47	48	49	Payoff
3	Selling price	$0.35	0.1	40	$6.00	$5.85	$5.70	$5.55	$5.40	$5.25	$5.10	$4.95	$4.80	$4.65	$ 6.00
4	Cost	$0.20	0.1	41	$6.00	$6.15	$6.00	$5.85	$5.70	$5.55	$5.40	$5.25	$5.10	$4.95	$ 6.15
5	Discount price	$0.05	0.1	42	$6.00	$6.15	$6.30	$6.15	$6.00	$5.85	$5.70	$5.55	$5.40	$5.25	$ 6.30
6	Demand	40	0.1	43	$6.00	$6.15	$6.30	$6.45	$6.30	$6.15	$6.00	$5.85	$5.70	$5.55	$ 6.45
7	Order Quantity	43	0.1	44	$6.00	$6.15	$6.30	$6.45	$6.60	$6.45	$6.30	$6.15	$6.00	$5.85	$ 6.60
8			0.1	45	$6.00	$6.15	$6.30	$6.45	$6.60	$6.75	$6.60	$6.45	$6.30	$6.15	$ 6.75
9	Profit	$5.55	0.1	46	$6.00	$6.15	$6.30	$6.45	$6.60	$6.75	$6.90	$6.75	$6.60	$6.45	$ 6.90
10		.	0.1	47	$6.00	$6.15	$6.30	$6.45	$6.60	$6.75	$6.90	$7.05	$6.90	$6.75	$ 7.05
11			0.1	48	$6.00	$6.15	$6.30	$6.45	$6.60	$6.75	$6.90	$7.05	$7.20	$7.05	$ 7.20
12	Key cell formulas		0.1	49	$6.00	$6.15	$6.30	$6.45	$6.60	$6.75	$6.90	$7.05	$7.20	$7.35	$ 7.35
13															
14			Expected profit	$6.00	$6.12	$6.21	$6.27	$6.30	$6.30	$6.27	$6.21	$6.12	$6.00	$ 6.68	
15			Standard deviation	$ -	$0.09	$0.19	$0.31	$0.42	$0.54	$0.65	$0.74	$0.82	$0.86		
16			Return-to-risk	#DIV/0!	68.00	32.33	20.49	14.85	11.65	9.65	8.34	7.48	6.96		
17														EVPI	$ 0.37
18															
19			This table is used to compute the weighted standard deviation based on probabilities in C3:C12												
20			0.1	40	0	0.0729	0.2601	0.5184	0.81	1.1025	1.3689	1.5876	1.7424	1.8225	
21			0.1	41	0	0.0009	0.0441	0.1764	0.36	0.5625	0.7569	0.9216	1.0404	1.1025	
22			0.1	42	0	0.0009	0.0081	0.0144	0.09	0.2025	0.3249	0.4356	0.5184	0.5625	
23			0.1	43	0	0.0009	0.0081	0.0324	3E-30	0.0225	0.0729	0.1296	0.1764	0.2025	
24			0.1	44	0	0.0009	0.0081	0.0324	0.09	0.0225	0.0009	0.0036	0.0144	0.0225	
25			0.1	45	0	0.0009	0.0081	0.0324	0.09	0.2025	0.1089	0.0576	0.0324	0.0225	
26			0.1	46	0	0.0009	0.0081	0.0324	0.09	0.2025	0.3969	0.2916	0.2304	0.2025	
27			0.1	47	0	0.0009	0.0081	0.0324	0.09	0.2025	0.3969	0.7056	0.6084	0.5625	
28			0.1	48	0	0.0009	0.0081	0.0324	0.09	0.2025	0.3969	0.7056	1.1664	1.1025	
29			0.1	49	0	0.0009	0.0081	0.0324	0.09	0.2025	0.3969	0.7056	1.1664	1.8225	

Excel Exercise 6.2

Open the Excel file *newsvendor.xls* and change the probability distribution of demand to the following:

Demand   40 41 42 43 44 45 46 47 48 49

Probability   .05 .10 .15 .20 .15 .10 .08 .07 .06 .04

What is the best order quantity?

Expected Value of Perfect Information

By *perfect information*, we mean knowing in advance what event will occur. Although we never have perfect information in practice, it is worth knowing how much we could improve the value of our decision if we had such information. This is called the *expected value of perfect information*, or EVPI. EVPI is the difference between the expected payoff under perfect information and the expected payoff of the optimal decision without perfect information. We compute EVPI by asking the following question: If each event occurs, what would be the best decision and payoff? Then we weight these payoffs by the probabilities associated with the events to obtain the expected payoff under perfect information.

The last column in Figure 6.8 shows the best payoffs for each level of demand and the expected payoff under perfect information. By subtracting the optimal payoff without perfect information, $6.30, from this value, we obtain in cell O17

$$\text{EVPI} = \$6.68 - 6.30 \approx \$0.37 \text{ (difference due to rounding)}$$

EVPI is the *maximum* amount that we could improve our average outcome by having better information about the future. Therefore, we would never want to pay any more than this amount for better information.

Another way of computing EVPI is to compute the *expected opportunity loss* for the best decision using the expected value criterion. The opportunity loss tells us how much we gain if we could switch our decision based on perfect information. Given a decision from the expected value criterion, the opportunity loss for that decision is the potential gain from perfect information. Let $OL(D_i, S_j)$ be the opportunity loss for decision alternative D_i under event S_j, and let n be the number of events. The expected opportunity loss for D_i, denoted $EOL(D_i)$, is

$$EOL(D_i) = \sum_{j=1}^{n} P(S_j)OL(D_i, S_j)$$

If D^* is the optimal decision using the expected value criterion, then $EVPI = EOL(D^*)$.

Excel Exercise 6.3

Modify the Excel file *newsvendor.xls* to compute the EVPI using the expected opportunity loss formula. You will first have to compute the opportunity loss for each combination of demand and order quantity.

OPTIMAL EXPECTED VALUE DECISION STRATEGIES

If we can estimate probabilities of events in a decision tree, we may determine an optimal strategy based on the expected value criterion. To do this, we "roll back" the tree. We always work *from right to left.* At each event node, the **rollback value** is the expected payoff obtained by multiplying the values at the end of the branches by their associated probabilities and summing. When we encounter a decision node, we select the best decision and set the rollback value of this node to the value associated with the best decision branch. We repeat this process until we arrive at the initial node in the tree. To find the optimal decision strategy, we work forward through the tree, selecting the best decision at each decision node that is conditional upon the event that occurs.

The CD-ROM accompanying this book contains an Excel add-in called TreePlan. TreePlan allows you to easily construct decision trees on an Excel worksheet, compute the rollback values for all nodes, and perform sensitivity analysis on the model data. Figure 6.9 shows the general structure of a TreePlan decision tree model. Decision nodes are shown as squares and event nodes as circles. Each decision branch can be assigned a name corresponding to the decision and a cash flow. The cash flows on branches leading to a terminal node are summed to determine the terminal value. You may, as an alternative, simply enter the terminal values, but using branch cash flows allows you to easily change the model assumptions and examine the sensitivity of the results to these changes. Event branches have a name and a probability of occurrence. The sum of the probabilities for the events immediately following an event node must equal 1.0. TreePlan automatically computes the rollback values and the optimal decision strategy. Complete documentation and a tutorial example for constructing decision trees using Treeplan are provided on the disk, so we will not present the details here. We use TreePlan in the following example and encourage you to use it for problems and exercises.

FIGURE 6.9
TreePlan
structure

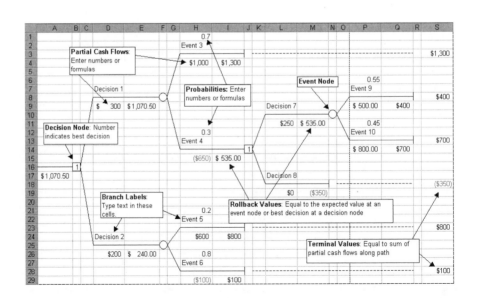

Example 6: An Optimal Decision Strategy for Introducing New Products

Figure 6.10 shows a TreePlan decision tree for the situation described in Example 2. The first number below each decision branch represents the discounted cash flow in thousands of dollars associated with that decision. For example, it costs $200,000 to introduce the product nationally (cell D6) and $30,000 to market the product regionally (cell D31). The cash flows at the terminal event branches represent the total revenue associated with the sequence of decisions and events from the beginning of the tree to the end of a path. For instance, a high response after a national introduction will yield revenues of $700,000; thus, the net payoff is $-\$200,000 + \$700,000 = \$500,000$. The terminal node values are summarized in column S of the worksheet.

The rollback values are shown just below and to the left of the nodes. For example, the rollback value associated with introducing the product nationally is computed by summing the probability of high response (0.6) multiplied by the associated payoff ($500) and the probability of low response (0.4) multiplied by the payoff $(-\$50)$, or $(0.6)(500) + (0.4)(-50) = 280$ (cell E6). At a decision node, the rollback value chosen is the one with the highest payoff among the following decisions. For example, consider the decision associated with cell J20. If a high regional response is found after introducing regionally, we must decide whether to remain regional (expected payoff = $170), market nationally (expected payoff = $415), or drop the product (expected payoff = $-\$30$). The best decision is to market nationally, so the payoff associated with this decision is selected as the rollback value for this decision node. Similarly, the rollback value in cell A18 is the highest between cells E6 and E31.

The numbers within the decision nodes provide the decision branch associated with the optimal decision based on the rollback values. These define the decision strategy. For example, the number "2" in cell B17 states that the second decision (introduce regionally) should be chosen. If the regional response is high, the second decision at node J20, market nationally, should be selected; if the regional response is low, the first decision at node J40, remain regional, is best. Note that we cannot choose which event will occur; we simply have to let nature take its course.

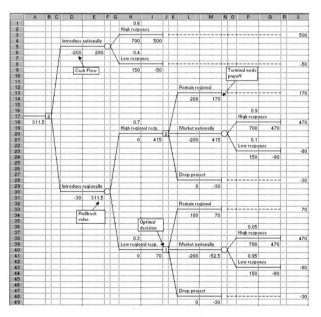

FIGURE 6.10

Decision tree for marketing example created using TreePlan

Sensitivity Analysis of Decision Strategies

In some cases, such as in the airline decision to sell discount tickets, the probabilities used to find optimal decision strategies can be estimated quite accurately because extensive historical data are available. In most cases, however, these probabilities are at best crude estimates. This suggests that we should examine carefully the impact of our estimates on the decision strategies we choose. This can be done rather easily on spreadsheets.

Figure 6.11 shows a data table created to examine the sensitivity of the high regional response probability assumption (cell H18) in the decision tree Figure 6.10. (You must first change the low regional response probability, cell H38, to be = 1 − H18 so that the probabilities add to 1 in the data table calculations.) The values in the data table represent the rollback values for the decision branch "Introduce regionally." You can see that if the probability is 0.6 or less (actually somewhere between 0.6 and 0.7), then the optimal decision strategy would be to market nationally initially. Because the current estimate is close to this threshold, the company might want to acquire better information to assess these probabilities.

FIGURE 6.11

Data table to examine sensitivity to high regional response probability

	U	V
1		
2	High regional	Regional rollback value
3	response probability	311.50
4	0	70.00
5	0.1	104.50
6	0.2	139.00
7	0.3	173.50
8	0.4	208.00
9	0.5	242.50
10	0.6	277.00
11	0.7	311.50
12	0.8	346.00
13	0.9	380.50
14	1	415.00

Excel Exercise 6.4

Open the file *Marketing decision tree.xls* and create a two-way data table to study the sensitivity on the regional introduction rollback value of both the high regional response probability and the high response probability if the regional response is high and the company decides to market nationally.

TECHNOLOGY FOR DECISION ANALYSIS

TreePlan was developed primarily for academic instruction; many commercial software packages are available for more complex decision analysis. Table 6.2 summarizes

Product	Company/Web Site	Description	TABLE 6.3
DPL	Applied Decision Analysis, 2710 Sand Hill Rd., Menlo Park, Calif.	Tool for combining influence diagrams, decision trees, and spreadsheets to produce a variety of outputs	Some popular decision analysis software technology
Expert Choice Professional	Expert Choice, Inc. **www.expertchoice.com**	A multicriteria decision support tool based on a theory called the analytic hierarchy process	
Decision Analysis by TreeAge (DATA)	TreeAge Software, Inc. **www.treeage.com**	Tool for building influence diagrams or trees, and converting influence diagrams to fully configured trees	
Precision Tree	Palisade Corporation **www.palisade.com**	Decision analysis add-in to Excel using influence diagrams and decision trees	
Sensitivity/ Supertree	Strategic Decisions Group— Decision Systems, 2440 Sand Hill Rd., Menlo Park, Calif.	Decision analysis tool with full range of tree manipulation	

some of these packages. More complete software surveys are regularly published in *OR/MS Today,* a publication of INFORMS **(www.informs.org)**.

RISK TRADE-OFFS AND MULTIOBJECTIVE DECISIONS

Many practical decisions involve multiple objectives. Usually the decision maker must make trade-offs among objectives that reflect risk and economic or other factors when selecting a decision. For example, in Chapter 1 we developed the economic order quantity (EOQ) model for minimizing the total cost of placing orders and holding inventory. We assumed that demand was deterministic. With this assumption, the demand rate follows a smooth curve, and we could place the order at a time that would ensure that no shortages would occur before the order was received (you might wish to refer back to Figure 1.17). In many practical situations it is unreasonable to assume that demand is deterministic. In such cases, the inventory pattern would resemble that shown in Figure 1.15. Demand is erratic, and inventory managers face the risk of either not being able to satisfy a customer's demand or carrying excess inventory. Thus, they must make a trade-off between the higher costs of carrying inventory and of not satisfying customer demand. Models can provide insight to help make these decisions.

In the EOQ model developed in Chapter 1, we saw that the total annual cost is composed of the cost of placing orders and the cost of holding inventory:

$$Q = \text{order quantity}$$
$$D = \text{annual demand}$$
$$C = \text{unit cost of the item}$$
$$C_0 = \text{ordering cost}$$
$$i = \text{inventory carrying charge}$$

The order quantity that minimizes total annual cost is

$$Q^* = \sqrt{\frac{2DC_0}{(iC)}}$$

Suppose that the annual demand is a random variable with some probability distribution. We might consider using the average demand, μ_D, in computing Q^*. In Chapter 1 we saw that the economic order quantity is relatively insensitive to changes in the parameters of the model. Therefore, even if the actual value of D would vary from the average, this order quantity should provide a reasonable solution.

A more critical decision is *when* to place an order. In the EOQ situation, we assumed a continuous review inventory system in which an order is triggered when the inventory position reaches the reorder point r. In the deterministic case, we computed $r = Dt$, where t is the lead time (time from which the order is placed until it is received). Again, we might consider using the average demand, μ_D, to compute r. Then $\mu_D t$ represents the *average demand during the lead time*. This presents a problem. If the distribution of the lead time demand is symmetric about the mean (for example, normally distributed), then about half the time the actual lead time demand will be less than $\mu_D t$ and half the time it will be greater than $\mu_D t$. This means that 50 percent of the time we will not be able to satisfy customers' demands, an unacceptable situation.

One way to resolve this is to define a service level. A **service level** is a constraint that represents the probability that a routine demand can be satisfied. For example, we might wish to ensure that demand can be satisfied 95 percent of the time. If we know the probability distribution of the lead time demand, then we can select r to meet the service level constraint.

Example 7: Setting a Service Level for Inventory Management

Data Management Supplies (DMS) is a small company that sells various computer peripheral and photocopier supplies to small businesses. One popular item is a toner cartridge for laser printers. Historical data suggest that demand is normally distributed with a mean of 150 every four weeks and a standard deviation of 25. Each cartridge costs $40 from the manufacturer. An order costs $10, and DMS estimates its carrying cost rate to be 20 percent. Lead time is one week.

The expected annual demand, D, is 150 (13) = 1,950. Applying the EOQ formula, we find that the economic order quantity is

$$Q^* = \sqrt{\frac{2DC_0}{(iC)}}$$
$$= \sqrt{\frac{2(1950)(10)}{(.2)(40)}} = 69.82 \text{ or about 70 units}$$

Using $Q^* = 70$, DMS will place about $D/Q^* = 1,950 / 70 = 27.86$, or about 28, orders each year.

If X represents the demand over a four-week period, then $0.25X$ represents the demand during the lead time. Since X is normally distributed with mean $\mu = 150$ and $\sigma = 25$, then $0.25X$ is also normally distributed with

$$\text{Mean} = 0.25\mu = 0.25(150) = 37.5$$

and

$$\text{Variance} = (0.25)^2\sigma^2 = (0.25)^2(25)^2 = 39.0625$$

Therefore, the standard deviation during the lead time is 6.25.

Suppose that the owner of DMS is willing to allow the lead time demand to exceed the reorder point 5 percent of the time at most. (With 27 orders each year, this will happen only once or twice each year.) This is shown in Figure 6.12. Therefore, the reorder point must satisfy

$$r = \mu + z\sigma$$

where z is the number of standard deviations associated with the service level. From Appendix A, we see that the probability of exceeding $z = 1.645$ in a standard normal probability distribution is 0.05. Therefore, the reorder point that meets the required service level is

$$r = 37.5 + 1.645(6.25)$$
$$= 47.78, \text{ or about 48 units}$$

Thus, DMS should place an order for 70 units whenever the inventory position drops to 48 units.

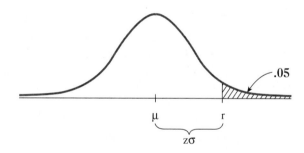

FIGURE 6.12
Computing the
reorder point
and safety
stock level

Since the average demand during the lead time is only 37.5 units, DMS will carry an average of $48 - 37.5 = 10.5$ units as safety stock. This increases the annual cost for operating this inventory system. The expected annual cost is as follows:

$$\text{Order cost: } (D/Q)C_0 = (1950/70)(10) = \$278.57$$
$$\text{Normal inventory holding: } iC(Q/2) = .2(40)(70/2) = \$280$$
$$\text{Safety stock holding: } iC(10.5) = .2(40)(10.5) = \$84$$
$$\text{Total annual cost} = \$642.57$$

DMS is paying about a 13 percent premium to better meet customer demands. (Note that order cost and normal inventory holding costs are not exactly equal—as would be expected—since we rounded the economic order quantity up to 70 units.)

Figure 6.13 shows an Excel worksheet (*service level analysis.xls*) for evaluating this model for various service levels. As the service level decreases, so does the additional

FIGURE 6.13
Spreadsheet
model for
Example 7

	A	B	C	D	E	F
1	Service Level Analysis			Key cell formulas		
2						
3	*Parameters and Uncontrollable Inputs*					
4						
5	Annual Demand	1950				
6	Ordering Cost	$ 10.00				
7	Unit Cost	$ 40.00				
8	Carrying Charge	20%				
9						
10	Lead time demand					
11	Mean	37.5				
12	Standard deviation	6.25				
13						
14	*Model*					
15						
16	Optimal order quantity		69.82			
17	Annual holding cost		$ 279.28			
18	Annual ordering cost		$ 279.28			
19	Total annual cost		$ 558.57			
20						
21					Safety	Additional
22	Service level	P(stockout)	z-value	Reorder point	stock	cost
23	0.99	0.01	2.326342	52.0	14.5	$ 116.32
24	0.95	0.05	1.644853	47.8	10.3	$ 82.24
25	0.90	0.10	1.281551	45.5	8.0	$ 64.08
26	0.85	0.15	1.036433	44.0	6.5	$ 51.82
27	0.80	0.20	0.841621	42.8	5.3	$ 42.08

cost. With multiple objectives, there is no "optimal" answer; the decision simply depends on the trade-off the company is willing to make between service level and cost.

UTILITY AND DECISION MAKING

A typical charity raffle involves selling one thousand $100 tickets to win a $35,000 automobile. A decision analysis (see Figure 6.14) based on expected value clearly shows this to be a poor gamble. Nevertheless, many people would take this chance (even ignoring the charitable issues). On the other hand, if the ticket cost $5,000, and the probability of winning was increased to 0.50, yielding a much higher expected value, most people would not take the chance. An approach for assessing risk attitudes quantitatively is called *utility theory*. This approach quantifies a decision maker's relative preferences for particular outcomes.

We can determine an individual's utility function by posing a series of decision scenarios. This is best illustrated with an example; we will use Rick's investment problem.

FIGURE 6.14
Decision tree
for the car raffle

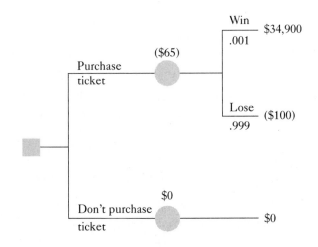

Example 8: Determining a Utility
Function for Rick's Investment Outcomes

In Example 3, Rick Martin is faced with selecting a decision that could result in a variety of payoffs, from a profit of $1,700 to a loss of $900. The first step in determining Rick's utility function is to rank-order the payoffs from highest to lowest. We arbitrarily assign a utility of 1.0 to the highest payoff and a utility of 0 to the lowest:

Payoff x	Utility $U(x)$
$1,700	1.0
$1,000	
$840	
$600	
$400	
($500)	
($900)	0.0

Next, for each payoff between highest and lowest, we present Rick with the following situation: suppose you have the opportunity of achieving a *guaranteed return of R* or taking a chance of receiving $1,700 with probability p or losing $900 with probability $1 - p$. What value of p would make you indifferent to these two choices? Let us choose $R = $1,000$. This is illustrated in Figure 6.15. Because this is a relatively high value, Rick decides that p would have to be at least .9 to take this risk. This represents the utility of a payoff of $1,000, denoted as $U($1,000)$. We repeat this process for each payoff. The probabilities p that Rick selects for each scenario form his utility function. Suppose this process results in the following:

Payoff x	Utility $U(x)$
$1,700	1.00
$1,000	.90
$840	.85
$600	.80
$400	.75
($500)	.35
($900)	0.00

If we compute the expected value of each of the gambles for the chosen values of p, we see that they are higher than the corresponding payoffs. For example, if $p = .9$, the expected value of taking the gamble is

$$.9($1,700) + .1(-$900) = $1,440$$

This is larger than accepting $1,000 outright. We can interpret this to mean that Rick requires a risk premium of $1,440 - $1,000 = $440 to feel comfortable enough to risk losing $900 if he takes the gamble. In general, the **risk premium** is the amount an individual is willing to forgo to avoid risk. This indicates that Rick is a *risk-averse individual;* that is, he is relatively conservative.

Another way of viewing this is to find the *break-even probability* at which Rick would be indifferent to receiving the guaranteed return and taking the gamble. This probability is found by solving the equation:

$$1,700p - 900(1 - p) = 1,000$$

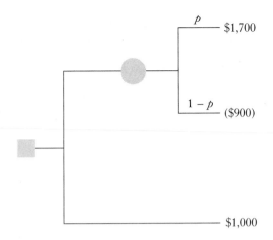

resulting in $p = 19/26 = 0.73$. Because Rick requires a higher probability of winning the gamble, it is clear that he is uncomfortable taking the risk.

If we graph the utility versus the payoffs, we can sketch Rick's utility function as shown in Figure 6.16. Rick's utility function is generally *concave downward*. This type of curve is characteristic of risk-averse individuals. Such decision makers avoid risk, choose conservative strategies and those with high return-to-risk value. Thus, a gamble must have a higher expected value than a given payoff to be preferable, or equivalently, a higher probability of winning than the break-even value.

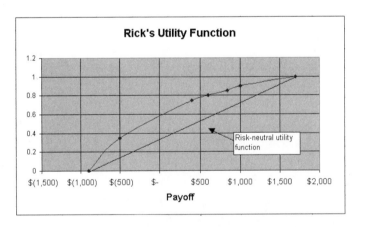

Other individuals might be risk taking. What would their utility functions look like? As you might suspect, they are *concave upward*. These individuals would take a gamble that offers higher rewards even if the expected value is less than a certain payoff. An example of a utility function for a risk-taking individual in Rick's situation would be as follows:

Payoff x	Utility $U(x)$
$1,700	1.00
$1,000	0.60
$840	0.55
$600	0.45
$400	0.40
($500)	0.10
($900)	0.00

We suggest you graph this to see its shape. For the payoff of $1,000, this individual would be indifferent between (1) receiving $1,000 and (2) taking a chance at $1,700 with probability .6 and losing $900 with probability 0.4. The expected value of this gamble is

$$.6(\$1,700) + .4(-\$900) = \$660$$

Since this is considerably less than $1,000, the individual is taking a larger risk to try to receive $1,700. Note that the probability of winning is less than the break-even value. Risk takers generally prefer more aggressive strategies.

Finally, some individuals are risk neutral; they prefer neither taking risks nor avoiding them. Their utility function would be linear and would correspond to the break-even probabilities for each gamble. For example, a payoff of $600 would be equivalent to the gamble if

$$\$600 = p(\$1,700) + (1 - p)(-\$900)$$

Solving for p, we obtain $p = 15/26$, or .58, which represents the utility of this payoff. The decision of accepting $600 outright or taking the gamble could be made by flipping a coin. These individuals tend to ignore risk measures and base their decisions on the average payoffs.

A utility function may be used instead of the actual monetary payoffs in a decision analysis by simply replacing the payoffs by their equivalent utilities and then computing expected values. The expected utilities and the corresponding optimal decision strategy then reflect the decision maker's preferences toward risk. For example, if we replace the payoffs in Table 6.1 with the (risk-averse) utilities that Rick defined, and assume that the events are equally likely (average payoff strategy), the optimal decision now is to choose the bank CD as opposed to the stock fund, as shown in the next table.

Decision/Event	Rates Rise	Rates Stable	Rates Fall	Average Utility
Bank CD	0.75	0.75	0.75	0.75
Bond fund	0.35	0.85	0.90	0.70
Stock fund	0.00	0.80	1.00	0.60

If other assessments of event probabilities are available, these can be used to compute the expected utility and identify the best decision.

Exponential Utility Functions

It can be rather difficult to compute a utility function, especially for situations involving a large number of payoffs. Because most decision makers are risk averse, we may use an exponential utility function to approximate the true utility function. The exponential utility function is

$$U(x) = 1 - e^{-x/R}$$

where e is the base of the natural logarithm (2.71828...) and R is a shape parameter. Figure 6.17 shows several examples of $U(x)$ for different values of R. Notice that all of these functions are concave, and that as R increases, the functions become flatter, indicating more tendency toward risk neutrality.

One approach to estimating a reasonable value of R is to find the maximum payoff $P for which the decision maker is willing to take an equal chance on winning $P or losing $P/2. The smaller the value of P, the more risk averse is the individual. For Rick's problem, suppose that P = $400. This results in the following utility values:

Payoff x	Utility $U(x)$
$1,700	.9857
$1,000	.9179
$840	.8775
$600	.7769
$400	.6321
($500)	−2.4903
($900)	−8.4877

Using these values, we find that the bank CD remains the optimal decision, as shown in the next table.

Decision/Event	Rates Rise	Rates Stable	Rates Fall	Average Utility
Bank CD	0.6321	0.6321	0.6321	0.6321
Bond fund	−2.4903	0.8775	0.9179	−0.2316
Stock fund	−8.4877	0.7769	0.9857	−2.2417

FIGURE 6.17
Exponential utility functions

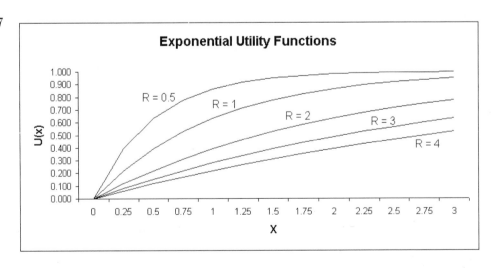

PROBLEMS

1. A patient arrives at an emergency room complaining of abdominal pain. The ER physician must decide whether to operate or to place the patient under observation for a non-appendix-related condition. If an appendectomy is performed immediately, the doctor runs the risk that the patient does not have appendicitis. If it is delayed and the patient does indeed have appendicitis, the appendix might become perforated, leading to a more severe case and possible complications. However, the patient might recover without the operation.
   a. Construct a decision tree for the doctor's dilemma.
   b. How might payoffs be determined?
   c. Would utility be a better measure of payoff than actual cost? If so, how might utilities be derived for each path in the tree?

2. Slaggert Systems is considering becoming certified to the ISO 9000 series of quality standards. Becoming certified is expensive, but the company could lose a substantial amount of business if its major customers suddenly demand ISO certification and the company does not have it. At a management retreat, the senior executives of the firm developed the following payoff table, indicating the net present value of profits over the next five years. What decision should they make under these decision strategies: average payoff, aggressive, conservative, opportunity loss, and return-to-risk?

	Customer Response	
	Standards Required	Standards not Required
Become certified	$550,000	$480,000
Stay uncertified	$300,000	$520,000

3. The Durr family is considering purchasing a new home and would like to finance $150,000. Three mortgage options are available, a one-year ARM at 6.125 percent, a three-year ARM at 7 percent, and a 30-year fixed mortgage at 7.875 percent. Both adjustable rate mortgages are sensitive to interest rate changes. As the family anticipates staying in the home for at least five years, they are interested in the interest costs over this time. A spreadsheet analysis indicates that the total interest costs over five years for possible interest rate scenarios are

Type of mortgage	Rates Rise	Rates Stable	Rates Fall
1-year ARM	$59,134	$44,443	$37,161
3-year ARM	$56,901	$51,075	$46,721
30-year fixed	$57,658	$57,658	$57,658

   a. What decisions should the Durr family make under the average payoff, aggressive, conservative, and opportunity loss strategies?
   b. If they estimate the probabilities P(rates rise) = 0.4, P(rates stable) = 0.5, and P(rates fall) = 0.1, what decision should they make?
   c. How much is perfect information worth?

4. Big Bob of Bob's Bagels is trying to decide if he should expand his business. He can choose not to expand, expand by buying an existing business, or expand by opening at another (new) location. The net present values in thousands of dollars for each alternative under various events are shown in the following table:

	S_1	S_2	S_3	S_4
Do not expand	$290	$290	$290	$290
Buy existing	$510	$130	$580	$150
Open new	$580	$100	$520	$160

a. What is the optimal decision for Bob under the average payoff, aggressive, conservative, opportunity loss, and return-to-risk decision strategies?

b. Bob has determined the following probabilities for events: $P(S_1) = .4$, $P(S_2) = .2$, $P(S_3) = .3$, $P(S_4) = .1$. Draw a decision tree of his problem and find the optimal decision using the expected value criterion. Show your calculations on the decision tree.

c. What is the return-to-risk value for each decision using the probabilities in part (b)?

d. Calculate the expected value of perfect information (EVPI).

5. The Doorco Corporation is a leading manufacturer of garage doors. All doors are manufactured in the firm's plant in Carmel, Indiana, and shipped to distribution centers or major customers. Doorco recently acquired another manufacturer of garage doors, Wisconsin Door, and is considering moving its wood door operations to the Wisconsin plant. Key considerations in this decision are the transportation and production costs at the two plants and the new construction and relocation costs. Complicating matters is the fact that marketing is predicting a decline in the demand for wood doors. The company has developed three scenarios:

i.   Demand falls slightly, with no noticeable effect on production.
ii. Demand and production decline 20 percent.
iii. Demand and production decline 40 percent.

The next table shows the total costs under each decision and scenario.

	Slight Decline	20% Decline	40% Decline
Stay in Carmel	$982,000	$830,000	$635,000
Move to Wisconsin	$993,000	$832,000	$629,000

a. What decision should Doorco make using each of the decision criteria discussed in this chapter?

b. Suppose the probabilities of the three scenarios are estimated to be .15, .45, and .40, respectively. What is the optimal expected value decision?

6. Mountain Ski Sports, a chain of ski equipment shops in Colorado, purchases skis from a manufacturer each summer for the coming winter season. The most popular intermediate model costs $150 and sells for $260. Any skis left over at the end of the winter are sold at the store's half-price sale (for $130). Sales over the years are quite stable. Gathering data from all its stores, Mountain Ski Sports has developed the following probability distribution for demand:

Demand	Probability
150	.10
175	.30
200	.40
225	.15
250	.05

The manufacturer will take orders only for multiples of 20, so Mountain Ski is considering the following order sizes: 160, 180, 200, 220, and 240.

a. Construct a payoff table for Mountain Ski's decision problem of how many pairs of skis to order. What is the best decision from an expected value basis?

b. Find the risk-to-return for each decision and the expected value of perfect information.

c. What is the expected demand? Is the optimal order quantity equal to the expected demand? Why?

7. Mad Marty runs a shirt concession at Riverview Music Center. He buys commemorative T-shirts for $15 each and sells them for $25. As people are leaving the music center after the concert, he reduces his price to $10 to sell any remaining shirts. Historically, shirt sales follow the distribution shown next:

Demand	Probability
350	.1
400	.2
450	.3
500	.2
550	.1
600	.1

Marty must purchase lots of 50, so he is considering orders of 350, 400, 450, 500, 550, or 600. Construct a payoff table and determine Mad Marty's optimal ordering decision using the expected value criterion.

8. Bev's Bakery specializes in sourdough bread. Early each morning, Bev must decide how many loaves to bake for the day. Each loaf costs $0.35 to make and sells for $1.15. Bread left over at the end of the day can be sold to a store selling day-old baked goods for $0.25. Past data indicate that demand is distributed as follows:

Number of Loaves	Probability
15	.05
16	.05
17	.10
18	.10
19	.20
20	.40
21	.05
22	.05

a. Construct a payoff table and determine the optimal quantity for Bev to bake each morning.

b. What is the optimal quantity for Bev to bake if the unsold loaves cannot be sold to the day-old store at the end of the day (so that unsold loaves are a total loss)?

9. Midwestern Hardware must decide how many snow shovels to order for the coming snow season. Each shovel costs $6 and is sold for $8.50. No inventory is carried from one snow season to the next. Shovels unsold after February are sold at a discount price of $5. Past data indicate that sales are highly dependent on the severity of the winter season. Past seasons have been classified as mild or harsh, and the following distributions of regular price demand have been tabulated:

Mild Winter		Harsh Winter	
No. of Shovels	Probability	No. of Shovels	Probability
250	.5	1,500	.2
300	.4	2,500	.4
350	.1	3,000	.4

Shovels must be ordered from the manufacturer in lots of 200. Construct a decision tree to illustrate the components of the decision model, and find the optimal quantity for Midwestern to order if the forecast calls for a 70 percent chance of a harsh winter.

10. Perform a sensitivity analysis of the Midwestern Hardware scenario (Problem 9). Find the optimal order quantity and optimal expected profit for probabilities of a harsh winter ranging from .2 to .8 in increments of .2. Plot optimal expected profit as a function of the probability of a harsh winter.

11. For the forest fire management problem (Example 1), find the expected value of perfect information by considering the best strategy for each possible event. Show that this is equal to the expected opportunity loss for the optimal expected value decision. How practical would it be to obtain better information about the events for this decision situation?

12. Perform a sensitivity analysis about the probability of good weather for the forest fire management problem (Example 1). For what range of probabilities would it be better to postpone the burn, all else being equal?

13. Figure 6.18 shows a real-time fire execution decision on a Gifford Pinchot National Forest site in the state of Washington.

FIGURE 6.18 Real-time fire execution decision

Commitment Decision	Conditions at Time of Burn	Initiation Decision	Fire Results	Resource Value	Burn Cost	Escape Cost	Total
			Success 0.70	$22,500	$4,000	$ 0	$18,500
			Problems 0.20	22,500	6,000	0	16,500
	Optimal 0.70	Burn	Poor burn 0.09	11,250	4,000	0	7,250
			Escape 0.01	22,500	4,000	200,000	−181,500
			Success 0.40	22,500	4,000	0	18,500
		$10,750	Problems 0.30	22,500	6,000	0	16,500
		Burn	Poor burn 0.28	11,250	4,000	0	7,250
Mobilize	Weather marginal		Escape 0.02	22,500	4,000	200,000	−181,500
$12,632	0.20	−$1,000		0	1,000	0	−1,000
		Don't burn					
	Poor weather or other constraints			0	800	0	−800
$0	0.10	Don't burn					
Don't mobilize				0	0	0	0

Source: Figure from David Cohan, Stephen M. Haas, David L. Radloff, and Richard F. Yancik, "Using Fire in Forest Management: Decision Making under Uncertainty," *Interfaces*, Vol. 14, No. 5, September–October 1984, p. 17.

a. Determine the optimal decision strategy.

b. Perform a sensitivity analysis to determine weather conditions under which it would not be preferable to mobilize resources for the burn.

14. A car rental agency offers you the option of purchasing insurance. Insurance for a week would cost $70. A relatively minor accident might cost you $1,500.

a. Construct a payoff table for this decision problem. Show the decisions under the decision criteria discussed in this chapter.

b. What decision would you choose and why? Sketch your utility function for this decision.

c. Statistics show that the chance of a car renter being involved in an accident during one week is only 0.3 percent. What is the best expected value decision? Would this change your decision from part *b*?

15. Dean Kuroff started a business of rehabbing old homes. He recently purchased a circa-1800 Victorian mansion and converted it into a three-family residence. Recently, one of his tenants complained that the refrigerator was not working properly. As Dean's cash flow was not extensive, he was not excited about purchasing a new refrigerator. He is considering two other options: purchase a used refrigerator or repair the current unit. He can purchase a new one for $400, and it will easily last three years. If he repairs the current one, he estimates a repair cost of $150, but he also believes that there is only a 30 percent chance that it will last a full three years and he will end up purchasing a new one anyway. If he buys a used refrigerator for $200, he estimates that there is a .6 probability that it will last at least three years. If it breaks down, he will still have the option of repairing it for $150 or buying a new one. Develop a decision tree for this situation and determine Dean's optimal strategy.

16. Drilling decisions by oil and gas operators involve intensive capital expenditures made in an environment characterized by limited information and high risk. A well site is dry, wet, or gushing. Historically, 50 percent of all wells have been dry, 30 percent wet, and 20 percent gushing. The value (net of drilling costs) for each type of well is as follows:

Dry	($80,000)
Wet	$100,000
Gushing	$200,000

Wildcat operators often investigate oil prospects in areas where deposits are thought to exist by making geological and geophysical examinations of the area before obtaining a lease and drilling permit. This often includes recording shock waves from detonations by a seismograph and using a magnetometer to measure the intensity of the earth's magnetic effect to detect rock formations below the surface. The cost of doing such studies is approximately $15,000. Of course, one may choose to drill in a location based on "gut feel" and avoid the cost of the study. The geological and geophysical examination classify an area into one of three categories: no structure (NS), which is a bad sign; open structure (OS), which is an "OK" sign; and closed structure (CS), which is hopeful. The following table gives probabilities that the well will actually be dry, wet, or gushing based on the classification provided by the examination (in essence, the examination cannot accurately predict the actual event): If a study is conducted, history shows that 41 percent of locations show NS, 35 percent show OS, and 24 percent show CS as a result.

	Dry	Wet	Gushing
NS	.73	.22	.05
OS	.43	.34	.23
CS	.21	.38	.41

 a. Construct a decision tree of this problem that includes the decision of whether or not to perform the geological tests.

 b. What is the optimal decision under expected value when no experimentation is conducted?

 c. Find the overall optimal strategy by rolling back the tree.

17. Statewise Auto Parts sells a variety of automobile parts. Its highest-volume item is a headlight replacement bulb. Historical data suggest that monthly demand is normally distributed with mean 120 and standard deviation 30. Each bulb costs $9.50 from the manufacturer. An order costs $12, and Statewise estimates that its inventory carrying cost is 22 percent. Lead time is 7 days (assume a 28-day month).

 a. Calculate the economic order quantity (EOQ).

 b. The manager of Statewise is willing to allow the lead time demand to exceed the reorder point at most 5 percent of the time. Calculate the reorder point.

 c. Calculate the expected annual cost based on the lot size and reorder points found in parts *a* and *b*.

18. Suppose the manager at Statewise Auto Parts (see Problem 17) decides to use his clout with his current supplier. He intends to threaten to go to a new supplier unless the lead time for headlight bulbs is reduced to one day. How much would such a lead time reduction save Statewise?

19. Use the risk-taking utility function developed for the scenario in Example 8 and compute the expected utility of the decisions. What is the best? How does this decision compare with the average utility decision for the risk-averse utility function?

20. If Rick uses an exponential utility function with $R = 1,000$, find the best decision using the probabilities in Problem 19.

21. A college football team is trailing 14–0 late in the game. The team is getting close to making a touchdown. If they can score now, hold the opponent, and score one more time, they can tie or win the game. The coach is wondering whether to go for an extra-point kick or a two-point conversion now, and what to do if they can score again.

 a. Develop a decision tree for the coach's decision. Develop a utility function to represent the final score for each path in the tree.

 b. Estimate probabilities for successful kick or two-point conversions. (You might want to do this by some group brainstorming or by calling on experts, such as your school's coach or a sports journalist.) Using these probabilities and utilities from part a, determine the optimal strategy.

 c. Perform a sensitivity analysis on the probabilities to evaluate alternative strategies (such as when the starting kicker is injured).

22. One of the major decisions that most individuals eventually face is buying a house or condominium and securing a mortgage. A mortgage contract obligates the borrower to make a series of payments over time that will fully amortize the loan. Lenders have available several different types of mortgage instruments, such as fixed-rate (15, 20, or 30 years) and adjustable-rate mortgages (ARMs). The key decisions that borrowers face are the type of mortgage and length of time. ARMs typically have a lower interest rate than fixed-rate mortgages, but many borrowers

are reluctant to accept ARMs because of the uncertainty associated with the actual payments they will have to make in the future, particularly if interest rates rise. From the data in Table 6.3 (you may wish to use current data available from a bank or other lender), develop a decision tree to help a would-be borrower make a decision. Use the current Treasury bill index for the margin, and make whatever assumptions you deem appropriate. You might want to use a spreadsheet to compute present values of your cash flows to support your analysis.

30-year fixed rate: 8.875%

15-year fixed rate: 8.25%

1-year ARM:
Current rate: 5.5%
Maximum yearly increase: 2%
Lifetime increase: 6%

1-year ARM:
Current rate: 6.75%
Maximum yearly increase: 1%
Lifetime increase: 4%

3-year ARM:
Current rate: 7.5%
Maximum yearly increase: 2%
Lifetime increase: 6%
After 3 years, this becomes a 1-year ARM

5-year ARM:
Current rate: 8.125%
Maximum yearly increase: 2%
Lifetime increase: 5%
After 5 years, this becomes a 1-year ARM

TABLE 6.4
Alternative
home mortgage
contracts

23. Many automobile dealers are advertising lease options for new cars. Suppose that you are considering three alternatives:
   i.   Purchase the car outright (assuming that you have the cash on hand).
   ii. Purchase the car with 20 percent down and a 48-month loan.
   iii. Lease the car.

   Select an automobile whose leasing contract is advertised in your newspaper. Using current interest rates and advertised leasing arrangements, perform a decision analysis of these options. Make, but clearly define, any assumptions that may be required.

NOTES

1.   Bruce F. Baird, *Managerial Decisions under Uncertainty* (New York: John Wiley & Sons, 1989), 6; and Ralph L. Keeney, "Decision Analysis: An Overview," *Operations Research*, Vol. 30, No. 5, September–October 1982, 803–838.
2.   Nancy A. Nichols, "Scientific Management at Merck: An Interview with CFO Judy Lewent," *Harvard Business Review*, January–February 1994, 89–99.

3. Adapted from David Cohan, Stephen M. Haas, David L. Radloff, and Richard F. Yancik, "Using Fire in Forest Management: Decision Making under Uncertainty," *Interfaces*, Vol. 14, No. 5, September–October 1984, 8–19.

4. David G. Luenberger, *Investment Science* (New York: Oxford University Press, 1998).

5. Matt Siegel, "Guard Your Stocks against a Market Drop!" *Fortune*, August 3, 1998, 276–278.

6. The standard deviation for decision D_i is computed by $\sqrt{\Sigma P(S_j)(V(D_i,S_j) - EV_i^2)}$. Because the probabilities may weight the values differently, you cannot use the *STDEVP* function in Excel for this case.

BIBLIOGRAPHY

Balson, William E., Justin L. Welsh, and Donald S. Wilson. 1992. "Using Decision Analysis and Risk Analysis to Manage Utility Environmental Risk." *Interfaces*, Vol. 22, No. 6, November–December, pp. 126–139.

Bunn, Derek W. 1984. *Applied Decision Analysis*. New York: McGraw-Hill.

Byrd, Jack, Jr., and L. Ted Moore. 1982. *Decision Models for Management*. New York: McGraw-Hill.

Carter, Virgil, and Robert E. Machol. 1971. "Operations Research on Football." *Operations Research*, Vol. 19, No. 2, pp. 541–544.

———. 1978. "Optimal Strategies on Fourth Down." *Management Science*, Vol. 24, No. 16, December, pp. 1758–1762.

Clarke, John R. 1987. "The Application of Decision Analysis to Clinical Medicine." *Interfaces*, Vol. 17, No. 2, March–April, pp. 27–34.

Cohan, David, Stephen M. Haas, David L. Radloff, and Richard F. Yancik. 1984. "Using Fire in Forest Management: Decision Making under Uncertainty." *Interfaces*, Vol. 14, No. 5, September–October, pp. 8–19.

Engemann, Kurt J., and Holmes E. Miller. 1992. "Operations Risk Management at a Major Bank." *Interfaces*, Vol. 22, No. 6, November–December, pp. 140–149.

Feinstein, Charles D. 1990. "Deciding Whether to Test Student Athletes for Drug Use." *Interfaces*, Vol. 20, No. 3, May–June, pp. 80–87.

Heian, Betty C., and James R. Gale. 1988. "Mortgage Selection Using a Decision Tree Approach: An Extension." *Interfaces*, Vol. 18, No. 4, July–August, pp. 72–83.

Hosseini, Jinoos. 1986. "Decision Analysis and Its Application in the Choice between Two Wildcat Oil Ventures." *Interfaces*, Vol. 16, No. 2, March–April, pp. 75–85.

Janssen, C. T. L., and T. E. Daniel. 1984. "A Decision Theory Example in Football." *Decision Sciences*, Vol. 15, No. 2, Spring, pp. 253–259.

CHAPTER 7

Monte Carlo Simulation

OUTLINE

Applications of Monte Carlo Simulation
New Venture Planning
Pharmaceutical Research
Project Management

Building and Implementing Monte Carlo Simulation Models

Sampling from Probability Distributions

Building Simulation Models with Crystal Ball
Interpreting Crystal Ball Output

Statistical Issues in Monte Carlo Simulation

Monte Carlo Simulation Examples
Newsvendor Problem
Management Science Practice: Simulating a CD Portfolio
Pricing Stock Options

Crystal Ball *Tornado Chart Extender*

Optimization and Simulation

Problems

Appendix: Additional Crystal Ball Features
Correlated Assumptions
Freezing Assumptions
Overlay Charts
Trend Charts
Sensitivity Charts

As we have seen in the previous chapter, many decision problems involve random variables and risk. Although analytical models, expected value, and decision analysis techniques can be useful approaches for dealing with these types of models, they can be difficult to apply or may not present a complete picture of risk. Understanding risk and expressing it in meaningful terms is crucial to making good decisions.

A powerful technique for analyzing models involving probabilistic assumptions is **simulation.** Simulation models are logical descriptions of the interrelationships among elements in a decision problem, or the sequence of events that occur in a system over time. Simulation models can be used for a wide variety of problems; in fact, simulation is the approach used most often in management science.

Simulation approaches can be divided into two major types: **Monte Carlo simulation**—which is based on repeated sampling from probability distributions of model inputs to characterize the distributions of model outputs, and **systems simulation**—which models the dynamics and behavior of interacting elements of a system, such as a manufacturing facility or a service function like a call center or a bank.

The term "Monte Carlo simulation" was coined by scientists who worked on the development of the atom bomb; it is taken from the random behavior of casino games at Monte Carlo. By randomly selecting model inputs and evaluating the outcomes, we may construct a distribution of potential outcomes of key model variables along with their likelihood of occurrence. This provides an assessment of the risk associated with a set of decisions that analytical methods generally cannot capture. By manipulating key decision variables and evaluating the risks associated with them, managers can use a simulation model to help identify good decisions. Monte Carlo simulation using spreadsheets has become increasingly popular in recent years because of the availability of powerful spreadsheet add-ins. It has been particularly useful in financial planning to study capital investment decisions, cash flow analysis, and corporate budgeting. Monte Carlo simulation is the focus of this chapter.

In contrast, systems simulation tends to focus on operational issues in production and service systems. Examples include evaluating plant designs, technology improvements, different scheduling rules, and material handling systems; analyzing air traffic control systems, truck terminal operations, and railroad system performance; managing inventories; scheduling operating rooms and planning staffing schedules, and many others. The approaches used in systems simulation are distinctly different from Monte Carlo simulation, and require different concepts and software to capture the complex interactions and the time-dependent behavior of such systems. Thus, we will discuss this topic separately in the next chapter.

APPLICATIONS OF MONTE CARLO SIMULATION

In this section we present several applications of Monte Carlo simulation. Many of these applications use Crystal Ball—which we will employ in this chapter—as a tool to implement the simulations on a spreadsheet.

New Venture Planning

ExperCorp is a business-planning consulting firm located in Naperville, Illinois. The company specializes in new venture strategy and marketing research for entrepreneurs designing entries for the fitness, recreation, and sporting goods markets. A key to small business planning is the development of good risk and reward estimates. Following a determination of the size of the available target market, realistic assumptions for unit sales, realized selling price, production costs, and operating expenses in the first year of operation, ExperCorp wanted to develop a pro forma income statement. The objective was to create a profitability distribution of gross revenues and profits/losses.

The most critical and difficult aspect of venture planning is developing estimates of cash flow. Developing realistic statements for the first and subsequent years of operation along with determination of cash reserves is a task that all diligent planners face. The objective was to increase the precision of cash flow forecasts. An income statement template was developed to create appropriate probability distributions for unit sales, production costs, operating expenses, and profits/losses. Using the relevant data from first-year income statements, ExperCorp made assumptions about the percentage of receivables collected in 30, 60, and 90 days. Point estimates in the cash flow models were replaced with probability distributions of cash flows for key variables and used as inputs for Monte Carlo simulation analysis.

This approach increased the precision of forecasting under the conditions of uncertainty that new business planners typically confront. Rather than running the standard set of sensitivity analyses, they produced robust forecasts in less time with more accuracy. It gave ExperCorp's clients and their financial sponsors a better picture of their venture landscape (Decisioneering, Inc., 1998).

Pharmaceutical Research

DuPont Merck is a research-focused pharmaceutical company. Pharmaceutical research has led to numerous advances in medical therapy. However, such research by its nature involves a tremendous amount of risk and uncertainty. Often a thousand or more compounds are screened for each one that ultimately is nominated for development. On average, for every ten compounds that reach development, only one gains FDA approval. An important aspect of pharmaceutical development is selecting the proper portfolio of compounds to develop—a portfolio that will provide the expected level of sales at the right level of cost and risk. Monte Carlo simulations are used to model the key risks and uncertainties for each individual project. In addition, simulations are used to estimate total portfolio value and risk—a probability distribution of potential R&D spending and sales by year. Often, being the first or second to the market can have a significant impact on total sales realized. In certain high-potential markets, there may be ten or more competitors with compounds in development. This information helps focus attention on the need to achieve key project milestones to ensure a reasonable level of certainty that a compound will be first to market (Decisioneering, Inc., 1998).

Project Management

Boeing, Inc., is widely known as one of the world's leading manufacturers of commercial aircraft. Today, Boeing develops the key technologies, including software, behind state-of-the-art space navigation systems, space labs, and guided missile systems built by NASA, the U.S. Department of Defense, and other Boeing divisions. The development of any technology entails some degree of risk. For leading-edge software projects, that risk is significantly higher. The accuracy of a software project estimate can make the difference between profitability and nonprofitability, bidding and not bidding on a contract, or getting the project done right and coming up short. Boeing develops forecasts for software development costs and schedules based on factors such as the number of lines of code that a project is expected to require, the average cost rate for each line of code, and the software development mode being used. However, common project management tools provide only single point estimates of the worker hours required for a given project. Monte Carlo simulation is used to incorporate uncertainty in forecasts of project requirements, allowing the estimation of the probability of meeting or exceeding key targets (Decisioneering, Inc., 1998).

BUILDING AND IMPLEMENTING MONTE CARLO SIMULATION MODELS

Performing Monte Carlo simulation involves the following activities:

1. *Formulating the problem* to determine the objectives of the simulation study, the model inputs and outputs, and the scenarios to be evaluated. A *scenario* is any specification of the controllable inputs, that is, a particular variation of the problem that we wish to study. By comparing the performance of different scenarios, we expect to be able to draw conclusions as to which scenario is the best, if indeed there is a significant difference.

2. *Specifying probabilistic assumptions* needed to "drive" the simulation, that is, the probability distributions of input variables. This is perhaps the most critical task in building valid and useful models. In Chapter 2 we discussed fundamentals of common probability distributions, their properties, and some typical applications. We suggest that you review this material at this time. If historical data are available, we can use the data fitting techniques in Crystal Ball described in Chapter 2 to select distributions to drive a simulation. If not, then a distribution must be selected judgmentally. Often, a uniform or triangular distribution is used in the absence of any information.

3. *Implementing the model* to perform the calculations required to evaluate the distribution of model outputs. These calculations generally consist of
   a. sampling input values from the assumed probability distributions,
   b. using the samples to compute the output variables and recording the results,
   c. repeating steps (a) and (b) until a sufficient number of **trials**—replications of the experiment—have been performed to generate useful distributions of the outputs.

To illustrate the process of Monte Carlo simulation, we will use the profit model introduced in Chapter 1.

Example 1: Profitability Simulation

We will assume that the profit model, shown again in Figure 7.1, represents a financial model for a new product, for which historical data are unavailable. In this model, demand is likely to be uncertain. In addition, financial analysts are unsure of the exact values of the unit cost and the fixed cost if this project is undertaken. Therefore, the objective of the simulation is to better understand the risks associated with deciding to pursue this venture given the uncertainty in these uncontrollable inputs and decide the best quantity to produce to maximize the expected profit.

FIGURE 7.1
Profit model
spreadsheet

	A	B	C
1	**Profit Model**		Key cell formulas
2			
3	*Parameters and Uncontrollable Inputs*		
4			
5	Unit Price	$40.00	
6	Unit Cost	$24.00	
7	Fixed Cost	$400,000.00	
8	Demand	50000	
9			
10			
11	*Model*		
12			
13	Unit Price	$40.00	
14	Quantity Sold	40000	
15	Revenue		$1,600,000.00
16	Unit Cost	$24.00	
17	Quantity Produced	40000	
18	Variable Cost		$960,000.00
19	Fixed Cost		$400,000.00
20	Profit		$240,000.00

Because no historical data are available, the uncertain variables can only be estimated judgmentally. The best estimates are that demand may range from 40,000 to 60,000 units. Financial analysts believe that the unit costs might be $22, $23, $24, or $25, and the fixed cost might be either $400,000 or $450,000. We will assume that the demand follows roughly a triangular distribution, unit price is uniform, and the two possibilities for fixed costs have equal probabilities (summarized in Figure 7.2). These distributions were conveniently (!) chosen to allow us to sample from them using dice, cards, and a coin, respectively (later in this chapter we will discuss how to do this in a spreadsheet using Crystal Ball, as well as how to sample from more realistic distributions).

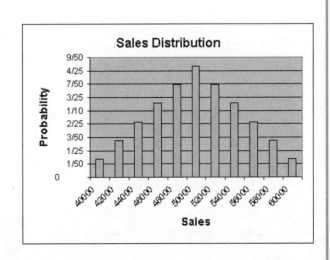

FIGURE 7.2
Probabilistic
assumptions
for the profit
model

Sales	Probability
40000	1/36
42000	1/18
44000	1/12
46000	1/9
48000	5/36
50000	1/6
52000	5/36
54000	1/9
56000	1/12
58000	1/18
60000	1/36

Unit Cost	Probability
$ 22.00	0.25
$ 23.00	0.25
$ 24.00	0.25
$ 25.00	0.25

Fixed Cost	Probability
$ 400,000	0.5
$ 450,000	0.5

The demand distribution corresponds to the probabilities of rolling a 2 through 12 on two dice. For example, if we roll a 2, the demand is 40,000; a 7 corresponds to 50,000, and a 9 corresponds to 54,000. Take a 2, 3, 4, and 5 from a deck of cards, shuffle these four cards, and select one. If a 2 is selected, the unit cost sampled is $22, a 3 corresponds to $23, and so on. Finally, let heads represent a fixed cost of $400,000 and tails represent $450,000.

To simulate one trial for this simulation requires us to roll a pair of dice, draw a card, and flip a coin. The corresponding outcomes can be substituted in the spreadsheet model, and the resulting profit can be recorded. For example, suppose we select a production quantity of 50,000 and obtain the following sequence of outcomes:

Dice	Card	Coin
4	5	H
8	2	H
7	3	T

These correspond to the following demands, unit cost, and fixed costs, with profit shown in the last column:

Demand	Unit Cost	Fixed Cost	Profit
44,000	$25	$400,000	$110,000
52,000	$22	$400,000	$500,000
50,000	$23	$450,000	$400,000

By repeating this process, we can generate enough outcomes to construct a distribution of profit for each potential value of the quantity produced.

Figure 7.3 shows a portion of the results for 50 trials of the simulation with a production quantity of 50,000. The average profit is $309,231 with a standard deviation of $134,317, yielding a coefficient of variation of .434 and a return-to-risk of 2.302. The frequency distribution and histogram in Figure 7.4 shows that profit might

FIGURE 7.3
Profit simulation and summary results

Trial	Demand	Qty. Sold	Revenue	Unit Cost	Var. Cost	Fixed Cost	Profit
1	54,000	50,000	$2,000,000	$ 24.00	$1,200,000	$ 400,000	$ 400,000
2	48,000	48,000	$1,920,000	$ 25.00	$1,250,000	$ 450,000	$ 220,000
3	54,000	50,000	$2,000,000	$ 23.00	$1,150,000	$ 400,000	$ 450,000
4	40,000	40,000	$1,600,000	$ 24.00	$1,200,000	$ 450,000	$ (50,000)
5	54,000	50,000	$2,000,000	$ 25.00	$1,250,000	$ 400,000	$ 350,000
6	52,000	50,000	$2,000,000	$ 24.00	$1,200,000	$ 400,000	$ 400,000
7	52,000	50,000	$2,000,000	$ 24.00	$1,200,000	$ 400,000	$ 400,000
8	54,000	50,000	$2,000,000	$ 25.00	$1,250,000	$ 450,000	$ 300,000
9	50,000	50,000	$2,000,000	$ 23.00	$1,150,000	$ 450,000	$ 400,000
10	46,000	46,000	$1,840,000	$ 22.00	$1,100,000	$ 400,000	$ 340,000
...	...	...	...	...	...	...	...
48	50,000	50,000	$2,000,000	$ 25.00	$1,250,000	$ 400,000	$ 350,000
49	56,000	50,000	$2,000,000	$ 25.00	$1,250,000	$ 450,000	$ 300,000
50	44,000	44,000	$1,760,000	$ 24.00	$1,200,000	$ 400,000	$ 160,000
Average	50,308	48,308	$1,932,308	$ 24.08	$1,203,846	$ 419,231	$ 309,231
Std. Dev.	4679.47	3146.0191	$ 125,841	$ 0.95	$ 47,704	$ 25,318	$ 134,317
C.V.							0.434
Return to risk							2.302

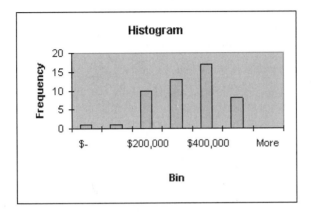

		Bin	Frequency
$	-	$ -	1
$	100,000	$ 100,000	1
$	200,000	$ 200,000	10
$	300,000	$ 300,000	13
$	400,000	$ 400,000	17
$	500,000	$ 500,000	8
		More	0

FIGURE 7.4
Summary results of profit simulation for quantity produced = 50,000, sample size = 50

range to over $400,000, with a fairly high probability of exceeding $200,000. While we can never know exactly what will happen, the simulation provides a profile (often called a **risk profile**) of what we might expect and an assessment of the risks involved with making the decision to purchase 50,000 units.

Excel Exercise 7.1

Modify the spreadsheet *profit model.xls* to include a table similar to Figure 7.3. Using dice, a coin, and cards, simulate 25 trials of this model, entering your values in the spreadsheet, and computing summary measures. How close do your results agree with Figure 7.3? Why do they differ?

To implement Monte Carlo simulation models on spreadsheets, we need to be able to sample from probability distributions and aggregate the results efficiently. Although this is possible within a general Excel environment using data tables and Excel's statistical analysis tools, it can be tedious. Fortunately, add-ins such as Crystal Ball and @Risk make this an easy task. @Risk, a product of the Palisade Corporation, is a spreadsheet add-in that provides similar features as Crystal Ball. We will use the student version of Crystal Ball,[1] which we introduced in Chapter 2 for distribution fitting, as the principal tool. The basic procedures for using Crystal Ball for Monte Carlo simulation are described later in this chapter; the appendix to this chapter provides some additional details and more advanced features of the software.

SAMPLING FROM PROBABILITY DISTRIBUTIONS

The engine that drives Monte Carlo simulation is the process of sampling from probability distributions. To gain a better understanding of how this is accomplished by Excel and Crystal Ball, we can propose another experiment. Suppose that we wish to sample from the unit cost distribution in Example 1. Take 100 business cards, numbered 1 through 100, and sort the cards into four groups. Cards numbered 1 through 25 correspond to a unit cost of $22; cards numbered 26 through 50 correspond to a unit cost of $23; cards numbered 51 through 75 correspond to a unit cost of $24; and finally, cards numbered 76 through 100 correspond to a unit cost of $25. We then shuffle the cards and draw one. All cards have an equal chance of being drawn, but 25 percent of the time we expect to draw a number corresponding to each of the outcomes we wish to generate. This experiment is more appealing from a quantitative perspective since we are dealing with numbers rather than four individual playing cards. Conceptually, it provides a means of sampling from *any* discrete probability distribution by simply grouping the numbers in proportion to the probabilities. The only thing we must ensure is that we can draw a number *uniformly*, that is, with equal probability from the entire population of possible numbers.

In simulation, probabilistic outcomes are generated using "random numbers." A **random number** is a number drawn from a uniform probability distribution between 0 and 1. Nearly every computer language and spreadsheet package has the ability to generate random numbers quite easily. In Excel, the function *RAND( )* generates a uniform random number between 0 and 1. Note that no arguments are included between the parentheses.

Excel Exercise 7.2

Copy the function *RAND()* into 100 cells of a spreadsheet. Use the *Histogram* tool in the Analysis Toolpak to create a histogram with 10 equally spaced cells between 0 and 1 for these data. Do the random numbers appear to be uniform? Press the F9 key several times to examine how the distribution changes as new random numbers are generated.

To use random numbers to simulate outcomes from a probability distribution for the unit cost, we first find the cumulative probability distribution and then assign random number intervals according to this distribution.

Example 2: Simulating Outcomes from a Discrete Distribution

The cumulative distribution of the unit costs is shown next, along with the corresponding random number interval.

Unit Cost	Probability	Cumulative Probability	Random Number Interval
$22	.25	.25	0–.25
$23	.25	.50	.25–.50
$24	.25	.75	.50–.75
$25	.25	1.00	.75–1.00

We generate a random number and determine the interval into which it falls. Thus, if a random number falls in the interval 0 and 0.25 (not including the upper limit), the sample outcome is \$22; if it falls between 0.25 and 0.50, the sample outcome is \$23, and so on. We may draw samples from discrete distributions in Excel by using the *VLOOKUP* function (see the Appendix to Chapter 1) as illustrated in Figure 7.5. Although only the first column of the lookup table is actually used in the function computations, the full random number intervals are listed to enhance understanding.

While this approach can be used for any discrete distribution, many simulation applications require the generation of sample outcomes (technically called **random variates**) from other probability distributions. We will present some formulas for some common distributions. While the formulas might seem like black magic, they actually derive from the same principle used in the discrete case. That is, we set the cumulative probability distribution function equal to a random number (since any cumulative probability must be between 0 and 1), and solve for the value of the random variate, either analytically or numerically. The details are beyond the scope of this book, but they can be found in many books devoted exclusively to simulation and statistics. Essentially, these formulas transform random numbers to random variates. In each of the following, R represents a random number.

Uniform Distribution. To generate uniform random variates over the interval $[a, b]$, use the formula

$$U = a + (b - a)R$$

This makes sense if you observe that if $R = 0$, $U = a$, and if $R = 1$, $U = b$. For any random number between 0 and 1, the formula takes a proportional amount of the interval.

Normal Distribution. To generate a normal random variate x with mean μ and standard deviation σ, use the Excel function:

NORMINV(RAND(),mean, standard_deviation)

In this case, *RAND()* represents a random value for the cumulative area between negative infinity and the random variate x under the normal distribution. The *NORMINV* function in Excel finds the value of x that corresponds to a cumulative probability of *RAND()*.

FIGURE 7.5

Sampling from a discrete distribution in Excel

	A	B	C	D	E	F
1	Sampling from a Discrete Distribution					
2						
3	Random Number Range		Unit Cost			
4	0.00	0.25	22			
5	0.25	0.50	23			
6	0.50	0.75	24			
7	0.75	1.00	25			
8						
9	Trial	Sample				
10	1	24	=VLOOKUP(RAND(),\$A\$4:\$C\$7,3)			
11	2	25				
12	3	22				
13	4	25				
14	5	23				
15	6	22				
16	7	24				
17	8	24				
18	9	24				
19	10	23				

Exponential Distribution. To generate a random variate from an exponential distribution having a mean μ, we multiply the negative of the mean by the natural logarithm of a random number:

$$E = -\mu \ln(R)$$

Triangular Distribution. For a triangular random variate T with a lower limit a, mode m, and upper limit b, first generate a random number R. Then,

$$\text{If } R \leq \frac{m-a}{b-a}, \text{ then } T = a + (b-a)\sqrt{R(m-a)/(b-a)}$$

$$\text{If } R > \frac{m-a}{b-a}, \text{ then } T = a + (b-a)\left[1 - \sqrt{(1-R)\left(1-\frac{m-a}{b-a}\right)}\right]$$

Crystal Ball provides a collection of functions that generate random variates and can be used as ordinary Excel functions in spreadsheets. For example, *CB.Normal (mean, standard_deviation)* generates a normal random variate. A list of these can be found by selecting a blank spreadsheet cell, clicking on the *Paste Function* button [*fx*] on the Excel toolbar, and going to the Crystal Ball category.

Excel Exercise 7.3

Copy the function *CB.Normal(0,1)* to a matrix of 500 cells in a spreadsheet. Create a histogram of the data. Does the histogram look like a standard normal distribution? Press the F9 key to examine random variation in the samples.

BUILDING SIMULATION MODELS WITH CRYSTAL BALL[2]

Crystal Ball automates the complex tasks required in Monte Carlo simulations, such as generating random variates, replicating the spreadsheet, aggregating results, and computing statistics. When Crystal Ball is loaded, the Excel screen will have two new menu items (*Cell* and *Run*) and an additional toolbar, as shown in Figure 7.6. The key toolbar buttons and menu location are described next:

- *Define Assumption* (*Cell* menu)—define a probability distribution for a spreadsheet cell value
- *Define Forecast* (*Cell* menu)—define a cell as a simulation output for which statistics will be generated

FIGURE 7.6
Crystal Ball
toolbar

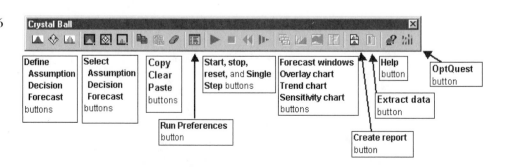

- *Select Assumptions* (*Cell* menu)—show all assumption cells in the worksheet
- *Select Forecasts* (*Cell* menu)—show all forecast cells in the worksheet
- *Run Preferences* (*Run* menu)—choose run preferences, such as number of trials, sampling method, and other features
- *Start Simulation* (*Run* menu)—start simulation run
- *Stop Simulation* (*Run* menu)—stop simulation run
- *Single Step* (*Run* menu)—perform a single trial of the simulation
- *Create Report* (*Run* menu)—generate a report of output statistics based on the simulation run

Crystal Ball requires the specification of two sets of cells in a spreadsheet model:

1. *Assumption Cells*—cells that represent uncertain inputs in the simulation model, and that are defined by some probability distribution. In the profit model in Figure 7.1 we will define cells B6, B7, and B8 as assumption cells.
2. *Forecast Cells*—cells that represent the outputs of the simulation model. In the profit model, cell C20 is the only forecast cell.

Selecting *Define Assumption* will invoke the Crystal Ball *Distribution Gallery*, which will allow you to select the appropriate probability distribution and parameter settings. A variety of known distributions are available as well as the capability of defining a distribution from empirical data. The *Distribution Gallery* is shown in Figure 7.7. Only the following distributions are available in the student version: normal, triangular, Poisson, uniform, exponential, and custom.

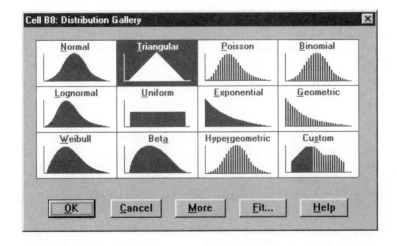

FIGURE 7.7
Crystal Ball Distribution gallery

Example 3: Defining an Assumption in Crystal Ball

Let us define the distribution for the demand in the profit model. We will assume that the distribution is a triangular with a minimum value of 40,000, likeliest value of 50,000, and maximum value of 60,000. (By assuming the distribution to be continuous we allow a larger range of possible values, instead of only those corresponding to dice rolls!) First click on cell B8. Then select *Define Assumption*. From the distribution gallery, select the triangular distribution and then click the *OK* button. In the dialog box that is displayed (Figure 7.8), enter the minimum value, likeliest value, and maximum value of the distribution. Clicking on *Enter* fixes these values

FIGURE 7.8

Triangular
distribution
assumption
dialog box

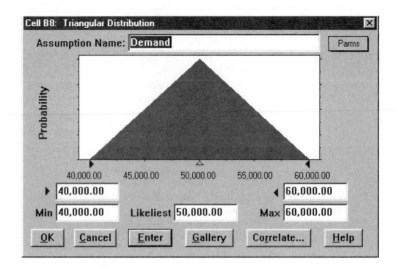

and rescales the picture to allow you to see what the distribution looks like. Clicking on *OK* accepts your choice and returns to the main screen. You may change the appearance of assumption cells in the spreadsheet (for example, with color or shading) using the *Cell Preferences* option in the *Cell* menu.

In Figure 7.8 you may notice two small black triangles and one open diamond just below the horizontal axis. These are called *grabbers* because you may click and hold the left mouse button on them and move them to change the distribution. For the triangular distribution shown, moving the left or right triangle grabber will change the minimum and maximum values; moving the center diamond will change the likeliest value. You should experiment with moving the grabbers for this and other distributions to see how they work.

We repeat this process for each of the probabilistic assumptions in the model. Figure 7.9 shows the dialog box for a uniform distribution assumption for unit cost, and Figure 7.10 shows a custom distribution for the fixed cost, which we still assume

FIGURE 7.9

Uniform
distribution
assumption
dialog box

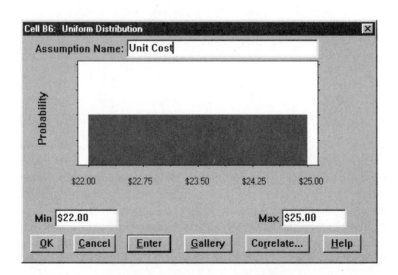

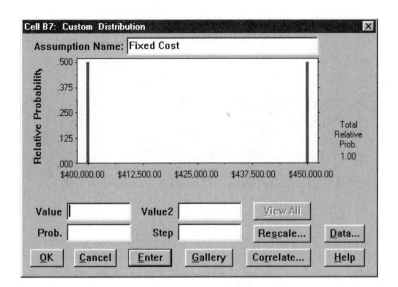

FIGURE 7.10

Custom distribution assumption dialog box

will be either \$400,000 or \$450,000. To create this distribution, enter each value in the *Value* box and the probability in the *Prob.* box and click *Enter*. Repeat for all discrete values of the distribution. The *Value2* box allows you to specify a uniform range having a fixed probability, for example, a range of \$390,000 to \$410,000 with probability 0.5.

After all assumptions are defined, you must define one or more forecast cells that define the output variables of interest. The student version is limited to six forecast cells. The output we are interested in is cell C20, profit. However, we would like to evaluate the profit for different values of the production quantity. Rather than simulate each case separately, we may construct a data table as shown in Figure 7.11, and define each of the data table values as forecast cells. In Figure 7.11, highlight the range of cells E4:E7, and select *Define Forecast*. Crystal Ball prompts you for the name of each cell and unit of measure (see Figure 7.12). If *Display Window Automatically* is checked, you will see the output distribution being built as the simulation is performed by Crystal Ball.

Next, select *Run Preferences*. This dialog box allows you to choose the number of trials (limited to 1,000 in the student version) and select other options such as

	A	B	C	D	E
1	**Profit Model**		Key cell formulas		
2				Forecast Cell Data Table	
3	*Parameters and Uncontrollable Inputs*			Production	\$240,000.00
4				30000	\$ 80,000.00
5	Unit Price	\$40.00		40000	\$240,000.00
6	Unit Cost	\$24.00		50000	\$400,000.00
7	Fixed Cost	\$400,000.00		60000	\$160,000.00
8	Demand	50000			
9					
10					
11	*Model*				
12					
13	Unit Price	\$40.00			
14	Quantity Sold	40000			
15	Revenue		\$1,600,000.00		
16	Unit Cost	\$24.00			
17	Quantity Produced	40000			
18	Variable Cost		\$960,000.00		
19	Fixed Cost		\$400,000.00		
20	Profit		\$240,000.00		

FIGURE 7.11

Data table for evaluating multiple forecast cells with Crystal Ball

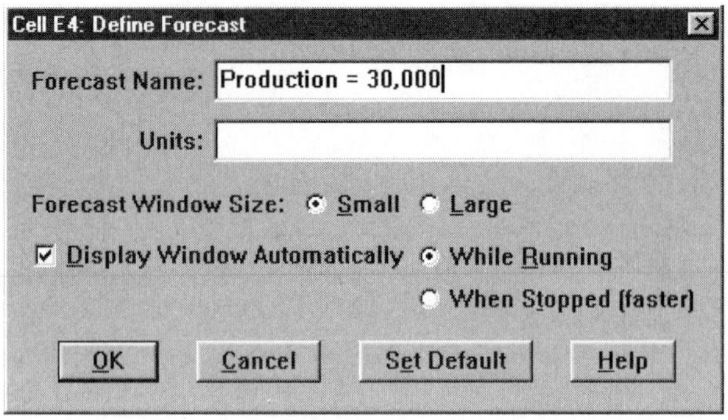

FIGURE 7.12
Define forecast
dialog box

controlling the random number generation process. If you click on the *Sampling* button, you may set the *initial seed value*. This seed value is used to start the process of generating random numbers. By clicking on *Use Same Sequence of Random Numbers* and using the same seed value, you can generate the same sequence of random numbers over again. This is sometimes useful, for example, in debugging models or in comparing different decisions (to make sure that the variation in results is from the different decisions and not because of differences in random numbers). You may also choose between two sampling methods: Monte Carlo or Latin Hypercube. Monte Carlo sampling generates random variates randomly over the entire range of possible values. With Latin Hypercube sampling, an assumption's probability distribution is divided into intervals of equal probability, and Crystal Ball generates an assumption value for each interval. The number of intervals is determined by the *Minimum Sample Size* option in the *Run Preferences* dialog box. Latin Hypercube sampling is more precise because it samples the entire range of the distribution in a more consistent manner. However, it requires additional memory requirements.

   If you select the *Options* button, another dialog box opens, and you may choose to have the assumption cells retain their original values or the estimated means from sampling after the simulation is completed. The latter option is useful if you wish to perform additional spreadsheet calculations using the estimated means after the simulation is completed and to leave the assumption cells at their original values. The *Speed* and *Macro* buttons have options for advanced users, and generally should be left alone. Finally, select *Run* and watch Crystal Ball perform!

Excel Exercise 7.4

Open the spreadsheet *profit model CB.xls* after Crystal Ball is loaded. Verify the assumptions of each uncertain variable and the run preferences and options described above. Then run the simulation.

Interpreting Crystal Ball Output

The principal outputs provided by Crystal Ball are the *forecast chart, percentiles summary,* and *statistics summary*. Figure 7.13 shows the forecast chart for the profit corresponding to a production quantity of 60,000 after 1,000 replications. The default forecast chart is a histogram of the outcome variable that includes all values within 2.6 standard deviations of the mean, which represents approximately 99 percent of the data. (This may be

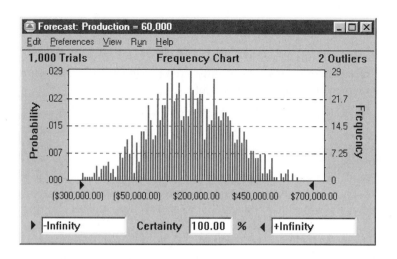

FIGURE 7.13

Forecast chart for cell E7

changed in the *Preferences/Choose Display Range* menu to include all the observations.) Just below the horizontal axis at the extremes of the distribution are two small triangles, called *end-point grabbers*. The range values of the variables at these positions are given in the boxes at the bottom left and right corners of the chart. The percentage of data values between the grabbers is displayed in the *Certainty* box at the lower center of the chart.

Questions involving risk can be answered by manipulating the end-point grabbers or by changing the range and certainty values in the boxes. Several options exist.

1. You may move an end-point grabber by clicking and holding the left mouse button on the grabber and moving it. As you do, the distribution outside of the middle range changes color, the range value corresponding to the grabber changes to reflect its current position, and the certainty level changes to reflect the new percentage between the grabbers. Figure 7.14 shows the result of moving the left grabber to the value $300,000. The dark portion of the histogram represents 24.1 percent of the distribution (as given in the *Certainty* box).
2. You may type specific values in the range boxes. When you do, the grabbers automatically move to the appropriate positions and the certainty level changes to reflect the new percentage of values between the range values. For example,

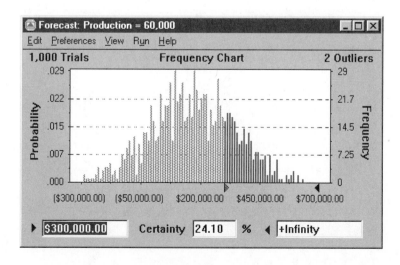

FIGURE 7.14

Finding the probability that profit exceeds $300,000 on the forecast chart suppose you wanted to determine the probability of realizing a positive profit. If you enter 0 in the left range box, the grabber will automatically move to that position, and the certainty level will change to reflect the percentage of the distribution between these values. This is illustrated in Figure 7.15, which shows an 83.3 percent chance of having a positive profit.

3. You may specify a certainty level. If the end-point grabbers are free (as indicated by a black color), the certainty range will be centered around the mean (or median, as specified in the *Preferences/Statistics* menu option). For example, Figure 7.16 shows the result of changing the certainty level to 90 percent. The range centered about the mean is from $122,470 to $448,327. Note that this is not a confidence interval, but simply a probability interval. You may anchor an end-point grabber by clicking on it. When anchored, the grabber will be a lighter color. (To free an anchored grabber, click anywhere in the chart area.) If a grabber is anchored and you specify a certainty level, the free grabber moves to a position corresponding to this level. Finally, you may move each grabber to the other end to determine certainty levels for the ends of the distribution.

FIGURE 7.15
Probability of a
positive profit

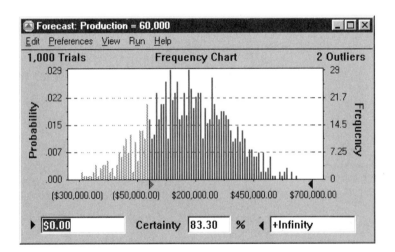

FIGURE 7.16
90 percent
certainty range
for profit

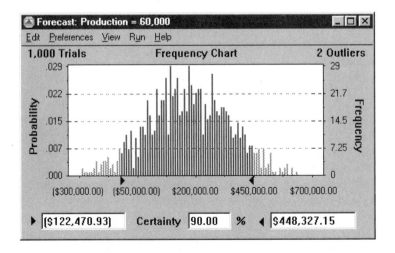

The forecast chart may be customized to change its appearance through the *Preferences...Chart Preferences* dialog box. The chart may be displayed as a column (bar) chart (default), area, or outline; and may be displayed as a frequency distribution (the default), cumulative distribution, or reverse cumulative distribution. The number of groups determines the granularity of the chart; a smaller number of groups provides less detail. The number format may be changed in the *Preferences...Format Preferences* dialog box.

The *percentiles chart* can be displayed from the *View* menu. An example is shown in Figure 7.17. This provides cumulative distribution values of the results. From the *View* menu you may also select a *statistics report*. This report, shown in Figure 7.18, provides a summary of key descriptive statistical measures (which we reviewed in Chapter 2).

Forecast: Production = 60,000

Edit   Preferences   View   Run   Help

Cell E7   Percentiles

Percentile	Value
0%	($329,944.96)
10%	($54,396.51)
20%	$25,245.16
30%	$76,184.20
40%	$120,844.61
50%	$170,910.64
60%	$215,930.50
70%	$271,806.69
80%	$323,411.31
90%	$396,973.59
100%	$630,381.40

FIGURE 7.17
Percentiles chart

Forecast: Production = 60,000

Edit   Preferences   View   Run   Help

Cell E7   Statistics

Statistic	Value
Trials	1,000
Mean	$170,654.94
Median	$170,910.64
Mode	---
Standard Deviation	$174,014.43
Variance	$30,281,020,269.14
Skewness	-0.06
Kurtosis	2.70
Coeff. of Variability	1.02
Range Minimum	($329,944.96)
Range Maximum	$630,381.40
Range Width	$960,326.37
Mean Std. Error	$5,502.82

FIGURE 7.18
Statistics report (1,000 trials)

Customized reports can be created from the *Run* menu by choosing *Create Report*. This option allows you to select a summary of assumptions and output information that we described. These are created in a separate Excel worksheet and may be printed.

Excel Exercise 7.5

Open the Excel file *profit model CB.xls* (Crystal Ball must be running). Run the simulation and compare the results of the four levels of production. Compute a return-to-risk metric for each and make a recommendation.

STATISTICAL ISSUES IN
MONTE CARLO SIMULATION

Because simulation is simply a sampling experiment, we need to be able to characterize the population of outcomes in order to draw statistical conclusions about the results. The random numbers used in a simulation are a *sample* from an infinite population of possible random number sequences. Hence, the output of a simulation model is also a sample from some infinite population of outcomes. Each time we run the simulation with a different set of random numbers, we would expect different results.

We may characterize the variability we can expect from a simulation by the *standard error of the mean*. The standard error is the standard deviation of the *sampling distribution* of the mean. That is, nearly all means for a simulation of n trials will fall within three standard errors of the true population mean. The standard error of the mean for a simulation with n trials is $s/\sqrt{n}$, where s is the sample standard deviation of the individual observations. For example, the standard error associated with the results in Figure 7.18 is 5,502.82, which is the standard deviation, 174,014.43 divided by the square root of 1,000.

From statistics, we may compute a confidence interval for the population mean by

$$\bar{x} \pm t_{n-1,\alpha/2}(s.e.)$$

In this formula, $\bar{x}$ represents the sample average and *s.e.* is the standard error. The value for the *t*-distribution can be found using the Excel function *TINV(two-tailed probability, degrees of freedom)*. For a large number of trials (> 30), we can use *z*-values from a normal distribution. Thus, a 95 percent confidence interval would be

$$\bar{x} \pm 1.96\, s.e.$$

Example 4: A Confidence Interval for Mean Profit

From the Crystal Ball statistics report in Figure 7.18, we find that the mean profit is $170,654.94, with a standard error of $5,502.82. Therefore, a 95 percent confidence interval for the mean profit for a production quantity of 60,000 is

$$\$170,654.94 \pm 1.96(\$5,502.82) = [\$159,871, \$181,439]$$

Excel Exercise 7.6

Run the profit model five times using Crystal Ball, each for 1,000 trials, each time changing the seed value in the *Sampling* dialog box of *Run Preferences*. Record the means, and determine if they fall within this confidence interval.

From statistics you should recall that as the sample size increases (for a single replication), the estimate of the mean becomes more precise. Small sample sizes have large standard errors, making it difficult to estimate the true population mean with a high level of confidence. For example, Figure 7.19 shows the results of the profit model simulation for only 250 trials. Note that the standard error is much larger that that obtained for 1,000 trials. Thus, to reduce the size of a confidence interval and obtain a better estimate of the true population mean, we can run the simulation for a larger number of trials.

We can use the information from a simulation run to determine the sample size required to estimate the population mean with a specified precision. For example, if we want the half-width of the confidence interval to be E, we solve the following equation for n:

$$E = 1.96\, s/\sqrt{n}$$

yielding

$$n = (1.96)^2 s^2/E^2$$

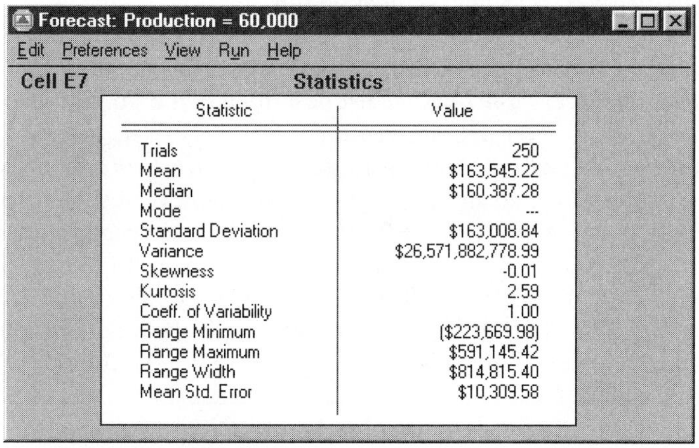

Statistic	Value
Trials	250
Mean	$163,545.22
Median	$160,387.28
Mode	---
Standard Deviation	$163,008.84
Variance	$26,571,882,778.99
Skewness	-0.01
Kurtosis	2.59
Coeff. of Variability	1.00
Range Minimum	($223,669.98)
Range Maximum	$591,145.42
Range Width	$814,815.40
Mean Std. Error	$10,309.58

FIGURE 7.19
Statistics report (250 trials)

Example 5: Computing a Sample Size

In the profit model, if we want the half-width of the confidence interval to be no more than $E = 5{,}000$, then, using the results from the 1,000-trial simulation, we have $s = 174{,}014$, and therefore, the minimum sample size required would be

$$n = (1.96)^2(174{,}014)^2/(5{,}000)^2 = 4{,}653$$

If the new confidence interval does not meet the required precision, it is because the standard deviation used in the formula was not precise enough, due to sampling error. In this case, we simply repeat the procedure with the new value of the standard deviation.

MONTE CARLO SIMULATION EXAMPLES

Monte Carlo simulation can be applied to many different types of problems (one practical example is described in the Management Science Practice box about CD portfolios). In this section we present two examples of using Crystal Ball, one for a variation of the newsvendor problem, introduced in the previous chapter, and a second involving the pricing of stock options.

Newsvendor Problem

In Chapter 6 we discussed the application of decision analysis to the newsvendor problem, assuming a discrete probability distribution of demand. Monte Carlo simulation provides an easier approach to analyzing this problem, particularly with respect to changes in the distributional assumptions. This example also shows how to use data tables to evaluate alternatives and how to use special Crystal Ball functions to create customized output reports.

Figure 7.20 shows a spreadsheet for the newsvendor problem we described in Chapter 6. The model in columns A and B is the same as in Figure 6.8. Cell B6 represents the demand. We have assumed that demand is triangular, with a minimum of 40, likely value of 47, and maximum value of 50. This assumption can easily be changed using the *Define Assumption* command in Crystal Ball. In columns D and E we constructed a data table to evaluate the profit for selected order quantities. Each entry in the data table (cells E3:E8) is defined as a forecast cell in Crystal Ball (the student version is limited to six forecast cells). Thus, we can obtain results for multiple decision alternatives in a single simulation run.

Crystal Ball has available a variety of functions for accessing information used in the simulation process or in the output reports (which can be found in the Crystal Ball category in the Excel *Paste Function* list described earlier in this chapter). These are particularly useful if you wish to create Excel Visual Basic macros to automate particular aspects of an analysis. One useful function is *CB.GetForeStatFN(ForeReference, Index)*. *ForeReference* is a valid forecast cell, and *Index* corresponds to one of the statistics in the *Statistics Report* (Figure 7.18). For instance, setting Index = 1 returns the number of trials; Index = 2 returns the mean, and so on. In Figure 7.20, the entry in cell G3 is = CB.GetForeStatFN(E3,2) and the entry in cell H3 is = CB.GetForeStatFN(E3,5). This allows you to automatically report output statistics within your spreadsheet without having to copy them from the forecast windows or a separate Crystal Ball report.

Figure 7.21 shows the forecast chart for the best order quantity $X = 46$. You can see that the risk of obtaining low-profit values is quite small. Figures 7.22 and 7.23 show the results for an alternative scenario, in which the likeliest value for the triangular demand assumption is set to 43, skewing the distribution to the right. In this case, the risk increases if the same order quantity is chosen. In this fashion, Monte Carlo simulation allows you to easily investigate the implications of errors or poor estimates of distributional assumptions of uncertain model inputs.

FIGURE 7.20
Spreadsheet for Crystal Ball analysis of newsvendor problem (base case)

	A	B	C	D	E	F	G	H	I
1	**Newsvendor Problem**			**Data Table**			**Forecast Cell Statistics**		
2				*Quantity*	$ 6.45		Mean	Std. Dev.	Return to risk
3	*Selling price*	$0.35		40	$ 6.00		$ 6.00	$ -	#DIV/0!
4	*Cost*	$0.20		42	$ 6.30		$ 6.29	$ 0.06	$ 96.95
5	*Discount price*	$0.05		44	$ 6.60		$ 6.50	$ 0.23	$ 28.81
6	*Demand*	44.894		46	$ 6.57		$ 6.59	$ 0.44	$ 15.07
7	*Order Quantity*	43		48	$ 6.27		$ 6.47	$ 0.60	$ 10.74
8				50	$ 5.97		$ 6.21	$ 0.65	$ 9.62
9	*Profit*	$6.45							
10									
11									
12	*Key cell formulas*								

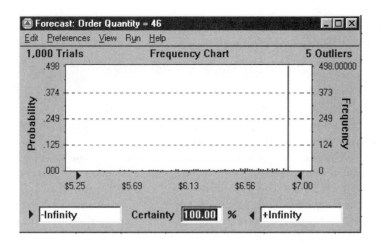

FIGURE 7.21

Forecast chart for order quantity = 46 (base case)

	A	B	C	D	E	F	G	H	I
1	**Newsvendor Problem**			**Data Table**			**Forecast Cell Statistics**		
2				*Quantity*	$ 6.45		Mean	Std. Dev.	Return to risk
3	*Selling price*	$0.35		40	$ 6.00		$ 6.00	$ -	#DIV/0!
4	*Cost*	$0.20		42	$ 6.30		$ 6.28	$ 0.08	$ 75.59
5	*Discount price*	$0.05		44	$ 6.60		$ 6.40	$ 0.28	$ 22.74
6	*Demand*	44.894		46	$ 6.57		$ 6.31	$ 0.48	$ 13.22
7	*Order Quantity*	43		48	$ 6.27		$ 6.09	$ 0.59	$ 10.32
8				50	$ 5.97		$ 5.80	$ 0.61	$ 9.50
9	*Profit*	$6.45							

FIGURE 7.22

Newsvendor results for alternative scenario

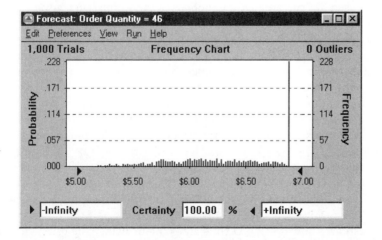

FIGURE 7.23

Forecast chart for order quantity = 46 (alternate scenario)

Excel Exercise 7.7

Open the file *newsvendor CB.xls*. Run the simulation for scenarios in which demand assumptions are (1) uniform between 40 and 50; (2) normal, with a mean of 45 and standard deviation of 5; and (3) normal, with a mean of 45 and standard deviation of 10. Compare your results with the two scenarios described above.

Management Science Practice: Simulating a CD Portfolio

At the First National Bank and Trust Company, managers wanted to assess the impact of various interest-rate scenarios of certificates of deposit (CDs) on portfolio yield. The mean portfolio yield reflects the cost to the bank of offering the CDs. By calculating the mean portfolio yield under a variety of future interest-rate scenarios, the bank can more effectively estimate risk and plan for profitability. The specific statistic calculated in the simulation analysis is mean dollar weighted yield, where the interest rates of larger CDs are weighted more heavily in calculating the overall portfolio yield.

The "jumbo" CD portfolio of the bank was composed primarily of corporate CDs with a minimum investment of $100,000. The total portfolio consisted of approximately $270 million. Typically 10 to 30 CDs mature on any given day and are "rolled over" for a specified length of time. This maturity period depends on two factors: the cash flow requirements of the corporation (which buys the CD) and anticipated return on investment. The bank has no way of knowing what maturity periods will be chosen for rollovers.

To determine the probability distribution of maturity periods, data were collected on CD renewals for three months. The data consisted of the dollar amount, rate, and time to maturity of all CDs renewing on a given day. Table 7.1 shows a distribution of 908 observations. This distribution, which follows no common probability distribution (notice the spikes at 14, 30, 60, 90, and 120 days), was used to generate maturity periods in the simulation analysis.

In constructing the simulation model, several assumptions were made. Since 95 percent of the CDs are renewed, a 100 percent renewal rate was assumed to simplify the analysis. Also, all CDs were assumed to roll over at their original dollar value. Finally, interest rates were assumed to be a function of maturity period and are input to the model.

The final input is the CD portfolio itself. Aggregated data were used instead of updated data on each individual CD, because the portfolio manager did not have easy access to complete data for individual CDs, but received aggregate daily reports.

In using the model, the bank simulated various interest rate scenarios, summarized in Table 7.2. Table 7.3 shows the results of a 60-day simulation of 20 replications. The simulation can help predict the impact on the spread between prime earning rates and CD funding costs. Also, if the portfolio manager can estimate how much CD costs are going to change, he can plan a more effective strategy for hedging against these costs in the CD futures market. From Table 7.2, the manager concluded that a half-point rise in interest rates after 30 days would cause the cost of the portfolio to increase by 33 basis points. Similarly, a half-point decline after 30 days would cause the cost of the portfolio to decrease 32.6 basis points. The cost of offering more competitive rates (Scenario A) for longer-term CDs appears to cost 13 basis points relative to the level approach of Scenario B.

continued

TABLE 7.1 Distribution of three months' data on maturity periods

Maturity Period (Days)	Probability	Maturity Period (Days)	Probability	Maturity Period (Days)	Probability
7	0.119	33	0.013	69	0.001
8	0.035	34	0.002	70	0.001
9	0.004	35	0.005	77	0.001
10	0.010	36	0.001	79	0.001
11	0.015	38	0.001	87	0.001
12	0.022	40	0.002	89	0.001
13	0.011	41	0.001	90	0.032
14	0.206	42	0.001	91	0.012
15	0.032	43	0.002	92	0.004
16	0.003	44	0.001	98	0.001
17	0.014	45	0.004	99	0.001
18	0.020	46	0.001	112	0.001
19	0.003	47	0.001	116	0.001
20	0.007	49	0.001	119	0.003
21	0.009	50	0.001	120	0.009
22	0.002	52	0.001	122	0.001
23	0.002	53	0.001	123	0.002
24	0.004	55	0.002	133	0.001
25	0.001	57	0.001	147	0.001
27	0.005	60	0.044	180	0.003
28	0.032	61	0.012	181	0.003
29	0.007	62	0.005	182	0.003
30	0.152	63	0.004	184	0.001
31	0.055	64	0.001	272	0.001
32	0.024	65	0.002		

TABLE 7.2 Interest-rate scenarios

Scenario	Days to Maturity							
	7	14	30	60	90	120	150	180 (or more)
A (1 to 60 days)	9.85	10.00	10.15	10.30	10.45	10.60	10.75	10.75
B (1 to 60 days)	9.85	10.00	10.15	10.15	10.15	10.15	10.15	10.15
C (1 to 30 days)	9.85	10.00	10.15	10.30	10.45	10.60	10.75	10.75
(31 to 60 days)	10.35	10.50	10.65	10.80	10.95	11.10	11.25	11.25
D (1 to 30 days)	9.85	10.00	10.15	10.30	10.45	10.60	10.75	10.75
(31 to 60 days)	9.35	9.50	9.65	9.80	9.95	10.10	10.25	10.25
E	Same as C, except issue $42 million of new 60-day CDs at 10.30 percent. This block is in addition to existing portfolio.							

TABLE 7.3 Simulation results for CD portfolio

Scenario	Mean Yield	Standard Deviation	Minimum Yield	Maximum Yield
A	10.25	0.0407	10.16	10.32
B	10.12	0.0124	10.08	10.13
C	10.58	0.0399	10.53	10.65
D	9.92	0.0729	9.78	10.11
E	10.54	0.0346	10.50	10.60

Source: Adapted from Robert A. Russell and Regina Hickle, "Simulation of a CD Portfolio," *Interfaces*, Vol. 16, No. 3, May–June 1986, pp. 49–54.

Pricing Stock Options

We discussed decisions involving stock options in Chapter 6. In this example, we illustrate how to use simulation to price European call options. A European call option gives you the right to purchase a given stock at a given price (the strike price) only on the expiration date of the option (whereas an American option allows you to purchase it any time prior to the expiration date). Call options trade approximately at their net present values. We may use simulation to estimate the net present value of such an option. First we need a method of simulating future stock prices.

Finance researchers have developed various models for simulating the movement of stock prices. One such model [3] is

$$P_{k+1} = P_k [1 + \mu t + z \sigma \sqrt{t}]$$

where

P_k = stock price in time period k

P_{k+1} = stock price in time period $k + 1$

v = expected annual growth rate

σ = annual standard deviation

$\mu = v + 0.5\sigma^2$

t = time period interval (expressed in years)

z is a standard normal random variable (mean 0, standard deviation 1)

To simulate stock prices, all we need is to determine the initial price P_0, and then apply this formula repeatedly as the time period is increased. Figure 7.24 shows a portion of a spreadsheet designed to simulate stock prices for a year over 1-week intervals. We have assumed a 20 percent annual growth rate, a 25 percent annual standard deviation, and a $50 initial price. The actual simulation is shown in columns D and E, applying the formula just given.

The spreadsheet allows you to enter any strike price and expiration date (up to 52 weeks). We assumed a strike price of $55 that can be exercised in six months (26 weeks). The Excel function *VLOOKUP* is used to find the stock price on the expiration date from the simulated prices. Profit is simply the maximum of the difference

FIGURE 7.24

Stock price and option evaluation simulation spreadsheet

	A	B	C	D	E
1	**Stock Price Simulation**			Week	Price
2	**European Call Option Analysis**			0	$50
3				1	$ 48.96
4	*Parameters and Uncontrollable Inputs*			2	$ 47.04
5				3	$ 48.13
6	Expected annual growth rate	20%		4	$ 49.01
7	Annual standard deviation	25%		5	$ 49.06
8	Time period	1/52		6	$ 51.95
9	Initial price	$50		7	$ 54.16
10	Option strike price	$ 55.00		8	$ 53.42
11	Expiration date (week)	26		9	$ 51.81
12	Risk-free rate	8%		10	$ 51.76
13				11	$ 53.60
14	*Model Results*			12	$ 53.29
15				13	$ 53.17
16	Stock price at expiration date	$ 56.35		14	$ 55.46
17	Profit	$ 1.35		15	$ 56.35
18	Net present value of profit	$1.30		16	$ 57.01
19				17	$ 57.14
20	Key cell formulas			18	$ 57.67
21				19	$ 55.52

between (1) the stock price on the expiration date and the strike price and (2) zero. That is, if the stock price on the expiration date is less than the strike price, the option is worthless. To find the net present value of the profit in cell B18, we use the risk-free rate in cell B12 and the *NPV* function in Excel.

To estimate the expected net present value of the profit—the value of the option—we may simulate this spreadsheet using Crystal Ball, with cell B18 defined as the forecast cell. No assumption cells are necessary. Figure 7.25 shows the results. From the forecast chart we see that about two-thirds of the time the option is worthless, although it can assume a value of as high as $25.47 (from the *Range Maximum* in the *Statistics* chart). The expected net present value is $2.12; this represents the fair price of the option.

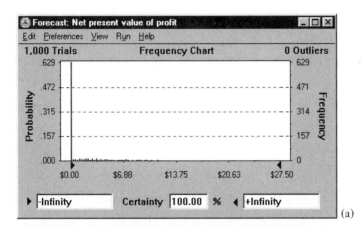

FIGURE 7.25
Forecast chart and statistics report for option pricing simulation

(a)

(b)

Excel Exercise 7.8

Open the file *stock option pricing.xls*.

a. Construct a line graph of the simulated stock prices by week.

b. Suppose an investor purchases an option to buy 100 shares of this stock. Modify the spreadsheet to compute his or her total return (including the cost of the option) on the expiration date of the option and use Crystal Ball to find the distribution of the total return.

CRYSTAL BALL *TORNADO CHART EXTENDER*

The student Crystal Ball Pro package contains two "extenders" that extend the functionality of Crystal Ball—*Tornado Chart* and *Decision Table* (others are available in the commercial version). The *Tornado Chart Extender* displays tornado and spider charts (which we introduced in Chapter 1). *Decision Tables* will be described in the next chapter.

The *Tornado Chart Extender* is useful for measuring the sensitivity of variables that you have defined in Crystal Ball, or quickly prescreening them to determine which are good candidates to define as assumptions or decision variables. To run *Tornado Chart*, select *CB Tornado Chart* from the Excel *Tools* menu. A series of three dialog boxes will be displayed. In the first box, *Specify Target*, you may select the dependent variable from all available forecasts defined in your Crystal Ball simulation model. The second dialog box, *Specify Input Variables*, asks you to define the inputs. Typically these are the assumption cells defined in the simulation model, but they may also include cells defined as decision variables or precedents—all cells within the active spreadsheet that are referenced as part of the formula or a subformula of the target cell. The last dialog box prompts you for various options.

Example 6: Crystal Ball Tornado and Spider Charts for the Profit Model

We will apply the *Tornado Chart Extender* to the profit model that we have been using to illustrate Crystal Ball. After selecting *CB Tornado Chart* from the *Tools* menu, select the profit forecast in the *Specify Target* dialog box. In the second dialog box, click the *Add Assumptions* button to define these as input variables. The third dialog box is shown in Figure 7.26. We have selected *Percentiles of the variables* as the *Tornado Method* option, which indicates that the extender should test the variables using percentiles of the assumption distributions as defined with a *Testing range* from 10 to 90 percent. *Selecting Percentage Deviations from The Base Case* tests variables using small changes that are specified percentages away from the base case. We also selected *Use median values* for the base case.

FIGURE 7.26
Crystal Ball
tornado chart
extender
options

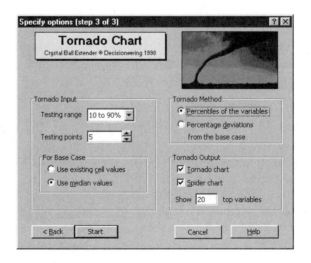

Figures 7.27 and 7.28 show the tornado and spider charts created by the extender. You can see that they are similar in structure to Figures 1.13 and 1.14, although they are based on different ranges of the variables because of the model assumptions. The top variables in the tornado chart—in this example, demand—have the most effect on the forecast, while the bottom ones have the least effect. The modeler might want to investigate this assumption further in hopes of reducing its uncertainty and therefore its effect on the target forecast. Similarly, the effect of the fixed cost is small and you might wish to eliminate this as an assumption in the simulation model. The slopes in the spider chart illustrate the relative effects of the assumptions on the forecast. Different colors are used to indicate the direction of the relationship between the variables and the forecasts.

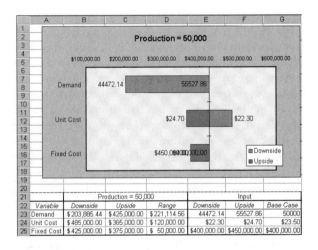

FIGURE 7.27
Crystal Ball
tornado chart

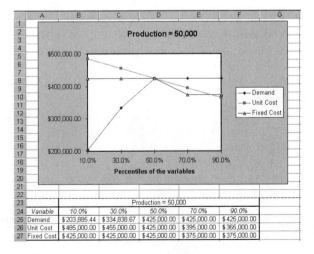

FIGURE 7.28
Crystal Ball
extender spider
chart

Excel Exercise 7.9

Open the workbook *Profit model CB.xls* and create the tornado and spider charts described in Example 6.

OPTIMIZATION AND SIMULATION

The purpose of many simulation models is to help a decision maker find an optimal solution to a problem. The newsvendor problem is one example. We used data tables to evaluate alternative values of the order quantity and Monte Carlo simulation to compute the expected profit and standard deviation to help assess risk. As with the optimization models we discussed in Chapters 4 and 5, many problems involve several decision variables. This makes it more difficult to identify optimal solutions, especially when uncertainty exists about some of the model inputs. Crystal Ball Pro includes an Excel add-in called OptQuest, which allows you to search for optimal solutions in the context of Monte Carlo simulation. We will illustrate the use of OptQuest in a groundwater cleanup problem.[4]

A small community gets its water from wells that tap into an old, large aquifer. Recently, an environmental impact study found toxic contamination in the groundwater due to improperly disposed-of chemicals from a nearby manufacturing plant. Since this is the community's only source of potable water and the health risk due to exposure to these chemicals is potentially large, the study recommends that the community reduce the overall risk to below a 1 in 10,000 cancer risk with 95 percent certainty; that is, the 95th percentile for the risk distribution should be less than .0001.

A task force narrowed down the number of appropriate treatment methods to three. It then requested bids from environmental remediation companies to reduce the level of contamination to recommended standards, using one of these methods. A company that wants to bid on the project has estimated possible cost distributions for the different cleanup methods, which vary according to the resources and time required for each (cleanup efficiency). Engineers have determined that the cost function is

$$Remediation\ cost = fixed\ cost * (1 + 2*(cleanup\ efficiency - 0.8))$$
$$+ variable\ cost * (1 + 3*(cleanup\ efficiency - 0.8))$$

With historical and site-specific data available, it would like to find the best process and efficiency level that minimizes cost and still meets the study's recommended standards with a 95 percent certainty.

The company has estimates of the contamination levels of the various chemicals. Each contaminant's concentration in the water is measured in micrograms per liter. Complicating the decision process is the fact that the cancer potency factor (CPF) for each chemical is uncertain. The CPF is the magnitude of the impact the chemical exhibits on humans; the higher the CPF, the more harmful the chemical is. The population risk assessment must also account for the body weights and volume of water consumed by the individuals in the community per day. All these factors lead to the following equation for population risk:

$$population\ risk = \frac{cancer\ potencies \times contaminant\ concentrations \times water\ consumed\ per\ day}{body\ weight \times conversion\ factor}$$

Figure 7.29 shows a spreadsheet for evaluating this situation. The conversion factor is 1000, and the uncertainties are modeled by the following:

- Cancer potency: normally distributed
- Concentration before: normally distributed
- Volume of water/day: normal, with a lower bound of 0
- Body weight: Normal, with a lower bound of 0

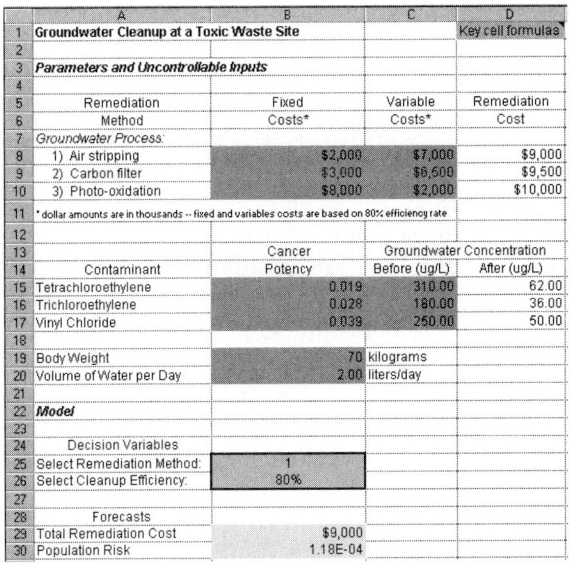

	A	B	C	D
1	Groundwater Cleanup at a Toxic Waste Site			Key cell formulas
2				
3	*Parameters and Uncontrollable Inputs*			
4				
5	Remediation	Fixed	Variable	Remediation
6	Method	Costs*	Costs*	Cost
7	*Groundwater Process:*			
8	1) Air stripping	$2,000	$7,000	$9,000
9	2) Carbon filter	$3,000	$6,500	$9,500
10	3) Photo-oxidation	$8,000	$2,000	$10,000
11	* dollar amounts are in thousands -- fixed and variables costs are based on 80% efficiency rate			
12				
13		Cancer	Groundwater Concentration	
14	Contaminant	Potency	Before (ug/L)	After (ug/L)
15	Tetrachloroethylene	0.019	310.00	62.00
16	Trichloroethylene	0.028	180.00	36.00
17	Vinyl Chloride	0.039	250.00	50.00
18				
19	Body Weight	70	kilograms	
20	Volume of Water per Day	2.00	liters/day	
21				
22	*Model*			
23				
24	Decision Variables			
25	Select Remediation Method:	1		
26	Select Cleanup Efficiency:	80%		
27				
28	Forecasts			
29	Total Remediation Cost	$9,000		
30	Population Risk	1.18E-04		

FIGURE 7.29
Groundwater cleanup spreadsheet model

- Fixed costs: triangular
- Variable costs: uniform

The first thing to do is to define the Crystal Ball assumption and forecast cells. In this case, both the total remediation cost and population risk are defined as forecast cells. With OptQuest, you must also define the decision variables. You can do this from the *Define Decision Variable* option under the *Cell* menu (or using the second icon on the Crystal Ball toolbar). This opens a *Decision Variable* selection window as shown in Figure 7.30. You must specify lower and upper bounds to limit the search process

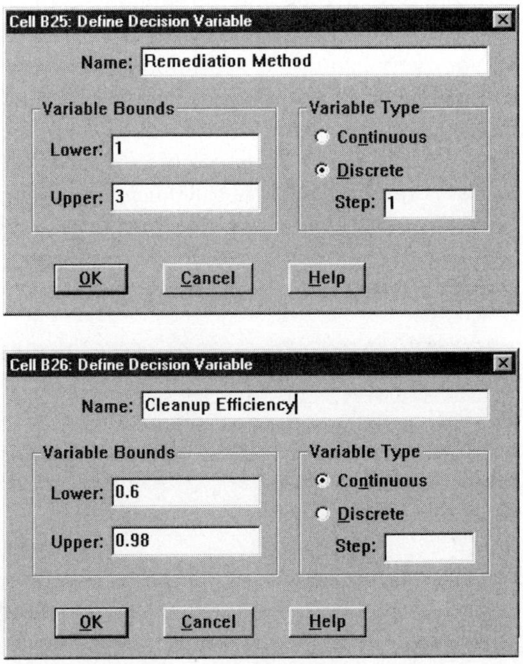

FIGURE 7.30
Decision variable definitions for groundwater cleanup model

and the type of variable—continuous or discrete. If the variable is discrete, specify a step size (the smaller the size, the quality of the solution will be better, but the process will be more time-consuming). In this example, the decision variables are

(1) the remediation method, which we define by the index 1, 2, or 3—thus, we define it as a discrete variable with a step size of 1, lower bound of 1, and upper bound of 3—and

(2) the cleanup efficiency, which company engineers have determined would be between 0.6 and 0.98 (a continuous variable).

To run OptQuest, select it from the *Tools* menu or by clicking on the OptQuest icon on the Crystal Ball toolbar (the bar chart with a yellow line graph running through it). Select *File...New*. This invokes the OptQuest *Wizard* to guide you in defining the problem (you may access any of the following windows from the *Tools* menu in the main OptQuest window). The *Decision Variable* selection window will appear with the variables you have defined. You may deselect or add new ones, or change the parameters as needed. After clicking *OK*, the *Constraints* window appears. This allows you to define any constraints that are functions of decision variables, similar to the types we discussed in Chapters 4 and 5. For this problem, no constraints are necessary. The next window is *Forecast Selection*, shown in Figure 7.31. We selected the total remediation cost as the objective to minimize (click the box under *Select Objective/Requirements* and choose from the drop-down menu). A **requirement** is a restriction on a forecast statistic; you may set upper and lower limits for any statistic of a forecast distribution. In this case, we want the 95th percentile of the population risk distribution to be no greater than 0; however, for numerical purposes, we choose a small tolerance value, such as 0.0001, or in scientific notation, 1E-4. This ensures that the upper tail of the distribution will not exceed 5 percent. Finally, the *Options* window allows you to specify various run settings for OptQuest. Computational efficiency will depend on the computer system you use; you should start with at least 10 minutes of running time, and you will be prompted to extend the search if you wish at the end of the process. We ran this problem for 15 minutes on a 266 Mhz laptop.

OptQuest uses various heuristics and intelligence gained during the search process to identify potential solutions. Each candidate is evaluated by running a Crystal Ball simulation. Not every candidate will improve the objective or be feasible; however, over time, the process generally finds very good solutions, although it cannot guarantee optimality. Figure 7.32 shows the results from one search. Over 50 solutions were evaluated. Notice that all the feasible solutions identified are virtually the same; the mean cost is approximately $11,000 using method 3 and a cleanup efficiency between 92 percent and 95 percent. Close examination reveals the tradeoffs—

FIGURE 7.31   Forecast selection window

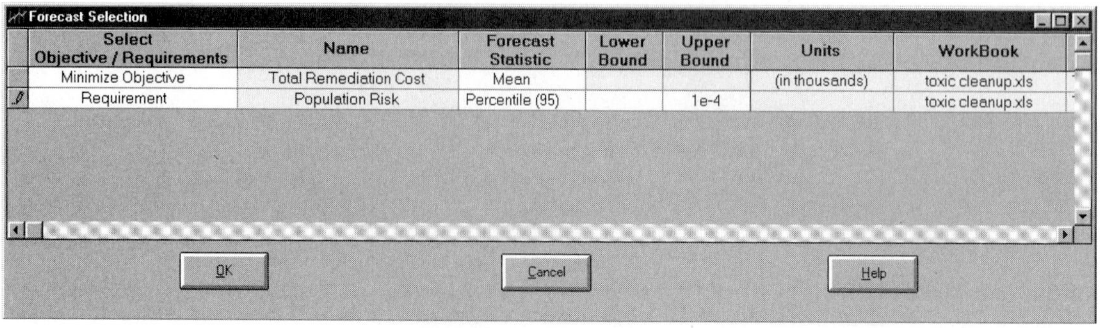

Select Objective / Requirements	Name	Forecast Statistic	Lower Bound	Upper Bound	Units	WorkBook
Minimize Objective	Total Remediation Cost	Mean			(in thousands)	toxic cleanup.xls
Requirement	Population Risk	Percentile (95)		1e-4		toxic cleanup.xls

OK   Cancel   Help

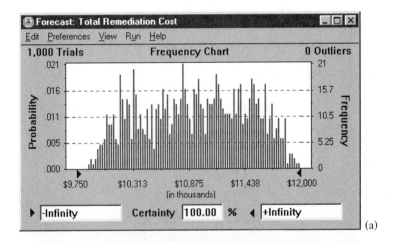

Simulation	Minimize Objective Total Remediation Cost Mean	Requirement Population Risk Percentile (95)	Cleanup Efficiency	Remediation Method
1	9009.31	2.2665E-04 - Infeasible	0.800000	1
2	11205.2	5.5590E-05	0.950679	3
18	11193.2	5.6338E-05	0.950770	3
19	11136.7	7.1208E-05	0.942518	3
20	11119.5	6.2098E-05	0.943950	3
24	11117.7	6.8310E-05	0.942054	3
33	11113.1	6.9670E-05	0.941604	3
44	11018.3	8.9245E-05	0.923133	3
45	11008.4	9.0392E-05	0.924342	3
Best: 50	10945.5	9.3742E-05	0.916997	3

FIGURE 7.32
OptQuest results for groundwater cleanup model

as cleanup efficiency is reduced, the cost decreases, but the requirement moves closer to its upper limit. A slightly better solution might be found by extending the search. Figure 7.33 shows the forecast windows associated with the best solution. In particular, we see that the considerable variation in the cost exists over 1,000 trials, and that the population risk requirement is met.

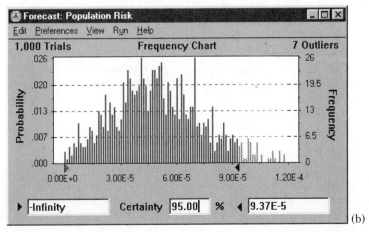

FIGURE 7.33
Forecast windows for best OptQuest solution

Excel Exercise 7.10

Open the file *toxic cleanup.xls*. Although the photo-oxidation method was the cheapest process for the required efficiency level, the community decides to relax the risk requirement to 3 out of 10,000. Use OptQuest to determine what method is best and how much the community would save.

PROBLEMS

1. Using a spreadsheet and the formulas in this chapter, generate 50, 100, and 500 outcomes from each of the following distributions.
   - Uniform between 0 and 1
   - Normal with mean 50 and standard deviation 10
   - Exponential with mean 1
   - Triangular with minimum = 0, likeliest = 3, and maximum = 10
   a. Construct histograms for each data set and explain the differences due to sample size.
   b. Use a Chi-square goodness-of-fit test with the Crystal Ball distribution fitting capability (see Chapter 2) to test the hypothesis that the data are drawn from the distribution from which they were generated.
2. Consider the simple profit model used in Figure 7.1. Develop a financial simulation model for a new product proposal and construct a distribution of profits under the following assumptions. Price is fixed at $1,000. Variable costs are unknown and follow the distribution:

Costs	Probability
$500	.20
$600	.40
$700	.25
$800	.15

Quantity sold is also variable and follows the following distribution:

Quantity Sold	Probability
120	.25
140	.50
160	.25

Fixed costs are estimated to follow the following distribution:

Fixed Costs	Probability
$45,000	.20
$50,000	.50
$55,000	.30

The company can sell all production.
   a. Describe how to use random numbers to generate random variates from these distributions.
   b. Generate three columns of 10 random numbers using the *RAND( )* function on an Excel worksheet. Simulate 10 trials of this model manually.

   c. Implement your model using Crystal Ball. Use 500 trials. Would you conclude that this product is a good investment?

   d. Construct a 90 percent confidence interval for the mean profit using your simulation results.

   e. Repeat the Crystal Ball simulation using 1,000 trials and recompute your confidence interval.

3. You are considering purchasing an apartment complex. Before deciding, you would like to do a risk analysis of the situation to assess whether or not the investment is a good idea for you. The complex has 40 units. Historical data indicate that the number of units rented in a given month is distributed uniformly between 30 and 40 units. The rent per unit, which you are not likely to change since demand is very price sensitive, is $600 per month. Monthly expenses for the entire complex seem to be distributed normally with mean $17,000 and standard deviation $1,500.

   a. Develop a Crystal Ball simulation for 1,000 months. What is the expected profit?

   b. According to the simulation results, what is the probability of making a profit in any given month?

   c. Suppose your debt payments on this complex will be $2,400 per month. Would you take this investment? Defend your decision using the simulation output data.

   d. How many trials would you need to estimate mean profit to within $50 for a 95 percent confidence level?

4. A company is considering two investments. Both cost $10 million and have a 10-year life. Investment A expects to generate an annual net cash flow of $1.3 million, and investment B expects to generate an annual net cash flow of $1.4 million. However, the distribution of annual net cash flow is triangular, with $a = $900,000$ and $b = 1.7 million for investment A, and $a = $600,000$ and $b = 3.4 million for investment B. Recommend which investment to choose, using a Crystal Ball simulation.

5. A plant manager is considering investing in a new $30,000 machine. Use of the new machine is expected to generate a cash flow of about $8,000 per year for each of the next five years. However, the cash flow is uncertain, and the manager estimates that the actual cash flow will be normally distributed with a mean of $8,000 and a standard deviation of $500. The interest rate is estimated to be 9 percent and assumed to remain constant over the next five years. The company evaluates capital investments using net present value and internal rate of return. How risky is this investment? Develop an appropriate simulation model and conduct experiments and statistical output analysis to answer this question. (Hint: use the functions NPV and IRR.)

6. MasterTech is a new software company that develops and markets productivity software for municipal government applications. In developing their income statement, the following formulas are used:

Gross profit = Net sales − Cost of sales

Net operating profit = Gross profit − Administrative expenses

Net income before taxes = Net operating profit − Interest expense

Net income = Net income before taxes − Taxes

    Net sales in the first year (in hundreds of thousands of dollars) is uniformly distributed between 6 and 12. The net sales are expected to increase each year by 5 percent plus or minus 1 percent. Cost of sales in year 1 (in hundreds of thousands of dollars) is normally distributed, with mean of 5.6 and standard deviation 0.2, and is expected to increase 5 percent per year. Selling expenses has a fixed component that is uniform between $75,000 and $110,000 in year 1, and will increase 4 percent per year. The variable component is estimated to be 7 percent of net sales. Administrative expenses are normal, with a mean of $50,000 and standard deviation of $4,000, and will increase 3 percent per year. Interest expenses are $10,000 in year 1 and will decline by $1,000 each year. The company is taxed at a 50 percent rate.

   a. Develop a risk profile of net income for each year using Crystal Ball and write a report to management. Hint: Use the *CB.Uniform* function for the annual increases in sales, and define assumptions for the base sales, cost of sales, fixed selling cost, and administrative expense in year 1.

   b. Create tornado and spider charts using the Crystal Ball *Tornado Chart Extender* and analyze the results.

7. Rattel Toys is considering introducing a new action toy based on an anticipated blockbuster movie to be released next summer. The toy is expected to have a marketable lifetime of only one year. The price proposed is $7.95, and marketing expects to sell 900,000 units. Fixed production costs are estimated to be $675,000, and per unit variable cost, $3.00. Selling costs are expected to be $875,000, and general and administrative costs, $300,000. All these estimates, however, are uncertain. Many movies have been surprise hits, and others have been enormous flops. Upon further analysis, the management team estimates that sales will be normal, with mean 900,000 and standard deviation 300,000; fixed costs will be uniform between $625,000 and $725,000; variable costs triangular, with $a = \$2.75$, $m = \$3$, and $b = \$3.25$; and selling costs normal, with mean $875,000 and standard deviation $50,000.

   a. How risky is this project?

   b. Which assumptions have the most impact on the results?

8. A small project network is shown in Figure 7.34. The total project completion time is the length of the longest path from the beginning of the project (point 1) to the end (point 4). The time for each activity, depicted by the arrows on the network, is triangular, with the parameters a, m, and b as shown in the figure. Construct a spreadsheet to model the project completion time.

   a. Use Crystal Ball to simulate the project and determine a distribution of project completion times as well as relevant statistical information. Translate your

FIGURE 7.34

A project network

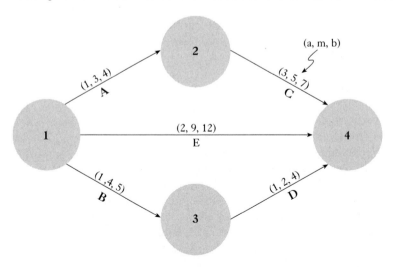

results into a memo to the project manager that describes risk associated with the project.

b. What activity has the largest impact on the project completion time?

9. A stockbroker calls on potential clients after referrals. For each call, there is a 15 percent chance that the client will decide to invest with the firm. Sixty percent of those interested are found not to be qualified based on the brokerage firm's screening criteria. The remaining are qualified. Of these, half will invest between $2,000 and $10,000, 25 percent will invest between $10,000 and $25,000, 15 percent will invest between $25,000 and $50,000, and the remainder will invest between $50,000 and $100,000. The commission schedule is as follows:

Transaction Amount	Commission
Up to $25,000	$60 + 0.5% of the amount
$25,001 to $50,000	$85 + 0.4% of the amount
$50,001 to $100,000	$135 + 0.3% of the amount

The broker keeps half the commission. How many calls per month must the broker make to have at least a 75 percent chance of making at least $5,000?

10. The director of a nonprofit ballet company in a medium-sized U.S. city is planning its next fund-raising campaign. In recent years, the program has found the following percentages of donors and gift levels:

Gift Level	Amount	Number of Gifts
Benefactor	$10,000	1–3
Philanthropist	$5,000	3–7
Producer's Circle	$1,000	16–25
Director's Circle	$500	31–40
Principal	$100	5–7% of solicitations
Soloist	$50	5–7% of solicitations

The company has set a financial goal of $125,000. How many prospective donors must they contact for donations at the $100 level or below to have a 90 percent chance of meeting this goal? Assume that the number of gifts at each level follow a uniform distribution (at the high levels, these must be discrete numbers!).

11. A textile supplier sells fabric to furniture manufacturers. A typical contract includes a financial contribution to "sampling programs" that customers use to promote their products. Most fabrics have market life cycles of three years, after which time they will need to be replaced by newer styles. Develop a three-year financial model to compute the total profit based on the following assumptions:

a. Unit cost of goods sold is $12 in year 1 and is expected to increase each year somewhere between 2 and 5 percent

b. Sampling costs (incurred only in the first year) are $1,600,000

c. Average selling price is $14.95 in year 1 and is expected to change each year between −2 and +3 percent.

d. Number of yards of fabric sold is 350,000 in year 1 and will increase between 0 and 25 percent, with a most likely increase of 15 percent (triangularly distributed)

Use Crystal Ball to evaluate the sensitivity of the assumptions to the total profit forecast, and determine the probabilities that total profit will exceed $1 million, $1.25 million, $1.5 million, and $1.75 million.

12. A local pharmacy orders 15 copies of a monthly magazine. Depending on the cover story, demand for the magazine varies. Historical records suggest that the probability distribution of demand is Poisson with a mean of 10. The pharmacy purchases the magazines for $1.25 and sells them for $3. Any magazines left over at the end of the month are donated to hospitals and other health care facilities. Is the 15-copy order quantity the most profitable? Conduct a simulation analysis using Crystal Ball to answer this question.

13. A buyer for a large department store must make a decision on the purchase of fall merchandise in May. Clearly, there is an element of risk. If not enough merchandise is purchased, the store will forgo the opportunity to gain additional profit. If too much is purchased, the store will have to sell the surplus at a loss at the end of the season to make room for the spring line. For a particular style of winter coat, marketing researchers have estimated a probability distribution for the sales of the merchandise, shown in the next table. The coats are purchased for $55 and sold for $135. However, any that are left over at the end of the season are sold for $40. What is the best quantity to order to maximize expected profit? Use a data table with Crystal Ball. Discuss the relative risks associated with other decisions.

Sales	Probability
50	.1
60	.2
70	.3
80	.2
90	.1
100	.1

14. Find the price of a European call option having a current price of $20, a strike price of $21.50, an annual growth rate of 7 percent, and an annual standard deviation of 25 percent with an exercise date of 12 weeks. Use a risk-free rate of 7 percent.

15. Find the price of a European call option for a stock having an annual growth rate of 35 percent, an annual standard deviation of 30 percent, an initial price of $7 and strike prices of $9 and $11 for expiration dates of 13 and 26 weeks. For each scenario, find the probability that the option will be exercised. Use a risk-free rate of 4.5 percent.

16. Develop a spreadsheet for evaluating a 26-week European put option (see the discussion in Chapter 6 for the concept of a put option) with a strike price of $72 for a stock that currently sells at $70, has an annual growth rate of 16 percent, and an annual standard deviation of 30 percent. Use a risk-free rate of 5 percent.

17. Consider the situation described in Problem 3. Suppose that you have determined that the demand for rental units is a linear function of the rent charged:

$$\text{Number of units rented} = -0.1(\text{rent per unit}) + 85$$

for rents between $400 and $600. However, the parameters in this regression model are uncertain. The slope is assumed to be triangular, with a minimum of -0.11, likeliest of -0.1, and maximum of -0.08, and the intercept is also triangular, with parameters 75, 85, and 90. In addition, monthly operating cost will average about $15,000, and is normally distributed with a standard deviation of $1,000. The manager wants to determine the optimal amount of rent to charge to maximize the expected profit.

   a. Develop a Crystal Ball model and evaluate the risks associated with charging a rent of $500 per month.

   b. Apply OptQuest to find a good solution to this problem.

18. Apply OptQuest to the newsvendor problem discussed in this chapter to find the best solution for the order quantity to maximize the expected profit.

NOTES

1. The student version of Crystal Ball is limited to a maximum of six assumption cells, six forecast cells, 1,000 trials, and one pairwise correlation. In addition, the *Overlay Chart* option is disabled, and only the normal, triangular, uniform, Poisson, exponential, and custom distributions are available in the *Distribution Gallery*. The *Decision Table* extender is limited to two decision variables.

2. Much of this section has been adopted from Crystal Ball *Version 4.0 User Manual*, © 1988–1996 Decisioneering, Inc. The *User Manual* provides other details about Crystal Ball that are not described here. Copies of the *User Manual* can be purchased from the Technical Services or Sales department of Decisioneering, at (303) 534-1515; toll-free sales: 1-800-289-2550; however, this is not necessary to work the problems in this book. The Decisioneering web site is **http://www.decisioneering.com**

3. See David G. Luenberger, *Investment Science* (New York: Oxford University Press, 1998).

4. Adapted from OptQuest *Version 1.0 User Manual* (Denver, Colo.: Decisioneering, Inc., 1998).

BIBLIOGRAPHY

Decisioneering, Inc. 1998. Company web site: **http://www.decisioneering.com.**

Evans, James R. 1992. "A Little Knowledge Can Be Dangerous: Handle Simulation with Care." *Production and Inventory Management*, Vol. 33, No. 2, pp. 51–54.

Hertz, David B. 1983. *Risk Analysis and Its Applications*. New York: John Wiley & Sons.

Kelton, W. David. 1986 "Statistical Analysis Methods Enhance Usefulness, Reliability, of Simulation Models." *Industrial Engineering*, Vol. 28, No. 9, September, pp. 74–84.

Law, Averill M., and W. David Kelton. 1991. *Simulation Modeling and Analysis*. New York: McGraw-Hill.

Thesen, Arne, and Laurel E. Travis. 1992. *Simulation for Decision Making*. St. Paul, Minn.: West Publishing Co.

Watson, Hugh J., and John H. Blackstone, Jr. 1989. *Computer Simulation*. 2d edition. New York: John Wiley & Sons.

APPENDIX: ADDITIONAL CRYSTAL BALL FEATURES

Crystal Ball includes several options that provide additional modeling and analysis capabilities. We illustrate these using a simple cash balance model, shown in Figure 7A.1 (*cash balance CB.xls* in the textbook example files folder). In this model, we assume that monthly cash inflows and outflows are normally distributed with means and standard deviations shown in the range C6:E9. The model assumes a beginning cash balance of $50,000. All inflows and outflows are defined as assumption cells, and the ending cash balances for each month are forecast cells.

Correlated Assumptions

Normally, each assumption is independent of the others. In many situations, we might wish to explicitly model dependencies between variables. Crystal Ball allows

**FIGURE
7A.1**
Cash balance
model

	A	B	C	D	E
1	**Cash Balance Model**				
2					
3	*Parameters and Uncontrollable Inputs*				
4					
5	Month		Jan	Feb	Mar
6	Cash inflow mean		$12,000.00	$15,000.00	$20,000.00
7	Cash inflow std. dev.		$ 1,000.00	$ 2,500.00	$ 4,000.00
8	Cash outflow mean		$10,500.00	$17,300.00	$13,000.00
9	Cash outflow std. dev.		$ 1,500.00	$ 3,000.00	$ 1,500.00
10					
11	*Model*				
12					
13	Month	Dec	Jan	Feb	Mar
14	Cash Inflows		$12,000.00	$15,000.00	$20,000.00
15	Cash Outflows		$10,500.00	$17,300.00	$13,000.00
16	Net gain (loss)		$ 1,500.00	$ (2,300.00)	$ 7,000.00
17	Cash Balance	$50,000.00	$51,500.00	$49,200.00	$56,200.00

you to use correlation coefficients to define dependencies between assumptions.* This can be done only after assumptions have been defined. For example, suppose that we wish to correlate the January and February cash inflows. We select one of these cells, for instance, C14, then select *Define Assumption* from the *Cell* menu. When the distribution dialog box appears, click on the *Correlate...* button. The *Correlation Dialog Box*, shown in Figure 7A.2, appears. The *Select Assumption* button provides a list of the assumptions that you have defined. We then select the assumption to correlate with the first and enter a correlation coefficient.

You may enter a correlation coefficient in one of three ways. First, you can enter a value between −1 and 1 in the *Coefficient Value* box. Second, you may drag the slider control along the Correlation Coefficient scale; the specific value you select is displayed in the box to the left of the scale. Third, you may click on *Calc...* and enter ranges of cells in the spreadsheet that contain empirical values that should be used to calculate a correlation coefficient. After the correlation coefficient is specified, Crystal Ball displays a sample correlation chart, as shown in Figure 7A.3. The solid line in-

**FIGURE
7A.2**
Correlation
dialog box

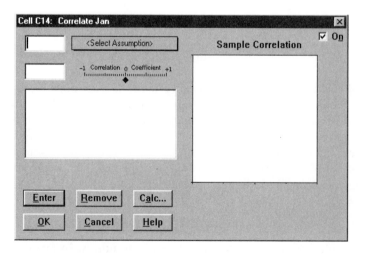

*In the student version of Crystal Ball, only one pairwise correlation can be defined for each model.

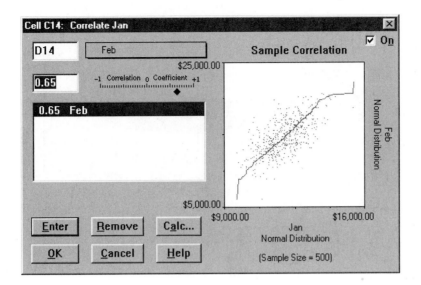

**FIGURE
7A.3**
Correlating
January and
February cash
inflows

dicates where values of a perfect correlation would fall; the points represent the actual pairing of assumption values that would occur during the simulation. When using this option you must be cautious, since it is possible that some correlations might conflict with others, preventing Crystal Ball from running.

Freezing Assumptions

The *Freeze Assumptions...* command under the *Cell* menu allows you to temporarily exclude certain assumptions from a simulation. This allows you to see the effect that certain assumptions might have on the forecast cells while holding others constant to their spreadsheet values.

Overlay Charts*

If a simulation has multiple, related forecasts, the overlay chart feature allows you to superimpose the frequency data from selected forecasts on one chart in order to compare differences and similarities that might not be apparent. This option is invoked from the *Run* menu by selecting *Open Overlay Chart.*

Trend Charts

If a simulation has multiple related forecasts, you can view the certainty ranges of all forecasts on a single chart, called a trend chart. Trend charts are particularly useful for time-related forecasts. The trend chart displays certainty ranges in a series of patterned bands. For example, the band representing the 90 percent certainty range shows the range of values into which a forecast has a 90 percent chance of falling. Figure 7A.4 shows a trend chart for the three monthly cash balances in our example. The chart clearly shows that the variance increases over time.

*The *Overlay Chart* option is not available in the student version of Crystal Ball. This section is included to provide a complete overview of the capabilities of the software.

FIGURE
7A.4
Trend chart

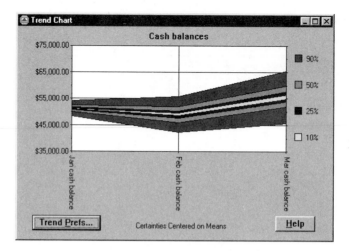

Clicking on the *Trend Preferences...* button displays the dialog box shown in Figure 7A.5. This allows you to customize the trend chart and select the number and type of certainty bands, as well as to change the chart type.

Sensitivity Charts

The uncertainty in a forecast is the result of the combined effect of the uncertainties of all assumptions as well as the formulas used in the model. An assumption might have a high degree of uncertainty yet have little effect on the forecast because it is not weighted heavily in the model formulas. On the other hand, an assumption with a relatively low degree of uncertainty might influence a forecast greatly. "Sensitivity" refers to the amount of uncertainty in a forecast that is caused by the uncertainty of an assumption as well as the model itself. The *Sensitivity Chart* feature of Crystal Ball allows you to determine the influence that each assumption cell has on a forecast cell.

FIGURE
7A.5
Trend
preferences
dialog box

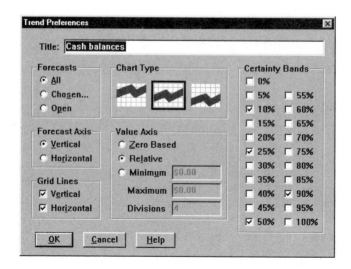

To create a sensitivity chart, you must ensure that the *Sensitivity Analysis* box is checked in the *Run Preferences* dialog box before running the simulation. After the simulation is completed, select *Open Sensitivity Chart* from the *Run* menu. The assumptions (and possible other forecasts) are listed on the left, beginning with the assumption having the highest sensitivity. The sensitivity chart for the example is shown in Figure 7A.6. We see that the March cash inflow assumption influences the March cash balance forecast the most, followed by the cash outflow assumption for February. The January cash inflow assumption has the least effect.

The sensitivities in Figure 7A.6 are measured by *rank correlation coefficients*, which describe how strongly forecasts move together. Positive coefficients indicate that an increase in the assumption is associated with an increase in the forecast; negative coefficients imply the reverse. The larger the value of the correlation coefficient, the stronger the relationship. You may also express sensitivities as a *percent of the contribution to the variance* of the forecast by opening the *Sensitivity Prefs...* dialog box, shown in Figure 7A.7. This dialog box also allows you to select different forecasts for the

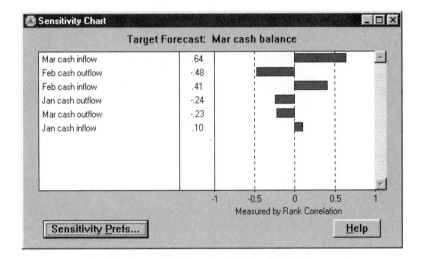

FIGURE 7A.6
Sensitivity chart (rank correlation)

FIGURE 7A.7
Sensitivity preferences dialog box

sensitivity chart. Figure 7A.8 shows the sensitivity chart for the March cash balance, ranked by contribution to variance. These results show that the variation in cash inflows for March causes about 44 percent of the variation in the forecast. Note the similarity to a tornado chart, except that this information is provided after assumptions are defined and the simulation is run. Since the January cash inflow assumption has little effect on the March cash balance, we might consider dropping it as an assumption and instead, using its mean value in the model.

FIGURE 7A.8
Sensitivity chart (contribution to variance)

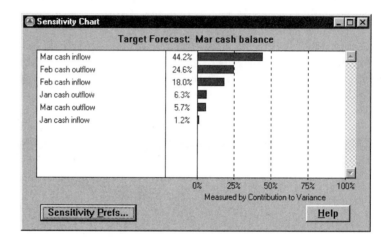

C H A P T E R   8

Systems Modeling and Simulation

O U T L I N E

Applications of Dynamic System Models
Toll Booth Improvement
Designing Security Checkpoints
Technology Evaluation
Dental Practice Management
Forest Fire Management

Modeling and Simulating Dynamic Systems

Queueing Systems
Customer Characteristics
Service Characteristics
Queue Characteristics
System Configuration
Management Science Practice: One Line or More?

Modeling and Simulating Queueing Systems

The Dynamics of Waiting Lines

Analytical Queueing Models
The M/M/1 Queueing Model
Other Queueing Models
Little's Law
Analytical Models vs. Simulation
Sensitivity Analysis
Management Science Practice: Queueing Models for Telemarketing at L.L. Bean

Modeling and Simulating Dynamic Inventory Systems
Using Simulation to Optimize Inventory Systems

Systems Simulation Technology
Management Science Practice: Designing an Air Force Repair Center Using Simulation
Arena Business Edition

Using Simulation Successfully
Verification and Validation
Statistical Issues

Problems

Dynamic **systems** involve processes that consist of interacting events occurring over time. For example, nearly everyone experiences waiting lines, or **queues** (the French word for "line")—at supermarkets, banks, toll booths, telephone call centers, restaurants, amusement parks—and in many other places. Experts have estimated that Americans wait an average of 30 minutes per day. This translates to 37 billion hours each year! Many other important types of waiting line systems involve "customers" other than people; for example, messages in communication systems, trucks waiting to be unloaded at warehouses, work-in-process at manufacturing plants, and photocopying machines awaiting repair by traveling technicians. In these systems, customers arrive at random times, and service times are rarely predictable. Managers of these systems would be interested in how long customers have to wait, the length of waiting lines, utilization of the server, and other measures of performance. Another example of a dynamic system is an inventory operation, like the type we described in Chapter 1. Managers would be interested in inventory levels, numbers of lost sales or backorders incurred, and the costs of operating the system. When demand and lead times are unpredictable, we cannot rely on the EOQ model to forecast these performance measures. More complex examples of dynamic systems include entire production systems, which might incorporate aspects of both waiting lines and inventory systems, as well as material movement, information flow, and so on.

Modeling dynamic systems is considerably more difficult than modeling other types of problems that you have studied in this book, primarily because the sequence of events over time must be explicitly taken into account. Modeling dynamic systems involves specifying the logical sequence of events as they take place. For instance, key events that occur in a waiting line system are the arrival of a customer, completion of the service, and occasionally line-switching or a customer's decision to leave a line because of excessive waiting. Events in an inventory system would include the demand for items, receipt of replenishment orders, and placement of new orders. Performance measures, such as the length of a waiting line and level of inventory, depend on the sequence of events and vary over time. Nevertheless, improving modern business processes depends on the ability to model and analyze such systems.

In this chapter we focus on modeling and analyzing dynamic systems, primarily queueing and inventory systems. Although we will discuss some analytical approaches, our major emphasis will be on using simulation as a modeling and analysis tool, drawing upon the concepts developed in the previous chapter. Dynamic simulation models allow us to draw conclusions about the behavior of a real system by studying the behavior of a model of the system, usually with random sequences of events similar to what we might observe in the real system. Simulation has the advantage of being able to incorporate nearly any practical assumption, and thus is the most flexible tool for dealing with dynamic systems.

APPLICATIONS OF DYNAMIC SYSTEM MODELS

Toll Booth Improvement

At Port Authority tunnels and bridges in New York, toll-plaza activities involve hundreds of traffic officers, with millions of dollars of payroll expenses. In a classic application of queueing analysis, the Port Authority undertook a study to balance expenses and employee and customer satisfaction. Specifically, the purposes of the study were to evaluate the grade of service given to customers as a function of traffic volume, to establish optimum standards of service, and to develop a more precise method of controlling expenses and service while providing for toll-collector rest periods.

Data collected included the traffic arrivals at the toll plaza, the number of cars in each toll-lane queue, and the toll-transaction count and times. The scheduling procedure used queueing models and saved a yearly equivalent of 10 times the cost of the study. In addition, the scheduling procedure offered better service to the public and benefits to toll-collection personnel (Edie, 1954).

Designing Security Checkpoints

Vehicles entering a secured work facility at the Westinghouse Hanford Company each morning pass through a security gate. Normally two security guards checked vehicles, drivers, and passengers. Upon approaching the gate, vehicles formed one line that extended past the available queue space onto the adjacent highway, resulting in a major traffic safety problem. A study was conducted to minimize the queue length while also minimizing the number of security guards at the gate. Using a simulation model, the study showed that a significant improvement could be achieved using the same number of guards in two parallel traffic lines rather than the obvious solution of increasing the number of guards. The study was so successful that several other queueing studies were initiated shortly thereafter (Landauer and Becker, 1989).

Technology Evaluation

The United States Postal Service delivers over 500 million pieces of mail each day to over 100 million delivery locations. The Postal Service uses a comprehensive simulation model for capacity planning and the evaluation of technology investment. The model has three major components: a mail operations component that models the flow of each stream of mail through sorting operations; a workforce component that assigns staff to each sorting method; and a financial model to calculate yearly costs. The system can identify bottlenecks and cost impacts of various automation alternatives. The annual labor savings from the automation program is over $4 billion per year (Cebry et al., 1992).

Dental Practice Management

Business schools have used computerized simulations as teaching and learning tools for many years. Computerized business simulations are games in which the participants make management decisions about a business based on past performance and the strategic direction they have chosen. A computer program analyzes the decisions based on the actual competitive and simulated economic environments and then determines the players' outcomes, which are used to make a new set of decisions. Few dental students have managed a small business or made any managerial decisions, yet they face important management issues when they venture into private practice. A simulation of the business decisions required by dental practitioners was developed to raise dental students' knowledge of business and financial issues. The simulation allows participants to make operational decisions, such as choosing hours of operation and making credit and collection policy decisions; staffing decisions, including hiring, termination, raises, and benefits; marketing decisions; financial decisions, such as loans and investments; and other decisions, such as incident response (Willis et al., 1997).

Forest Fire Management

The forest industry is an important sector of the economy of Ontario, Canada. As the value of forest resources increases, so does the need for fire managers to improve their operations. One of the most important components of a forest fire management system is the initial attack subsystem, which is designed to start fire-fighting action quickly on newly reported fires. The initial attack dispatch might be a three- to five-person crew that travels to the fire by truck, fixed-wing aircraft, or helicopter. It may also include one or more water-based or amphibious air tankers. The initial attack force constructs a fire line around the fire's perimeter, using digging tools, a flow of high-pressure water from a nearby water source, water dropped from air tankers, and natural barriers such as lakes and rivers.

To help forest fire managers of the Ontario Ministry of Natural Resources evaluate initial attack resources, a simple simulation model was developed of the kinds of fires that occur in Ontario and the resources used to fight them. Specifically, the model indicated that well-trained firefighters equipped with adequate transport would perform well on most fires without air tanker support. However, for a small but important class of fires, air tanker support is critical. The model was useful in planning the acquisition of air tankers and future use of air tankers, transport aircraft, and fire fighters (Martel et al., 1984).

MODELING AND SIMULATING DYNAMIC SYSTEMS

The process of modeling a dynamic system is usually performed by developing a flowchart of the logical sequence of events that occur as the system operates over time. Once such a flowchart is constructed, a simulation model can be developed by translating the logic into a spreadsheet, a general purpose computer language like BASIC or C++, or a special simulation language. Most events that "drive" dynamic systems, such as the arrivals at a waiting line or demands from an inventory system, cannot be predicted. Therefore, simulation models of dynamic systems generally rely on sampling from probability distributions as we described in the previous chapter to characterize the occurrences of such unpredictable events.

We will illustrate both the modeling and simulation process for a simple example.

Example 1: A Dynamic Production/Inventory Model

Murthy Manufacturing, Inc. (MMI), produces a variety of automobile aftermarket components. The demand for a certain type of rebuilt alternator fluctuates each month between 120 and 170, with any demand in this range being equally likely. MMI likes to keep its production level at 145 units per week, but this may also fluctuate because of variations in labor availability, defect rates, material delays, and so on. In approximately 4 of every 10 weeks, production falls to 130, and 10 percent of the time MMI can produce 150 units each week. MMI likes to maintain an inventory level of at least 200 units. If inventory falls below 200 units, MMI authorizes overtime to produce an extra run of 100 units. MMI management would like to know how often overtime is necessary or, equivalently, how long the inventory requirement can be maintained.

The fundamental equation that governs this system each week is

$$Ending\ Inventory = Beginning\ inventory + Production - Demand$$

Let us assume that we begin with an initial inventory of 250 units in the first week. Figure 8.1 shows a flowchart that models the logic of this system. If we know the demand and the production level for a given week, we may compute value for the ending inventory that week, which also becomes the beginning inventory for the next week. If at any time the inventory falls below 200, we add an extra 100 units to the production for that week.

Figure 8.2 shows a spreadsheet, *MMI.xls*, designed to simulate this model. In this situation, we see that both demand and production are random variables. We randomly generate a value for production and demand using Crystal Ball functions. For example, the function in cell C13 is *TRUNC(CB.Uniform(120,170))*. The *TRUNC* function truncates a uniform random demand between 120 and 170 to a whole number. We use the Crystal Ball *CUSTOM* function, *CB.Custom($A$6:$B$8)*, to generate production from the discrete probability distribution in the range A6:B8. In addition, we must check when the inventory falls below 200 units. This is done using an *IF* function to determine if an overtime production run is necessary. In cell H13, we use

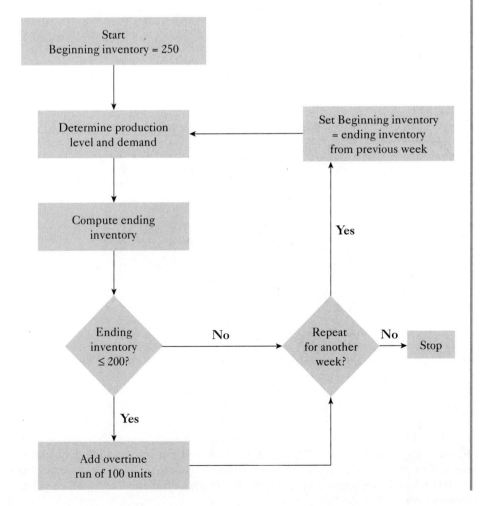

FIGURE 8.1
Logical flowchart for dynamic production/ inventory model

FIGURE 8.2

Spreadsheet for MMI production/inventory example

	A	B	C	D	E	F	G	H	I
1	Dynamic Production/Inventory Simulation					Key cell formulas			
2									
3	*Input Data*								
4									
5	Production	Probability							
6	130	0.4							
7	145	0.5							
8	150	0.1							
9									
10	*Simulation Results*								
11		Beginning			Ending				
12	Week	Inventory	Demand	Production	Inventory		Additional production runs		
13	1	250	123	145	272		6		
14	2	272	157	145	260				
15	3	260	139	150	271				
16	4	271	149	145	267				
17	5	267	161	130	236				
18	6	236	139	145	242				
19	7	242	129	130	243				
20	8	243	166	150	227				
21	9	227	141	130	216				
22	10	216	159	145	202				

the *COUNTIF* function to count the number of times production exceeds 200 to determine the number of additional production runs.

Figure 8.3 shows a graph of the ending inventory and production levels each week. The spikes clearly show when additional production is used. In this simulation, we see that the inventory level fell below the required 200 units 6 times in a 100-week period, or, on average, every 16 to 17 weeks.

FIGURE 8.3

Graph of ending inventory and production levels

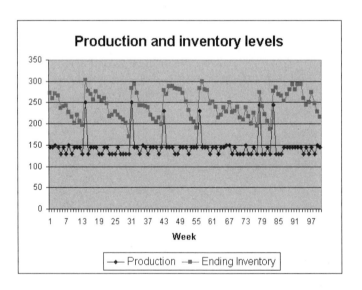

As we observed in the previous chapter, the output of a simulation model is simply a sample from an infinite population. Clearly, a 100-week sample is rather small to draw a good conclusion about the frequency of overtime runs. We have two ways of addressing this. First, the length of the simulation can be extended, for example, to 1,000 or more weeks. This is rather cumbersome, particularly for a spreadsheet model. A second approach is to replicate the 100-day simulation many times. This is done easily with Crystal Ball.

Example 2: Crystal Ball Replication of the MMI Model

Figure 8.4 shows the results of 1,000 trials of a Crystal Ball run for the MMI model
using the number of additional production runs (cell H13) as the forecast cell. These
results show that the number of additional runs in the 100-week simulation might
range from as low as 0 to as high as 10, with a median and approximate mean of 5.
Such information can help the company to determine an appropriate budget as well
as the risk of exceeding the budget that is chosen.

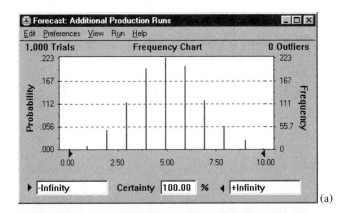

(a)

(b)

FIGURE 8.4
Crystal Ball
results for
1,000 trials of
the MMI
simulation

Excel Exercise 8.1

Open the file *MMI.xls* and copy the formulas so that 1,000 weeks are simulated. On the basis of a single simulation run, how close does the frequency of
overtime runs compare with the results from Example 2?

Example 3: A Simulation Model for Cash Management

In Chapter 1 we introduced the Miller-Orr model for cash management. Recall that
this model specifies a decision rule on which to base the sale or purchase of securities to minimize the expected costs of transactions and opportunity costs associated
with accrued interest by computing an upper limit on the cash balance and return
point that defines how many dollars of securities to buy or sell. Figure 8.5 shows a

FIGURE 8.5

Spreadsheet implementation and simulation of Miller-Orr Model

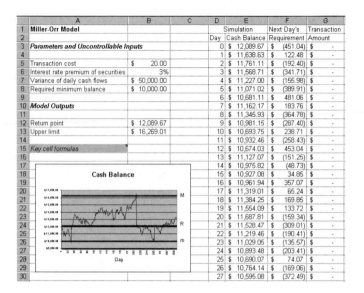

	A	B	C	D	E	F	G
1	Miller-Orr Model				Simulation	Next Day's	Transaction
2				Day	Cash Balance	Requirement	Amount
3	*Parameters and Uncontrollable Inputs*			0	$ 12,089.67	$ (451.04)	$ -
4				1	$ 11,638.63	$ 122.48	$ -
5	Transaction cost	$ 20.00		2	$ 11,761.11	$ (192.40)	$ -
6	Interest rate premium of securities	3%		3	$ 11,568.71	$ (341.71)	$ -
7	Variance of daily cash flows	$ 50,000.00		4	$ 11,227.00	$ (155.98)	$ -
8	Required minimum balance	$ 10,000.00		5	$ 11,071.02	$ (389.91)	$ -
9				6	$ 10,681.11	$ 481.06	$ -
10	*Model Outputs*			7	$ 11,162.17	$ 183.76	$ -
11				8	$ 11,345.93	$ (364.78)	$ -
12	Return point	$ 12,089.67		9	$ 10,981.15	$ (287.40)	$ -
13	Upper limit	$ 16,269.01		10	$ 10,693.75	$ 238.71	$ -
14				11	$ 10,932.46	$ (258.43)	$ -
15	*Key cell formulas*			12	$ 10,674.03	$ 453.04	$ -
16				13	$ 11,127.07	$ (151.25)	$ -
17				14	$ 10,975.82	$ (48.73)	$ -
18		Cash Balance		15	$ 10,927.08	$ 34.85	$ -
19				16	$ 10,961.94	$ 357.07	$ -
20			M	17	$ 11,319.01	$ 65.24	$ -
21				18	$ 11,384.25	$ 169.85	$ -
22				19	$ 11,554.09	$ 133.72	$ -
23				20	$ 11,687.81	$ (159.34)	$ -
24			R	21	$ 11,528.47	$ (309.01)	$ -
25				22	$ 11,219.46	$ (190.41)	$ -
26			m	23	$ 11,029.05	$ (135.57)	$ -
27				24	$ 10,893.48	$ (203.41)	$ -
28		Day		25	$ 10,690.07	$ 74.07	$ -
29				26	$ 10,764.14	$ (169.06)	$ -
30				27	$ 10,595.08	$ (372.49)	$ -

spreadsheet for implementing the model and for simulating the results. In the simulation model, we begin with a cash level equal to the return point. The next day's requirement is randomly generated as a normal variate with mean 0 and variance given in cell B7 using the Excel function =*NORMINV(RAND(),0,SQRT($B$7))*. The decision rule is applied in column G. For example, the formula in cell G4 for day 1 is =*IF(E4<=$B$8,$B$12−E4,IF(E4>=$B$13,−E4+$B$12,0))*. If the cash balance for the current day (cell E4) is less than or equal to the minimum level (cell B8), we sell B12-E4 dollars of securities to bring the balance up to the return point. Otherwise, if the cash balance exceeds the upper limit (B13), we buy enough securities (i.e., subtract an amount of cash) to bring the balance back down to the return point. If neither of these conditions hold, there is no transaction, and the balance for the next day is simply the current value plus the net requirement.

Excel Exercise 8.2

Open the worksheet *Miller-Orr Model.xls*. The model is designed to simulate 365 days. Modify the spreadsheet to compute the average daily cash balance, a count (hint: use the *COUNTIF* function) of the number of securities transactions, and the total annual cost of securities transactions and opportunity cost associated with maintaining the cash balance. Using Crystal Ball, define each of these new variables as forecasts, and simulate the model for 1,000 trials. What conclusions do you reach?

Note the similarity between the MMI and the Miller-Orr simulation models. Both increment time in equal intervals as the simulation progresses, and capture the activities that occur in each time period. Such simulation models are called **process models**. Clearly, the logic for these examples is quite simple. Most dynamic systems, however, involve a more complex set of interactions among the elements of the system, and therefore are more difficult to model. Many also involve waiting lines; therefore, we will discuss the characteristics of waiting line, or queueing, systems next.

QUEUEING SYSTEMS

The analysis of waiting lines, called *queueing theory*, applies to any situation in which customers arrive at a system, wait, and receive service. Queueing theory had its origins in 1908 with a Danish telephone engineer, A. K. Erlang, who began to study congestion in the telephone service of the Copenhagen Telephone Company. Erlang developed mathematical formulas to predict waiting times and line lengths. Over the years, queueing theory has found numerous applications in telecommunications and computer systems and has expanded to many other service systems. The objectives of queueing theory are to improve customer service and reduce operating costs. As consumers began to differentiate firms by their quality of service, reducing waiting times became an obsession with many firms. Many restaurant and department store chains take waiting seriously—some have dedicated staffs that study ways to speed up service.

All queueing systems have three elements in common:

1. *Customers* that wait for service. Customers need not be people but can be machines awaiting repair, airplanes waiting to take off, subassemblies waiting for a machine, computer programs waiting for processing, or telephone calls awaiting a customer service representative.
2. *Servers* that provide the service. Again, servers need not be only people, such as clerks, customer service representatives, or repairpersons; servers may be airport runways, machine tools, repair bays, ATMs, or computers.
3. A *waiting line* or *queue*. The queue is the set of customers waiting for service. In many cases, a queue is a physical line, like the one you join in a bank or grocery store. In other situations, a queue may not be visible and may not even be in one location, as with computer jobs waiting for processing or telephone calls waiting for an open line.

To understand the operation of a queueing system, we need to describe the characteristics of the customer, server, and queue, and how the system is configured.

Customer Characteristics

Customers arrive at the system according to some *arrival process*, which can be deterministic or probabilistic. Examples of deterministic arrivals would be parts feeding from an automated machine to an assembly line or patients arriving at appointed times to a medical facility. Most arrival processes, such as people arriving at a supermarket, are probabilistic. We can describe a probabilistic arrival process by a probability distribution representing the number of arrivals during a specific time interval, or by a distribution that represents the time between successive arrivals.

Many models assume that arrivals are governed by a **Poisson process**. This means that

1. Customers arrive one at a time, independently of each other and at random.
2. Past arrivals do not influence future arrivals; that is, the probability that a customer arrives at any point in time does not depend on when other customers arrived (sometimes we say that the system has no memory).
3. The probability of an arrival does not vary over time (the arrival rate is **stationary**).

One way to validate these assumptions is to collect empirical data about the pattern of arrivals. Not only can we observe and record the actual times of individual arrivals in order to determine the probability distribution and check if the arrival rate is constant over time, but we can also observe if customers arrive individually or in groups, and whether they exhibit any special behavior, such as not entering the system if the line is perceived as too long.

If the arrival pattern is described by a Poisson process, then the Poisson probability distribution with mean arrival rate λ (customers per unit time) can be used to describe the probability that a particular number of customers arrives during a specified time interval. For example, an arrival rate of two customers per minute means that on the average, customers arrive every half-minute, or every 30 seconds. Thus, an equivalent way of expressing arrivals is to state the *mean interarrival time* between successive customers. If λ is the mean arrival rate, the mean interarrival time is simply 1/λ. One useful result is that if the number of arrivals follows a Poisson process with mean λ, then the time between arrivals has an exponential distribution with a mean rate 1/λ. This fact will be very useful when simulating queueing systems later in this chapter.

Sometimes the arrival rate is not stationary (that is, customers arrive at different rates at different times). For instance, the demand for service at a quick-service restaurant is typically low in the mid-morning and mid-afternoon and peaks during the breakfast, lunch, and dinner hours. Individual customers may also arrive singly and independently (telephone calls to a mail order company) or in groups (a palletload of parts arriving at a machine center, or patrons at a movie theater).

The "calling population" is the set of potential customers. In many applications, the calling population is assumed to be infinite; that is, an unlimited number of possible customers can arrive at the system. This would be the case with telephone calls to a mail order company or shoppers at a supermarket. In other situations, the calling population is finite. One example would be a factory in which failed machines await repair.

Once in line, customers may not always stay in the same order as they arrived. It is common for customers to "renege," or leave a queue, before being served if they get tired of waiting. In queueing systems with multiple queues, customers may "jockey," or switch lines, if they perceive another to be moving faster. Some customers may arrive at the system, determine that the line is too long, and decide not to join the queue. This behavior is called "balking."

Service Characteristics

Service occurs according to some *service process*. The time it takes to serve a customer may be deterministic or probabilistic. In the probabilistic case, the service time is described by some probability distribution. In many queueing models, we make the assumption that service times follow an exponential distribution with a mean service rate μ, the average number of customers served per unit time. Thus, the average service time is 1/μ. One reason that the exponential distribution describes many realistic service phenomena is that it has a useful property—the probability of small service times is large. For example, the probability that an exponential random variable T exceeds the mean is only 0.368. This means that we see a large number of short service times and a few long ones. Think of your own experience in grocery stores. Most customers' service times are relatively short; however, every once in a while you see a shopper with a large number of groceries (who gets out the checkbook only *after* all of the groceries are scanned!).

The exponential distribution, however, does not seem to be as common in modeling service processes as the Poisson is in modeling arrival processes. Management scientists have found that many queueing systems have service time distributions that are not exponential, and may be constant, normal, or some other probability distribution.

Other service characteristics include nonstationary service times (taking orders and serving dinner might be longer than for breakfast), and service times that depend on the type of customer (patients at an emergency room). The service process may include one or several servers. The service characteristics of multiple servers may be identical or different. In some systems, certain servers may only service specific types of customers. In many systems, such as restaurants and department stores, managers vary the number of servers to adjust to busy and slack periods.

Queue Characteristics

The order in which customers are served is defined by the *queue discipline*. The most common queue discipline is first-come, first-served (FCFS). In some situations, a queue may be structured as last-come, first-served (LCFS); just think of the in-box on a clerk's desk. At an emergency room, patients are usually served according to a priority determined by triage, with the more critical patients served first.

System Configuration

The customers, servers, and queues in a queueing system can be arranged in various ways. Three common queueing configurations are as follows:

1. One or more parallel servers fed by a single queue (see Figure 8.6). This is the typical configuration used by many banks and airline ticket counters.
2. Several parallel servers fed by their own queues (see Figure 8.7). Most supermarkets and discount retailers use this type of system.

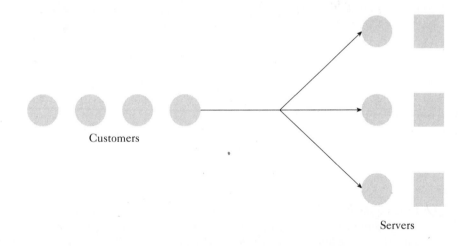

FIGURE 8.6
Single queue configuration

Customers

Servers

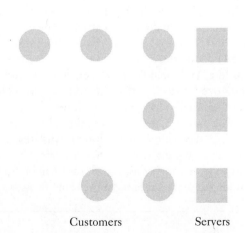

FIGURE 8.7
Parallel queue configuration

Customers   Servers

3. A combination of several queues in series. This structure is common when multiple processing operations exist, such as in manufacturing facilities and many service systems. The example of a typical voting facility is shown in Figure 8.8.

FIGURE 8.8
Queues in series in a typical voting facility

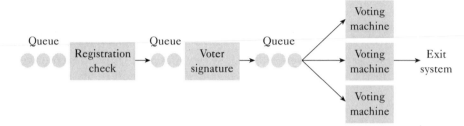

Source: Figure from Floyd H. Grant, III, " Reducing Voter Waiting Time," *Interface*, Vol. 10, No. 5, October 1980, pp. 19–25.

The choice of what system to use, particularly choosing between one or several queues, is not always easy, as shown in the next Management Science Practice Box.

Management Science Practice: One Line or More?

Customers become frustrated when a person enters a line next to them and receives service first. Of course, *that* customer feels a certain sense of satisfaction. People expect to be treated fairly; in queueing situations that means "first-come, first-served." In the mid-1960s, Chemical Bank was one of the first to switch from multiple parallel lines to a serpentine line (one line feeding into several servers). American Airlines copied this at their airport counters, and most others followed suit. Even the U.S. Postal Service has migrated to single lines. Studies have shown that customers are happier when they wait in a serpentine line, rather than in parallel lines, even if that type of line increases their wait.

Fast food franchises like Wendy's and Burger King have used single lines for many years.[1] Burger King found that multiple lines create stress and anxiety, whereas using a single line allows customers to focus on what they want to order and not be distracted by which line is shorter. However, critics call the single line system dehumanizing and make analogies to cattle corrals. Some are unfriendly to disabled customers. In fact, Wendy's reached a settlement with the Justice Department to remove or widen the roped lines to accommodate wheelchairs.

Other companies, like McDonald's, however, have stayed with multiple lines. Some McDonald's executives feel that the multiple line system accommodates higher volumes of customers more quickly, despite time studies that have proven a single line to be faster. But the perception of a long single line may cause customers to leave, and consumer perception is what really counts. Nevertheless, McDonald's has been testing whether to consolidate its multiple waiting lines into one. If the experiment succeeds, franchisees will be free to implement a single line system as long as all restaurants in the same market agree on it.

MODELING AND SIMULATING QUEUEING SYSTEMS

A queueing model provides measures of system performance, which typically are

1. The quality of the service provided to the customer.
2. The efficiency of the service operation and the cost of providing the service.

Various numerical measures of performance can be used to evaluate the quality of the service provided to the customer. These include the following:

*   Waiting time in the queue
*   Time in the system (waiting time plus service time)
*   Completion by a deadline

The efficiency of the service operation can be evaluated by computing such measures as the following:

*   Average queue length
*   Average number of customers in the system (queue plus those in service)
*   Throughput—the rate at which customers are served
*   Server utilization—percentage of time servers are busy
*   Percentage of customers who balk or renege

Usually we can use these measures to compute an operating cost in order to compare alternative system configurations. The most common measures of queueing system performance, called the "operating characteristics" of the queueing system, and the symbols used to denote them, are

$$L_q = \text{average number in the queue}$$

$$L = \text{average number in the system}$$

$$W_q = \text{average waiting time in the queue}$$

$$W = \text{average time in the system}$$

$$P_0 = \text{probability that the system is empty}$$

Let us examine a system with a single server, a single waiting line, and a FCFS queue discipline. A conceptual model of this system is shown in Figure 8.9. Customers arrive for service. They wait if the server is busy. They receive the service and finally leave the system. Performance measures that we would like to measure are the average waiting time per customer, the length of the queue, and the percentage of time that the server is idle.

To model this system and understand its behavior, we need to know the arrival pattern of customers and the time required for service. For arrivals, we could specify either the actual *times* that customers arrive for service or the *times between successive arrivals*. The second approach works better for simulation purposes because we need only know the time that the last customer arrived to generate the arrival time of the next customer.

Developing a detailed model requires a bit of logical thought. A key question to ask in developing any model is this: What do we need to know to compute the performance measures we want? In this example, we are interested in the *waiting time* of customers, the *length of the queue*, and the *idle time* of the server. We note the following:

FIGURE 8.9
A conceptual model of a single server queue

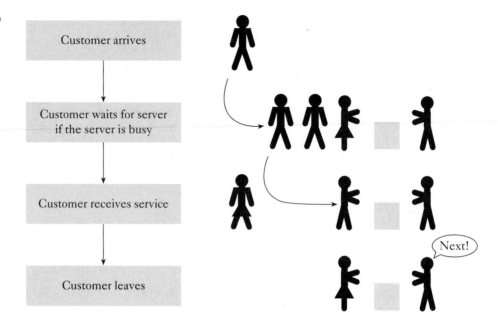

1. The waiting time of any customer is equal to the time at which the customer begins service minus the time the customer arrived.
2. The server is idle if the time at which a customer arrives is greater than the time at which the previous customer completed service. The idle time is the difference between these two times.
3. If a customer arrives at time t, then all prior customers who have not yet completed service by time t must still be in the system.

Thus, to compute the waiting time of a customer we need to know the *time that service begins* and the *arrival time* of the customer. To find the idle time of the server, we need to know the *time that the customer arrives* and the *time that the previous customer completed service*. The server remains busy when a customer arrives if the arrival time is less than the time that the previous customer will complete service. To determine the length of the queue upon the arrival of a customer, we need to know the number of customers whose scheduled completion time is later than the arrival time of the current customer minus one (the customer currently being served).

We now make the following additional observations:

4. If a customer arrives at time t and the server is not busy, then that customer can begin service immediately upon arrival.
5. If a customer arrives at time t and the server is busy, then that customer will begin service at the time that the *previous* customer completes service (which will be greater than t).
6. In either case, the time at which a customer completes service is computed as the time that the customer begins service plus the time it takes to perform the service.
7. Once the completion times are known, we may compare them to the arrival time of any customer to find the length of the queue.

Figure 8.10 translates these observations into a detailed flowchart. We may use this flowchart as a basis for constructing a spreadsheet or other computer simulation model. In this flowchart, *ST* represents the service time, and *TBA* the time between arrivals (from the previous customer to the current one).

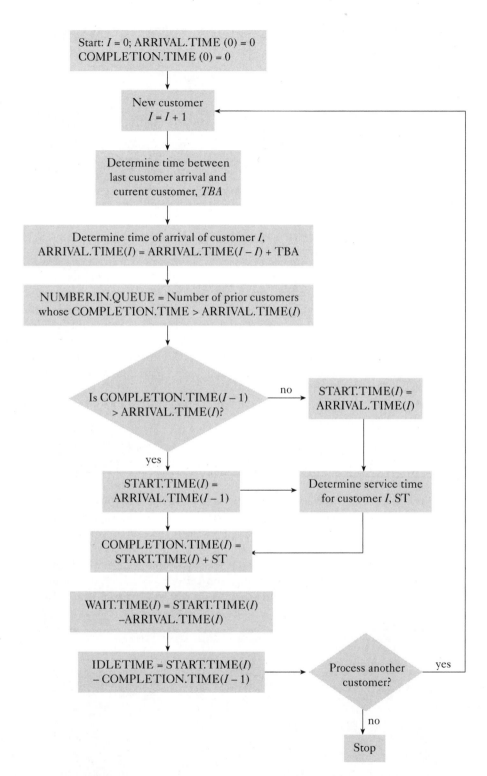

FIGURE 8.10
Simulation flow-
chart for a
single server
queue

Start: $I = 0$; ARRIVAL.TIME (0) = 0
COMPLETION.TIME (0) = 0

New customer
$I = I + 1$

Determine time between
last customer arrival and
current customer, *TBA*

Determine time of arrival of customer I,
ARRIVAL.TIME(I) = ARRIVAL.TIME($I - 1$) + TBA

NUMBER.IN.QUEUE = Number of prior customers
whose COMPLETION.TIME > ARRIVAL.TIME(I)

Is COMPLETION.TIME($I - 1$)
> ARRIVAL.TIME(I)?

no

START.TIME(I) =
ARRIVAL.TIME(I)

yes

START.TIME(I) =
ARRIVAL.TIME($I - 1$)

Determine service time
for customer I, ST

COMPLETION.TIME(I) =
START.TIME(I) + ST

WAIT.TIME(I) = START.TIME(I)
−ARRIVAL.TIME(I)

IDLETIME = START.TIME(I)
− COMPLETION.TIME($I - 1$)

Process another
customer?

yes

no

Stop

We can implement the logic and equations given in Figure 8.10 easily on a
spreadsheet, as shown in Figure 8.11. The start time for any customer is the maximum of the arrival time and the time of service of the previous customer. This is

	A	B	C	D	E	F	G	H	I
1	SINGLE SERVER QUEUE								
2									
3	CUSTOMER	TIME BETWEEN	ARRIVAL	NUMBER	START	SERVICE	COMPLETION	WAIT	IDLE
4		ARRIVALS	TIME	IN QUEUE	TIME	TIME	TIME	TIME	TIME
5							0		
6	1	1	=B6	0	=MAX(C6,G5)	1	=E6+F6	=E6-C6	=E6-G5
7	2	1	=C6+B7	=A7-MATCH(C7,$G$5:G6,1)	=MAX(C7,G6)	1	=E7+F7	=E7-C7	=E7-G6
8	3	1	=C7+B8	=A8-MATCH(C8,$G$5:G7,1)	=MAX(C8,G7)	1	=E8+F8	=E8-C8	=E8-G7
9	4	1	=C8+B9	=A9-MATCH(C9,$G$5:G8,1)	=MAX(C9,G8)	1	=E9+F9	=E9-C9	=E9-G8
10	5	1	=C9+B10	=A10-MATCH(C10,$G$5:G9,1)	=MAX(C10,G9)	1	=E10+F10	=E10-C10	=E10-G9
11	6	1	=C10+B11	=A11-MATCH(C11,$G$5:G10,1)	=MAX(C11,G10)	1	=E11+F11	=E11-C11	=E11-G10
12	7	1	=C11+B12	=A12-MATCH(C12,$G$5:G11,1)	=MAX(C12,G11)	1	=E12+F12	=E12-C12	=E12-G11
13	8	1	=C12+B13	=A13-MATCH(C13,$G$5:G12,1)	=MAX(C13,G12)	1	=E13+F13	=E13-C13	=E13-G12
14	9	1	=C13+B14	=A14-MATCH(C14,$G$5:G13,1)	=MAX(C14,G13)	1	=E14+F14	=E14-C14	=E14-G13
15	10	1	=C14+B15	=A15-MATCH(C15,$G$5:G14,1)	=MAX(C15,G14)	1	=E15+F15	=E15-C15	=E15-G14
16									
17	Average	=AVERAGE(B6:B15)				=AVERAGE(F6:F15)			

FIGURE 8.11   Spreadsheet template for queueing simulation

implemented using the *MAX* function. Waiting time is simply the difference between the start time and arrival time; idle time for the server is computed from the arrival time of the current customer and the completion time of the last customer. Finding the length of the queue when a customer arrives is a bit tricky. We use the MATCH function to determine the last customer, say customer 9, whose completion time is less than or equal to the arrival time of the current customer, say customer 15. Then all customers who arrived after customer 9 are still in the system (including, now, customer 15). Therefore, the system has 6 customers, one of whom is in service, so the length of the queue is 1 less, or 5. This is the number of the current customer minus the value of the MATCH function.

THE DYNAMICS OF WAITING LINES

Suppose we have a single server system with arrivals occurring at constant unit intervals that are perfectly matched to unit service times, as shown in Figure 8.12. Then clearly, each customer will start immediately upon arrival and will finish at the time of the next arrival, resulting in no waiting, no idle time, and no queue build-up.

Now suppose that we vary the arrival pattern by simply staggering the arrivals, but maintaining the same *average* time between arrivals, as shown in Figure 8.13. We see that this system now exhibits some idle time and waiting time. If we go

FIGURE 8.12
Queueing system with synchronized arrivals and services

	A	B	C	D	E	F	G	H	I
1	SINGLE SERVER QUEUE								
2									
3	CUSTOMER	TIME BETWEEN	ARRIVAL	NUMBER	START	SERVICE	COMPLETION	WAIT	IDLE
4		ARRIVALS	TIME	IN QUEUE	TIME	TIME	TIME	TIME	TIME
5							0.00		
6	1	1.0000	1.0000	0	1.0000	1.0000	2.0000	0.0000	1.0000
7	2	1.0000	2.0000	0	2.0000	1.0000	3.0000	0.0000	0.0000
8	3	1.0000	3.0000	0	3.0000	1.0000	4.0000	0.0000	0.0000
9	4	1.0000	4.0000	0	4.0000	1.0000	5.0000	0.0000	0.0000
10	5	1.0000	5.0000	0	5.0000	1.0000	6.0000	0.0000	0.0000
11	6	1.0000	6.0000	0	6.0000	1.0000	7.0000	0.0000	0.0000
12	7	1.0000	7.0000	0	7.0000	1.0000	8.0000	0.0000	0.0000
13	8	1.0000	8.0000	0	8.0000	1.0000	9.0000	0.0000	0.0000
14	9	1.0000	9.0000	0	9.0000	1.0000	10.0000	0.0000	0.0000
15	10	1.0000	10.0000	0	10.0000	1.0000	11.0000	0.0000	0.0000
16									
17	Average	1.0000				1.0000			

	A	B	C	D	E	F	G	H	I	J	K	L	M
1	SINGLE SERVER QUEUE												
2													
3	CUSTOMER	TIME BETWEEN	ARRIVAL	NUMBER	START	SERVICE	COMPLETION	WAIT	IDLE	CUMULATIVE	AVERAGE	CUMULATIVE	AVERAGE
4		ARRIVALS	TIME	IN QUEUE	TIME	TIME	TIME	TIME	TIME	WAIT TIME	WAIT TIME	IDLE TIME	IDLE TIME
5							0.00						
6	1	0.5000	0.5000	0	0.5000	1.0000	1.5000	0.0000	0.5000	0.0000	0.0000	0.5000	0.3333
7	2	1.5000	2.0000	0	2.0000	1.0000	3.0000	0.0000	0.5000	0.0000	0.0000	1.0000	0.3333
8	3	0.5000	2.5000	1	3.0000	1.0000	4.0000	0.5000	0.0000	0.5000	0.1667	1.0000	0.2500
9	4	1.5000	4.0000	0	4.0000	1.0000	5.0000	0.0000	0.0000	0.5000	0.1250	1.0000	0.2000
10	5	0.5000	4.5000	1	5.0000	1.0000	6.0000	0.5000	0.0000	1.0000	0.2000	1.0000	0.1667
11	6	1.5000	6.0000	0	6.0000	1.0000	7.0000	0.0000	0.0000	1.0000	0.1667	1.0000	0.1429
12	7	0.5000	6.5000	1	7.0000	1.0000	8.0000	0.5000	0.0000	1.5000	0.2143	1.0000	0.1250
13	8	1.5000	8.0000	0	8.0000	1.0000	9.0000	0.0000	0.0000	1.5000	0.1875	1.0000	0.1111
14	9	0.5000	8.5000	1	9.0000	1.0000	10.0000	0.5000	0.0000	2.0000	0.2222	1.0000	0.1000
15	10	1.5000	10.0000	0	10.0000	1.0000	11.0000	0.0000	0.0000	2.0000	0.2000	1.0000	0.0909
16													
17	Average	1.0000				1.0000							

FIGURE 8.13   Queueing system with unsynchronized arrivals and services

one step further, and introduce random arrival and service times (using the formula 0.5 + RAND() to maintain the same approximate average time between arrivals and service times), we obtain the results shown in Figure 8.14. We see that both the length of the queue and the waiting times appear to increase as more customers arrive. These experiments show that queueing behavior occurs because of the variation in arrival and service times. Because this is a characteristic of any real system, queueing is a fact of life!

Excel Exercise 8.3

Open the worksheet *single server queue.xls* and copy the formulas to model the arrival of 100 customers. Experiment with different random arrival and service distributions, specifically the following combinations:

Time between arrivals	Service time
1. 0.5 + *RAND()*	1.0 + *RAND()*
2. 0.5 + *RAND()*	*RAND()*

What differences do you observe in the performance of these systems?

	A	B	C	D	E	F	G	H	I	J	K	L	M
1	SINGLE SERVER QUEUE												
2													
3	CUSTOMER	TIME BETWEEN	ARRIVAL	NUMBER	START	SERVICE	COMPLETION	WAIT	IDLE	CUMULATIVE	AVERAGE	CUMULATIVE	AVERAGE
4		ARRIVALS	TIME	IN QUEUE	TIME	TIME	TIME	TIME	TIME	WAIT TIME	WAIT TIME	IDLE TIME	IDLE TIME
5							0.00						
6	1	1.1470	1.1470	0	1.1470	0.6481	1.7951	0.0000	1.1470	0.0000	0.0000	1.1470	0.6390
7	2	1.0183	2.1653	0	2.1653	1.1264	3.2917	0.0000	0.3702	0.0000	0.0000	1.5172	0.4609
8	3	0.7725	2.9378	1	3.2917	1.0181	4.3098	0.3539	0.0000	0.3539	0.1180	1.5172	0.3520
9	4	0.8047	3.7425	1	4.3098	1.4539	5.7637	0.5673	0.0000	0.9212	0.2303	1.5172	0.2632
10	5	1.1913	4.9338	1	5.7637	0.7313	6.4950	0.8299	0.0000	1.7511	0.3502	1.5172	0.2336
11	6	0.6489	5.5827	2	6.4950	1.1486	7.6436	0.9123	0.0000	2.6634	0.4439	1.5172	0.1985
12	7	0.7973	6.3800	2	7.6436	1.1272	8.7708	1.2636	0.0000	3.9270	0.5610	1.5172	0.1730
13	8	1.0636	7.4436	2	8.7708	0.8877	9.6585	1.3272	0.0000	5.2542	0.6568	1.5172	0.1571
14	9	1.1447	8.5883	2	9.6585	0.7427	10.4012	1.0702	0.0000	6.3244	0.7027	1.5172	0.1459
15	10	1.0502	9.6385	2	10.4012	0.9537	11.3549	0.7627	0.0000	7.0871	0.7087	1.5172	0.1336
16													
17	Average	0.9639				0.9838							

FIGURE 8.14   Queueing system with random arrivals and service times

We can gain more understanding about the dynamics of queues by conducting further simulation experiments and computing additional performance measures. Figure 8.15 shows a portion of a spreadsheet for a single server system with Poisson arrivals and exponential service times. The mean arrival and service rates are entered in cells B5 and B6. The expected values of key operating characteristics are computed in cells B8:B10; these formulas will be described in the next section. From the fact that Poisson arrivals correspond to exponential interarrival times, we may generate the times between arrivals (as well as the exponential service times) using the formula described in the previous chapter:

$$-(1/\lambda)*LN(RAND())$$

or using the Crystal Ball function CB.Exponential(mean). The calculations in columns E through K correspond to the formulas used in Figure 8.10. We suggest you work through the first few rows of the spreadsheet in conjunction with Figure 8.10 to verify them.

Columns L through Q compute operating characteristic measures using the simulation results. The average number in the queue is a **time-averaged statistic**. To understand this, think of the following analogy. Suppose you want to know the *average balance* in your checking account over a month. This statistic depends not only on how much money is in the account, but also on *how long* it is held. For instance, if you have a balance of $100 for 15 days and a balance of $200 for 15 days, your average balance is $150. However, if you have $100 for 10 days, and $200 for 20 days, your average balance is

$$\frac{(\$100)(10) + (\$200)(20)}{10 + 20} = \$166.67$$

	A	B	C	D	E	F	G	H	I	J	K
1	Single Server Queueing		CUSTOMER	TIME BET.	ARRIVAL	NUMBER	START		SERVICE	COMPLETION	WAIT	IDLE
2	Simulation Spreadsheet			ARRIVALS	TIME	IN QUEUE	TIME		TIME	TIME	TIME	TIME
3										0.00		
4			1	0.0061	0.0061	0.0000	0.0061	0.6976	0.7036	0.0000	0.0061
5	Mean arrival rate	2	2	0.1755	0.1816	1.0000	0.7036	0.1218	0.8255	0.5220	0.0000
6	Mean service rate	3	3	0.2688	0.4504	2.0000	0.8255	0.4295	1.2550	0.3751	0.0000
7			4	0.3867	0.8371	1.0000	1.2550	0.0217	1.2767	0.4179	0.0000
8	Expected No. in Queue	1.333	5	0.1092	0.9463	2.0000	1.2767	1.1207	2.3974	0.3304	0.0000
9	Expected Waiting Time	0.667	6	0.3789	1.3252	1.0000	2.3974	0.3522	2.7496	1.0722	0.0000
10	Expected % idle time	33%	7	0.2516	1.5769	2.0000	2.7496	0.0789	2.8285	1.1727	0.0000
11			8	0.4198	1.9967	3.0000	2.8285	0.3128	3.1413	0.8318	0.0000
12	Key cell formulas		9	1.1742	3.1708	0.0000	3.1708	0.1015	3.2723	0.0000	0.0295
13			10	0.3388	3.5096	0.0000	3.5096	0.1054	3.6150	0.0000	0.2373

	L	M	N	O	P	Q
1	CUMULATIVE	AVERAGE	CUMULATIVE	AVERAGE	CUMULATIVE	AVERAGE
2	NO.IN QUEUE	NO. IN QUEUE	WAIT TIME	WAIT TIME	IDLE TIME	% IDLE TIME
3						
4	0.0000	0.0000	0.0000	0.0000	0.0000	0%
5	0.1755	0.9666	0.5220	0.2610	0.0061	1%
6	0.7131	1.5833	0.8971	0.2990	0.0061	0%
7	1.0998	1.3138	1.3150	0.3288	0.0061	0%
8	1.3182	1.3930	1.6454	0.3291	0.0061	0%
9	1.6971	1.2807	2.7175	0.4529	0.0061	0%
10	2.2004	1.3955	3.8903	0.5558	0.0061	0%
11	3.4598	1.7328	4.7221	0.5903	0.0061	0%
12	3.4598	1.0911	4.7221	0.5247	0.0356	1%
13	3.4598	0.9858	4.7221	0.4722	0.2729	8%

FIGURE 8.15   Queueing simulation spreadsheet for Poisson arrivals and exponential service times

Essentially, you multiply each balance by the amount of time held, sum them up, and divide by the total number of days to compute the average over the time period. Applying this idea to the queueing simulation results, we multiply the number in the queue when each customer arrives by the time interval between arrivals, summing as we go along (column L) and then divide by the current time (column M), yielding an average value to that time. (This is only approximate because we are not considering the time that customers leave the system. This is too difficult to attempt on a spreadsheet; however, this approach does provide useful insights, as we will see later in this chapter.)

The average wait time per customer is not a time-averaged statistic; we simply sum the total wait times of all customers and compute a simple average. Finally, the average percent idle time is found by dividing the cumulative idle time accrued when a customer leaves by the completion time in column I and converting to a percentage.

Figures 8.16 through 8.19 show graphs of these performance measures and their expected values for a simulation of 500 customers. We observe several things. First, the number in the queue fluctuates considerably between 0 and 13 customers, and seems to be lower at the beginning of the simulation than later. It does not appear, however, that the line is building up indefinitely. Second, the average number in the queue, average wait time, and average percent idle time also vary as the simulation progresses; however, these values appear to be approaching their expected values.

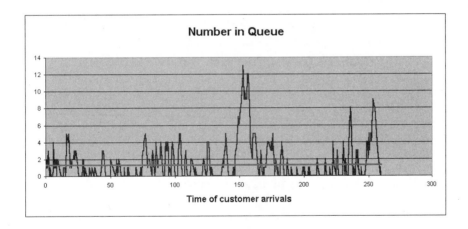

FIGURE 8.16
Graph of number in queue at time of customer arrivals

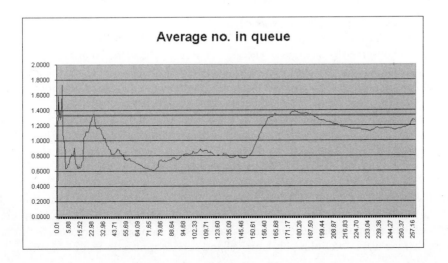

FIGURE 8.17
Graph of average number in queue over time

FIGURE 8.18
Graph of
average waiting
time/customer

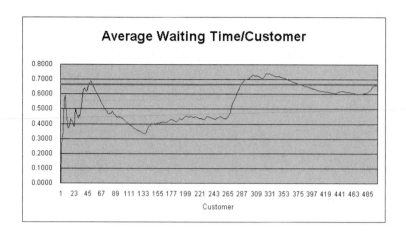

FIGURE 8.19
Graph of
average % idle
time for the
server

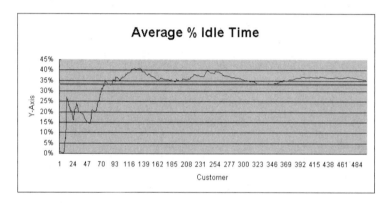

Replicating this simulation can result in dramatically different results, and we invite you to see this in the following Excel Exercise. We will explain these phenomena after we discuss analytical results in the next section.

Excel Exercise 8.4

Open the Excel workbook *queueing simulation model.xls*. Note that the graphs are stored in separate worksheets. Replicate the simulation by pressing the F9 key and observe the changes in the graphs. Record the values that average wait times and percent idle times seem to be approaching. How much do they vary? You might also wish to replicate the entire model using Crystal Ball, using cells M503, O503 and Q503 as forecast cells similar to Example 2.

ANALYTICAL QUEUEING MODELS

Analytical models of queueing behavior depend on some key assumptions about the arrival and service processes, the most common being Poisson arrivals and exponential services. In applying these models, as well as for simulation purposes, the unit of time that we use in modeling both arrival and service processes can be arbitrary. For example, a mean arrival rate of 2 customers per minute is equivalent to 120 customers per hour. We must be careful, however, to express the arrival rate and service rate in the same time units.

Because of the variety of assumptions that can be made, D. G. Kendall developed a taxonomy for queueing models that simplifies discussion and communication. We characterize a queueing model using the notation A/B/s, where

A = Arrival distribution

B = Service time distribution

s = Number of servers

Symbols used to specify the arrival and service time distributions include

M = Exponential distribution (Poisson process)

G = General (arbitrary) distribution of service times

GI = General (arbitrary) distribution of arrivals

D = Deterministic (constant) value

Thus, a queueing model with Poisson arrivals, exponential service times, and a single server is denoted as M/M/1; a model with Poisson arrivals, general service times, and two servers is denoted as M/G/2. Where does the M come from? Recall that a Poisson process has the *memoryless* property.

We will present analytical results for three basic queueing models. Except for special cases, queueing models in general are rather difficult to formulate and solve, even when the distribution of arrivals and departures is known. We will use simulation to better understand the implications of these analytical results.

The M/M/1 Queueing Model

The M/M/1 queueing model is the simplest. It assumes that arrivals follow a Poisson process with mean arrival rate λ, that service times are exponentially distributed with mean service rate μ, and that the queue discipline is FCFS. For this model, the operating characteristics are:

$$\text{Average number in the queue} = L_q = \frac{\lambda^2}{\mu(\mu - \lambda)}$$

$$\text{Average number in the system} = L = \frac{\lambda}{\mu - \lambda}$$

$$\text{Average waiting time in the queue} = W_q = \frac{\lambda}{\mu(\mu - \lambda)}$$

$$\text{Average time in the system} = W = \frac{1}{\mu - \lambda}$$

$$\text{Probability that the system is empty} = P_0 = 1 - \lambda/\mu$$

We note that these formulas are valid only if $\lambda < \mu$. If $\lambda \geq \mu$ (that is, the rate of arrivals is at least as great as the service rate), the numerical results become nonsensical. In practice, this means that the queue will never "average out" but will grow indefinitely (we will discuss this further in the section on simulation). It should be obvious that when $\lambda > \mu$, the server will not be able to keep up with the demand. However, it may seem a little strange that this will occur even when $\lambda = \mu$. You would think that an equal arrival rate and service rate should result in a "balanced" system.

This would be true in the deterministic case when both arrival and service rates are constant. However, when any variation exists in the arrival or service pattern, the queue will eventually build up indefinitely. The reason is that individual arrival times and service times vary in an unpredictable fashion even though their averages may be constant. As a result, there will be periods of time in which demand is low and the server is idle. This time is lost forever, and the server will not be able to make up for periods of heavy demand at other times. This also explains why queues form when $\lambda < \mu$.

Example 4: Computing M/M/1 Operating Characteristics

Customers arrive at an airline ticket counter at a rate of 2 customers per minute and can be served at a rate of 3 customers per minute. Note that the average time between arrivals is 1/2 minute per customer and the average service time is 1/3 minute per customer. Using the M/M/1 queueing formulas, we have

$$\text{Average number in the queue} = L_q = \frac{\lambda^2}{\mu(\mu - \lambda)} = \frac{2^2}{3(3 - 2)} = 1.33 \text{ customers}$$

$$\text{Average number in the system} = L = \frac{\lambda}{\mu - \lambda} = \frac{2}{3 - 2} = 2.00 \text{ customers}$$

$$\text{Average waiting time in the queue} = W_q = \frac{\lambda}{\mu(\mu - \lambda)} = \frac{2}{3(3 - 2)} = 0.67 \text{ minutes}$$

$$\text{Average time in the system} = W = \frac{1}{\mu - \lambda} = \frac{1}{3 - 2} = 1.00 \text{ minutes}$$

$$\text{Probability that the system is empty} = P_0 = 1 - \lambda/\mu = 0.33$$

These results indicate that on the average, 1.33 customers will be waiting in the queue. In other words, if we took photographs of the waiting line at random times, we would find an average of 1.33 customers waiting. If we include any customers in service, the average number of customers in the system is 2. Each customer can expect to wait an average of .67 minute in the queue, and spend an average of 1 minute in the system. About one-third of the time, we would expect to see the system empty and the server idle.

Other Queueing Models

Researchers have developed results for many other types of queueing situations. It is impossible to describe them all; however, we will discuss three common and useful models. The operating characteristic formulas are summarized in Table 8.1. All can easily be implemented on spreadsheets and are available in the Excel workbook *Queueing Models.xls*.

ARBITRARY SERVICE TIMES In some situations, we can assume that arrivals are Poisson, but cannot verify that service times are exponential. We may not even be able to fit a good distribution to service time data. With an M/G/1 model, all we need to know is the mean service time $1/\mu$ and the variance of the service time σ^2. In Table 8.1, notice that σ^2 appears explicitly in L_q and implicitly in L, W_q, and W. Thus, as the variance of the service time increases, the line lengths and waiting times increase. When the variance is zero, we have a constant service time and the formulas are somewhat simplified.

TABLE 8.1 Performance measures for three common queueing models

Model	M/G/1	M/M/s	Finite Calling Population
Assumptions	Poisson arrivals Arbitrary service times with variance $= \sigma^2$ Single server	Poisson arrivals Exponential service times s parallel servers	Poisson arrivals Exponential service times N customers in population
L_q	$\dfrac{\lambda^2\sigma^2 + (\lambda/\mu)^2}{2(1 - \lambda/\mu)}$	$P_0 \dfrac{(\lambda/\mu)^{s+1}}{(s-1)!(s-\lambda/\mu)^2}$	$N - \dfrac{\lambda+\mu}{\lambda}(1 - P_0)$
L	$L_q + \lambda/\mu$	λW	$L_q + (1 - P_0)$
W_q	L_q/λ	L_q/λ	$\dfrac{L_q}{(N-L)\lambda}$
W	$W_q + 1/\mu$	$W_q + 1/\mu$	$W_q + 1/\mu$
P_0	$1 - \lambda/\mu$	$\dfrac{1}{\left(\sum\limits_{n=0}^{s-1}\dfrac{(\lambda/\mu)^n}{n!}\right) + \dfrac{(\lambda/\mu)^s}{s!}\left(\dfrac{1}{1-\lambda/(s\mu)}\right)}$	$\dfrac{1}{\sum\limits_{n=0}^{N}\dfrac{N!}{(N-n)!}(\lambda/\mu)^n}$

MULTIPLE SERVERS You have probably experienced many situations, such as in banks or your college advising or cashier's office, where several servers are available. Customers form a single queue (behind the ubiquitous "Wait Here for Next Available Server" sign) and are served by the next available server. Many telephone-based customer service operations (call centers) operate in this manner. As with the M/M/1 model, steady state will occur only if the arrival rate is less than the service rate. However, since we have multiple servers, the combined service rate is $s\mu$. Therefore, steady state will occur when $\lambda < s\mu$; otherwise, the queue will grow indefinitely.

FINITE CALLING POPULATIONS All the models we have considered so far assume that the calling population is infinite; that is, an unlimited number of customers may seek service. With this assumption, the mean arrival rate remains constant regardless of how many customers are in the system. In many applications, though, the calling population is finite. One example would be a factory in which the "customers" are machines that have failed and are waiting for repair. Clearly, if all machines are down, then no further arrivals can take place. Thus, the mean arrival rate depends on the number of customers in the system. Another example is an office copy machine; the calling population consists only of the workers in the office.

Little's Law

MIT Professor John D. C. Little has made many contributions to the field of management science. He is most famous for recognizing a simple yet powerful relationship among operating characteristics in queueing systems. "Little's Law," as it has become known, is very simple: **For any steady-state queueing system, $L = \lambda W$.** This states that the average number of customers in a system is equal to the mean arrival rate times the average time in the system. An intuitive explanation of this result can be seen in the following way. Suppose that you arrive at a queue and spend W minutes in the system (waiting plus service). During this time, more customers will arrive at a rate λ. Thus, when you complete service, a total of λW customers will have arrived after you. This is precisely the number of customers that remain in the system when you leave, or L. Using similar arguments, we can also show that **for any steady-state queueing system, $L_q = \lambda W_q$.** This is similar to the first result and states that the average length of the queue equals the mean arrival rate times the average waiting time.

These results provide an alternative way of computing operating characteristics instead of using the formulas provided earlier. For example, if L is known, then we may compute W by L/λ. Also, W_q can be computed as L_q/λ. Two other general relationships that are useful are

$$L = L_q + \lambda/\mu$$

and

$$W = W_q + 1/\mu$$

The first relationship states that the average number in the system is equal to the average queue length plus λ/μ. This makes sense if you recall that the probability that the system is empty is $P_0 = 1 - \lambda/\mu$. Thus, λ/μ is the probability that at least one customer is in the system. If there is at least one customer in the system, then the server must be busy. The term λ/μ simply represents the expected number of customers in service.

The second relationship states that the average time in the system is equal to the average waiting time plus the average service time. This makes sense because the time spent in the system for any customer consists of the waiting time plus the time in service. As an exercise, you might show that these relationships are the same mathematically as the formulas presented earlier for the M/M/1 queueing system.

Analytical Models vs. Simulation

The M/M/1 queueing model (and all analytical queueing models) provides steady-state values of operating characteristics. By "steady state" we mean that the probability distribution of the operating characteristics does not vary with time. This means that no matter when we observe the system, we would expect to see the same average values of queue lengths, waiting times, and so on. This usually does not happen in practice, even if the average arrival rate and average service rate are constant over time.

To understand this, think of an amusement park that opens at 10 a.m. When it opens, there are no customers in the system and hence no queues at any of the rides. As customers arrive, it will take some time for queues to build up. For the first hour or so, the lines and waiting times for popular rides grow longer but then eventually decrease. This is called the "transient period." Eventually (if the arrival and service rates remain constant), the lines will stabilize and reach steady state. Mathematical queueing models provide only steady-state results. Thus, if we are interested in how long it takes to reach steady state, for example, we must resort to other methods of analysis, such as simulation.

We observed transient phenomena in Figures 8.17 to 8.19. The graphs appear to be approaching their steady state values, although the average number in queue and average wait time have not yet reached them. These differences are due to the relatively short simulation run (500 customers, or about 250 minutes). The large fluctuations during the transient period bias the averages.

Excel Exercise 8.5

Open the Excel workbook *queueing simulation model.xls* and copy the formulas to simulate 1,000 customers. You should also change the data series in the charts to reflect the new data. What do you observe about steady state?

Sensitivity Analysis

Spreadsheets can be useful in dealing with queueing models, since they allow managers to study the effects of variations in assumptions. For example, we might be uncertain of our estimate of the arrival rate and want to know how the performance measures would change if our estimate varied by as much as 20 percent. The spreadsheet in Figure 8.20 and its accompanying graph, shown in Figure 8.21,

	A	B	C	D	E	F	G	H	I	J
1	**M/M/1 Queueing Model**									
2										
3	Lambda	1.6	1.7	1.8	1.9	2	2.1	2.2	2.3	2.4
4	Mu	3.00	3.00	3.00	3.00	3.00	3.00	3.00	3.00	3.00
5										
6	Average number in queue	0.61	0.74	0.90	1.09	1.33	1.63	2.02	2.52	3.20
7	Average number in system	1.14	1.31	1.50	1.73	2.00	2.33	2.75	3.29	4.00
8	Average time in queue	0.38	0.44	0.50	0.58	0.67	0.78	0.92	1.10	1.33
9	Average waiting time in system	0.71	0.77	0.83	0.91	1.00	1.11	1.25	1.43	1.67
10	Probability system is empty	0.47	0.43	0.40	0.37	0.33	0.30	0.27	0.23	0.20
11										
12	*Key Cell Formulas*									
13										

FIGURE 8.20
Spreadsheet for M/M/1 Queueing Model analysis

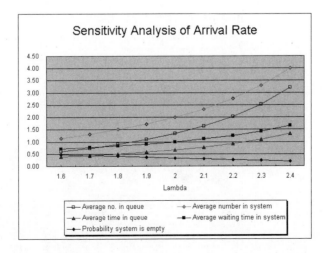

FIGURE 8.21
Graph of performance measures as arrival rate varies

are designed to answer this question. Each column B through J of the spreadsheet provides the operating characteristics for a constant service rate of 3 customers per minute, with the arrival rate varying from 1.6 to 2.4 customers per minute (a 20 percent variation from the original figure of 2 customers per minute). We see that the average numbers in the queue and system are the most sensitive to a 20 percent error in our estimate of λ. Waiting times are less sensitive. We also see that as the arrival rate increases, the average number in the queue and system increases at a growing rate, while the probability that the system is empty gradually declines. This makes sense, because the demand on the system is growing (more customers will have to wait), and the increased demand requires the server to be busy more often. What do you think happens as λ approaches the mean service rate of 3? Sophisticated models have been used in many practical settings to design systems that meet service objectives (see the Management Science Practice box on L.L. Bean).

Simulation models are an excellent means for performing sensitivity analysis. For example, in the single server model we developed in this chapter, suppose we would like to have a low risk of having a queue length of 8 or more customers in order to provide a high level of customer service. We might experiment with different service rates to find a value that achieves a certain risk level. Figure 8.22 shows a modification of the queueing simulation spreadsheet model to include a Crystal Ball forecast cell (B14) for the maximum queue length (=max(F4:F503) in the model). Figure 8.23 shows that for 1,000 trials, the chances of a length of 8 or more is about 11 percent.

Excel Exercise 8.6

Use the model in *queueing simulation.xls* and Crystal Ball to find the smallest service rate that provides less than a 5 percent probability of a queue length of 8 or more.

FIGURE 8.22
Crystal Ball
Modification to
Queueing
Simulation
Model

	A	B
1	**Single Server Queueing**	
2	**Simulation Spreadsheet**	
3		
4		
5	Mean arrival rate	2
6	Mean service rate	5
7		
8	Expected No. in Queue	0.267
9	Expected Waiting Time	0.133
10	Expected % idle time	60%
11		
12	*Key cell formulas*	
13		
14	Maximum No. in Queue	4
15		
16	=MAX(F4:F503)	
17		

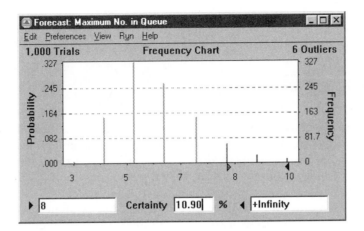

FIGURE 8.23
Risk of finding
8 or more
customers in
queue

Management Science Practice:
Queueing Models for Telemarketing at L.L. Bean

L.L. Bean is widely known for retailing high-quality outdoor goods and apparel, with more than 85 percent of sales generated through mail orders and telephone orders via 800-number service, which was introduced in 1986. About 65 percent of total annual sales volume is generated through orders taken at two telemarketing centers located in Maine. The telephone system at L.L. Bean is shown in Figure 8.24. An arriving call (point 1 in Figure 8.24) seizes one of the trunk lines if one is available; otherwise it is routed to a busy signal. The call arrives at an L.L. Bean switch (point 2); the switch checks the queue length; if the queue is full, the call is routed to a busy signal, freeing the trunk. Otherwise, the call enters the queue. Callers may abandon the call (hang up) while waiting for an agent, freeing the trunk and queue position (point 3). When one of the agents is available, the call is serviced, freeing a queue position (point 4). When the call is completed, a trunk and an agent are freed (point 5). From a strategic planning perspective, L.L. Bean can control the number of trunks, number of agent positions, and queue capacity. In the short term, decisions must be made to staff on a half-hourly basis, 24 hours a day, seven days a week to meet expected demand. In the intermediate term, the most critical decision is the number of agents to hire and train. These decisions affect the longer-term capacity issues as well as the short-term staffing issues.

L.L. Bean estimated that in 1988 it lost at least $10 million of profit by allocating telemarketing resources suboptimally. Customer service had become unacceptable; in some half-hours, 80 percent of the calls dialed received a busy signal because the trunks were full. Those customers who got through might have waited 10 minutes for an available agent. As a consequence, the company

continued

launched a project to better allocate resources and manage its queues. A mathematical model was developed for the key decision variables:

1. Number of trunks
2. Number of agents scheduled
3. Queue capacity (maximum number of wait positions)

The model searches for the combination of resources that minimizes the expected costs of trunking, labor, connect time, and lost-order profits.

A finite capacity queueing model (M/M/s/K, where K is the system capacity) was used to determine the rate of busy signals at the trunks. Caller abandonment was estimated by a simple linear regression model with the average queue time per caller as the independent variable. The most questionable assumption was the agent time distribution. A sample of 1,240 observations was tested using a Chi-square goodness-of-fit test to corroborate the validity of the exponential distribution assumption.

After the optimization effort, calls answered increased 24 percent, revenues increased 16.3 percent, the percent of calls spending less than 20 seconds in queue increased by 208 percent, the percent of abandoned callers fell by 81.3 percent, and the average time to answer fell from 93 seconds to 15 seconds.

FIGURE 8.24 L.L. Bean telephone system

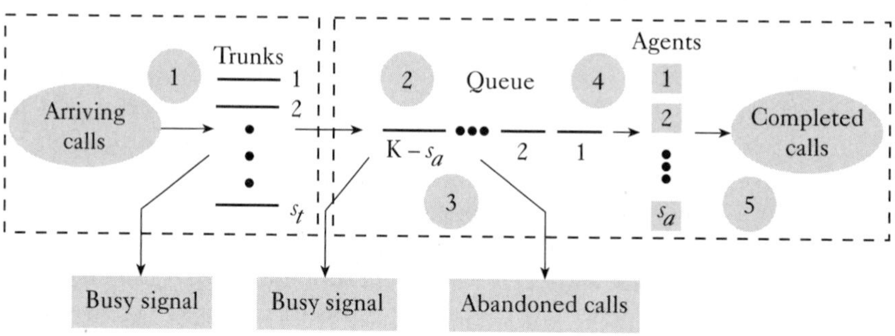

Source: Adaptation and figure from Quinn et al., "Allocating Telecommunications Resources at L.L. Bean, Inc.," *Interfaces*, Vol. 21, No. 1, January–February, 1991, 75–91.

MODELING AND SIMULATING DYNAMIC INVENTORY SYSTEMS

In Chapter 1 we developed a total cost model of a continuous review inventory system and discussed the economic order quantity. That model assumed a constant demand rate, a constant lead time, and no back orders. In realistic situations, neither the demand nor the lead time will be constant, and back orders may accumulate. Simulation provides a convenient way of evaluating the impact of different ordering policies. The decision maker can choose specific combinations of the order quantity Q and reorder point r and compare the costs of operating the inventory system under these different policies.

We can develop a simulation model for such a situation by translating the flowchart in Figure 1.16 into a more detailed, logical model. This is shown in Figure 8.25.

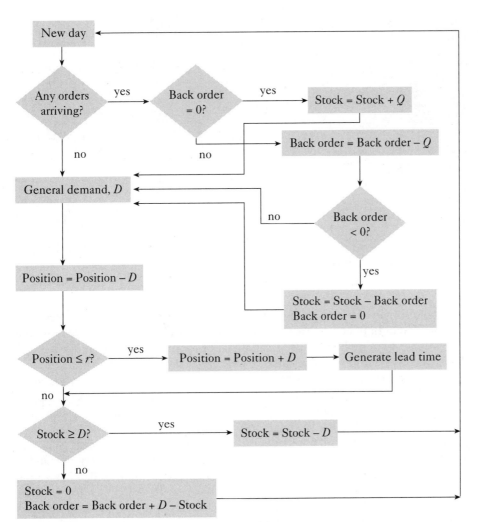

FIGURE 8.25

Simulation model of a continuous review inventory system

We begin each day by determining if any orders will arrive, assuming that all orders arrive at the beginning of the day. If the current level of back orders is zero, then the stock level is increased by the amount of the order, Q. If back orders have accumulated, we subtract Q from the current level of back orders. If the result is still positive, this means that we still have no inventory in stock and are carrying back orders until the next order arrives. If the result is negative, it means that we have filled all back orders and have some stock remaining. (Notice that in the box with "Stock = Stock − Back order," a negative value for back order will yield a positive level of stock.)

Next we generate the demand D for that day and decrease the inventory position by D. If the position is at or below the reorder point r, we place an order by increasing the position by Q and generating the lead time. If we have sufficient stock to meet the demand, we decrease the stock level by D. If not, the stock level drops to zero and the number of back orders increases by the difference. This completes the operation of the system for one day.

Figure 8.26 shows a portion of a spreadsheet designed to simulate a specific example (the actual spreadsheet simulates 100 days). Demand and lead time are

FIGURE 8.26

Spreadsheet model for inventory simulation with backorders

	A	B	C	D	E	F	G	H	I	J	K	L	M	N	O	P
1	Inventory Simulation With Backorders															
2				Order Quantity			5		Order Cost			$ 40				
3	Demand Distribution			Reorder Point			3		Holding Cost			$0.20				
4				Initial Inventory			5		Back Order Cost			$ 100		Average cost/day		$ 22.16
5	Demand	Probability														
6	0	0.50														
7	1	0.25		Beg	Order	Units		End	Back	Order	Inv	Lead	Hold	Order	Short	Total
8	2	0.15	Day	Inv	Rec'd	Rec'd	Dmd	Inv	Order	Placed?	Pos	time	Cost	Cost	Cost	Cost
9	3	0.05	1	5		0	1	4	0	NO	4		$0.80	$ -	$ -	$ 0.80
10	4	0.05	2	4		0	0	4	0	NO	4		$0.80	$ -	$ -	$ 0.80
11			3	4		0	3	1	0	YES	6	2	$0.20	$40.00	$ -	$ 40.20
12	Lead Time Distribution		4	1		0	1	0	0	NO	5		$ -	$ -	$ -	$ -
13			5	0	YES	5	0	5	0	NO	5		$1.00	$ -	$ -	$ 1.00
14	Lead Time	Probability	6	5		0	1	4	0	NO	4		$0.80	$ -	$ -	$ 0.80
15	1	0.20	7	4		0	1	3	0	YES	8	4	$0.60	$40.00	$ -	$ 40.60
16	2	0.10	8	3		0	4	-1	1	NO	4		$ -	$ -	$100.00	$100.00
17	3	0.40	9	-1		0	3	-4	3	YES	6	3	$ -	$40.00	$300.00	$340.00
18	4	0.20	10	-4		0	1	-5	1	NO	5		$ -	$ -	$100.00	$100.00
19	5	0.10	11	-5	YES	5	1	-1	1	NO	4		$ -	$ -	$100.00	$100.00
20			12	-1	YES	5	1	3	0	YES	8	4	$0.60	$40.00	$ -	$ 40.60
21			13	3		0	4	-1	1	NO	4		$ -	$ -	$100.00	$100.00
22			14	-1		0	0	-1	0	NO	4		$ -	$ -	$ -	$ -
23			15	-1		0	0	-1	0	NO	4		$ -	$ -	$ -	$ -

modeled using the Crystal Ball function *CB.Custom* for the distributions specified in columns A and B; of course these can be easily modified or replaced by some other type of distribution. The decision variables (order quantity and reorder point) are given in cells G2 and G3. The initial inventory at the start of the simulation is set equal to the order quantity. Costs are provided in L2:L4. Holding cost is the cost per unit, evaluated by the ending inventory each day. An arbitrarily high back order cost per unit is assumed as a penalty for shortages. The key performance measure is the average cost/day in cell P4. We might also compute the average stock level, average number of back orders, and number of orders, but we leave this to you as an exercise.

We will step through a part of the actual simulation. In the first day, demand is 1, resulting in an ending inventory (and inventory position) of 4. In day 2, a zero demand results in no changes. In day 3, a demand of 3 causes the inventory position to fall below the reorder point, triggering an order. At this time, the inventory position is increased to 6 to reflect the amount on order. A randomly generated lead time results in 2 days. (We assume that the order will arrive 2 days from the current day.) To explain how backorders are handled, look at row 16. On day 8, the beginning inventory is 3 and the demand is 4; thus 1 back order occurs. For accurate accounting purposes, the ending inventory is calculated as the *net* inventory (on hand minus backorders) so that when an order arrives, back orders may be fulfilled. If you wish to know the actual physical inventory, add a column to compute the maximum of zero or the ending inventory. We encourage you to examine the cell formulas to better understand how the simulation is implemented in Excel.

Example 5: A Crystal Ball Simulation of the Inventory System

We may use Crystal Ball to replicate the spreadsheet in Figure 8.26 easily and obtain a distribution of average cost for the chosen order quantity and reorder point. Figure 8.27 shows the result—an average daily cost of about $18. However, you can see that considerable variability exists in repeated trials of 100-day simulations. As with any Monte Carlo simulation, we may use the output statistics to assess risk and find a confidence interval for the average daily cost.

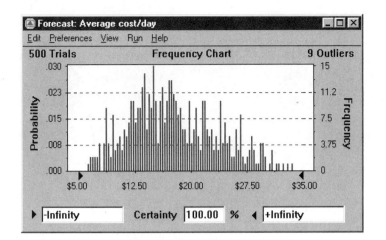

FIGURE 8.27
Crystal Ball
forecast chart
for average
daily cost

Excel Exercise 8.7

Modify the *inventory simulation.xls* spreadsheet to include a forecast cell for the average number of backorders per day. Using Crystal Ball, assess the risk of incurring an average of more than .20 backorders per day.

Using Simulation to Optimize Inventory Systems

The principal objective of developing a simulation model of an inventory system might be to find the optimal order quantity and reorder point that minimizes average daily cost or ensures a specified risk level of backorders or some other statistic. One simulation experiment simply provides information for a specific choice of order quantity and reorder point. We could run the simulation with new values for the order quantity and reorder point to determine the effects of these changes and move toward the best solution. Crystal Ball Pro includes a *Decision Table* extender that allows you to automatically run multiple simulations to test different values for one or two decision variables. The extender tests values across a specified range of the decision variables and puts the results in a table that you can analyze using Crystal Ball forecast or trend charts (and with overlay charts in the commercial version). We illustrate this using the inventory model.

Example 6: Optimizing an Inventory System

To apply the Decision Table extender to the inventory simulation model, first define the order quantity and reorder point cells (G2 and G3 in Figure 8.26) as decision variables by first selecting them, and then choosing *Define Decision...* from the *Cell* menu, or clicking on the *Define Decision* button on the *Crystal Ball* toolbar. Figure 8.28 shows the dialog box for the order quantity cell. We defined this variable to be discrete, with a lower bound of 3, upper bound of 7, and step size of 2. This causes the extender to evaluate the order quantity for values of 3, 5, and 7. In a similar fashion, we defined the reorder point to be evaluated at 1, 3, and 5. Next, select

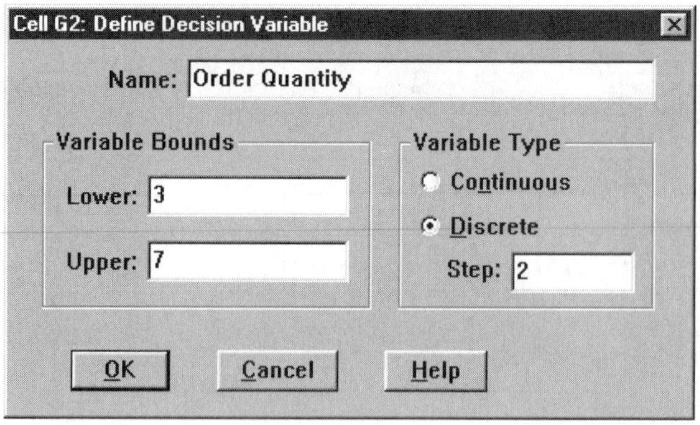

FIGURE 8.28
Define decision
dialog box

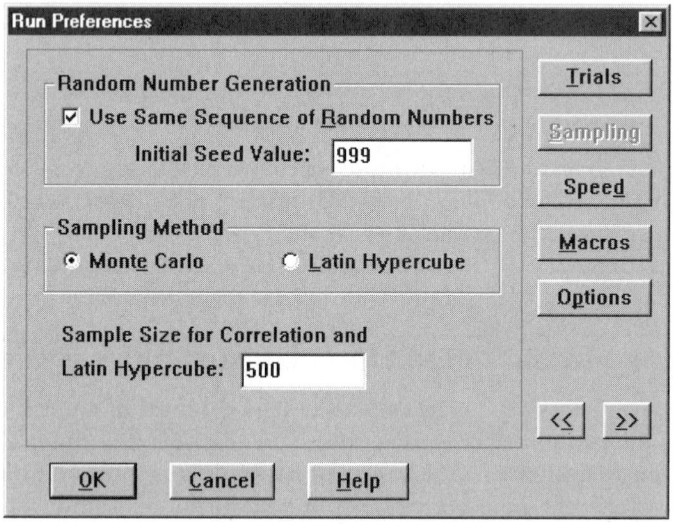

FIGURE 8.29
Run prefer-
ences sampling
dialog box

cell P4, the average cost/day, as the forecast cell. When using the *Decision Table* extender, you should check the *Use Same Sequence of Random Numbers* option from the *Sampling* option in the *Run Preferences* dialog, as shown in Figure 8.29. We selected an Initial Seed Value = 999. This makes the resulting simulations comparable by generating the same sequence of random variates for each experiment.

Next, select *CB Decision Table* from the *Tools* menu. This prompts you to select the appropriate forecast as the objective and the decision variables you wish to vary from among those you have defined. For this example, we selected both decisions. Finally, you are prompted to verify or change the options for the number of values to test for each variable and the number of trials per simulation. Clicking on *Start* begins the process.

Figure 8.30 shows the results. The buttons in cell A1 allow you to view *Forecast* charts for any of the output cells or *Trend* charts as a means of comparing, for example, a row or column of forecasts. (The *Overlay Chart* is disabled in the Student Version). It appears that average costs decrease as both the order quantity and reorder point increase from the base case values of 5 and 3. We may continue to experiment to find better values.

	A	B	C	D	E
1	Trend Chart / Overlay Chart / Forecast Charts	Order Quantity (3)	Order Quantity (5)	Order Quantity (7)	
2	Reorder Point (1)	$50.83	$33.74	$24.97	1
3	Reorder Point (3)	$27.90	$18.18	$13.34	2
4	Reorder Point (5)	$17.96	$11.43	$ 8.75	3
5		1	2	3	

FIGURE 8.30
Results from
Crystal Ball
Decision Table
Extender

Excel Exercise 8.8

Conduct a new set of experiments using the *inventory simulation.xls* spreadsheet and the *Decision Table* extender to find the best values for the order quantity and reorder point that minimize the average daily cost.

SYSTEMS SIMULATION TECHNOLOGY

Because the capabilities of spreadsheets are severely limited to simple situations, dynamic systems are usually simulated using special-purpose simulation software. This software technology automatically takes care of many details of simulation calculations and reporting, such as generating random numbers, advancing the simulation clock, collecting statistical information, managing files, and producing summary reports, thus allowing the user to concentrate on the modeling process itself. Many even have animation capabilities to allow users to observe the dynamics of a simulation in progress. They provide the basic building blocks to simulate nearly any type of situation (see the Management Science practice box). Table 8.2 summarizes some of the more popular products.

Arena Business Edition

The CD-ROM accompanying this book contains an evaluation copy of Arena Business Edition, a general-purpose simulation program designed to facilitate modeling and simulation of business problems. The evaluation copy is fully functional; however, the size of models that may be constructed with it is limited. Information on obtaining the full version of the software may be obtained from Systems Modeling Corporation.[2] In this section we will introduce you to some of the basic features of Arena Business Edition to give you a feel for the ease at which complex simulation models may be built using modern software technology. We will present the basic information needed to address the problems at the end of this chapter. However, much more information may be found in the help files, and the various examples and tutorials provided with the software. In addition, a PDF copy of the Guide to Arena Business Edition is available at the web site **www.sm.com/sim/ArenaBusiness/default.htm** (click on Getting Started Guide).

Management Science Practice: Designing an Air Force Repair Center Using Simulation

We describe some of the issues involved in developing a simulation model for Tinker Air Force Base. Tinker Air Force Base is one of five overhaul bases in the Air Force Logistics Command. It overhauls and repairs six types of jet engines and various aircraft and engine accessories, and it manages selected Air Force assets worldwide. Engines are received for periodic overhaul or modification. The engine is disassembled, and each part is inspected for wear and possible repair. Individual parts are repaired or replaced with new parts. The majority of parts are overhauled and returned to service for a fraction of the cost of a new part.

In 1984 a fire devastated one building at the base. The Air Force requested assistance to develop a simulation of the engine overhaul process to assist in redesign and layout of approximately 900,000 square feet of production floor space. A team from the University of Oklahoma received the contract.

Objectives of the study included a means to predict and forecast varying resource requirements as workload mixes changed. The baseline database consisted of 117 fields with over 2,500 records and provided information on each engine part, the annual requirement of each end item, the sequential routes of the part and machine/process time requirements, size and weight of the part (for storage estimation), and number of units per assembly required of each part.

The simulation model, called TIPS—Tinker Integrated Planning Simulation—was written in the SLAM language. The model included a three-shift operation; transfers to other operations such as painting, plating, and heat treatment; and storage facilities. The model included sick leave, training leave, and vacations for machine operators. Several assumptions had to be made because no data were available on arrival times of engines. Arrivals were assumed to be deterministic, based on the annual volume of demand. The model integrated both machine and labor resources to support bottleneck analysis, space analysis, and conveyor-routing analysis. They designed the model for *managers* and held two training sessions at the base to allow managers to use TIPS for decision making.

To ensure that the model worked correctly, the team developed it incrementally, which made debugging the programs easier. They also analyzed the outputs of each model component for reasonableness and consistency. To validate the model, they made a diagnostic check of how closely the simulation matched the actual system by cross-checking model assumptions and comparing results statistically to actual historical data.

The simulation output consisted of basic information about queueing behavior and a custom report at a level of detail suitable for prompt managerial decision making. The statistics in the output included the following:

- Machine availability by shift for each process
- Maximum number of waiting jobs in front of each process

continued

- Average processing time
- Average waiting time for each process
- Number of units of each type entering and leaving the system
- Time spent in storage waiting for a specific process
- Utilization level of each process per shift
- Total time in the system, including waiting, handling, and processing

The TIPS program proved valuable in aiding the transition to the new layout by estimating performance measures that helped determine the number of machines of each type to have in the facility.

Source: Adaptation from A. Ravindran et al., "An Application of Simulation and Network Analysis to Capacity Planning and Material Handling Systems at Tinker Air Force Base," *Interfaces*, Vol. 19, No. 1, January–February 1989, 102–115.

Product	Company/Web site	Description	
ALPHA/Sim	ALPHATECH, Inc. **www.alphatech.com**	General-purpose, discrete event simulation tool for business process reengineering, service delivery, manufacturing systems	**TABLE 8.2** Some popular systems simulation software
Arena	Systems Modeling Corporation **www.sm.com**	Process-driven software with animation for business process analysis, discrete event simulation	
Extend & BPR™	Imagine That, Inc. **www.imaginethatinc.com**	An integrated financial and operational approach for modeling and evaluating business processes	
FACTOR/AIM	SYMIX/Pritsker Division **www.symix.com**	Manufacturing decision support for process design and improvement, production planning, and production scheduling	
GPSS/H	Wolverine Software Corporation **www.wolverinesoftware.com**	General-purpose simulation language for a variety of applications, including manufacturing, transportation, computing networks, logistics, and queueing	
ProcessModel	PROMODEL Corporation **www.promodel.com**	General-purpose tool for business process improvement	
WITNESS 9.0	Lanner Group, Inc. **www.lanner.com**	A visual environment for business process reengineering, scheduling, queueing	

Source: Adapted from "1999 Simulation Software Product Listing," *OR/MS Today*, February 1999.

When Arena Business Edition is started, the modeling environment screen appears (Figure 8.31). The *Project Bar* on the left side of the screen contains panels with the primary types of objects needed for a simulation:

- *Basic Process* panel, which contains the modeling shapes, called modules, that define a process
- *Reports* panel, which contains the reports available for displaying results of simulation runs
- *Navigate* panel, which allows you to display different views of a model

In the model window, there are two main regions. The *Flowchart View* contains all of your model graphics, including the process flowchart, animation, and other drawing elements. The lower window, called the *Spreadsheet View*, displays model data, such as times, costs, and other parameters. Along the bottom of the *Spreadsheet View* are tabs with the module names. Clicking on a tab will display in the spreadsheet all modules of that type. Clicking the module icon on the Project Bar also activates the spreadsheet tab for that module type.

All information required to build and run simulation models is stored in modules. *Flowchart modules* are placed in the model window and connected to form a flowchart describing the logic of the process you are modeling. The basic flowchart modules and their functions are as follows:

- *Create*—the start of a process; the point at which entities—the items that move through the process—enter the simulation.
- *Dispose*—the end of process, at which entities are removed from the simulation.
- *Process*—an activity, usually performed by one or more resources and requiring some time to complete.
- *Decide*—a two-way branch in a process. Only one branch is taken.
- *Batch*—collection of a number of entities before they can continue processing.
- *Separate*—duplicate entities for concurrent or parallel processing, or separating a previously established batch of entities.
- *Assign*—change the value of some parameter during the process, such as the entity's type or a model variable.
- *Record*—collect a statistic, such as an entity count or cycle time.

FIGURE 8.31
Arena modeling
environment

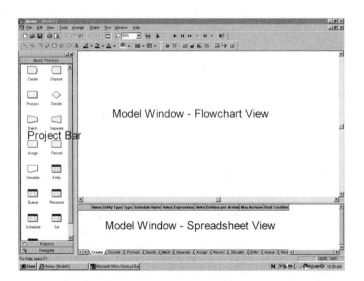

A separate *Simulate* module (not included in a process flowchart) defines the simulation run settings.

Data modules define the characteristics of various process elements, such as resources and queues, and are not placed in the model window. Instead, they are edited via a spreadsheet interface.

In full view, Arena will display the appropriate pane for the module you select. If you click on a flowchart module icon in the panel, the model workspace is displayed; and if you click on a data module icon in the panel, the spreadsheet section is displayed.

To illustrate the process of building and running simulation models with Arena Business Edition, we describe a simple example involving mortgage processing.

Example 7: A Simulation Model of Mortgage Processing[3]

Using a new electronic filing service, a banking office receives mortgage applications every two hours, on average. The application is reviewed by a mortgage clerk who earns $12 per hour. This review process generally takes between 1 and 3 hours, but most applications are reviewed in about 1.75 hours. About 12 percent of applications are not accepted either because they are incomplete or because the applicants do not meet the bank's financial requirements. Some of the questions the bank would like to answer are

1. On average, how long do applications spend in the review process?
2. What is the average cost of reviewing an application?
3. What is the longest time an application might spend in review?
4. What is the maximum number of applications waiting for review?
5. What proportion of time is the clerk busy?

Figure 8.32 shows a logical flowchart for this process. This logic is translated directly to Arena to build the simulation model. In Arena, flowcharts are constructed by clicking and dragging modules from the *Project Bar* to the *Flowchart View* in the model window. Thus, to begin the simulation, drag the *Create* module (which models the activity *Receive Application*) into the window. When a module is placed, a default name is provided; this can easily be changed to something more descriptive, as we will show later. The *Review Application* activity in Figure 8.32 is modeled by a *Process* module. If the *Auto-Connect* option is checked in the *Object* menu, then any new modules that are dragged to the *Flowchart View* will be connected automatically to the module that has been selected (to select a module, simply single-click on it). For example, if the *Create* module is selected, dragging the *Process* module to a position after the

RE 8.32

tgage
s logical
chart

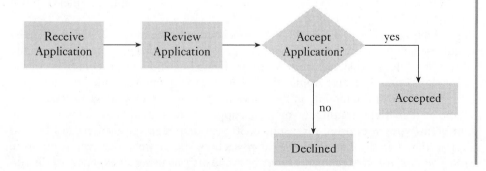

FIGURE 8.33
Building the mortgage process simulation model

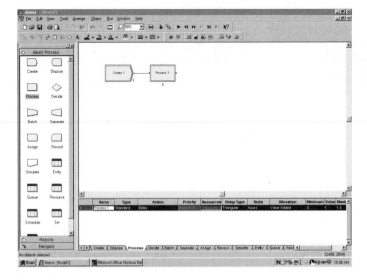

FIGURE 8.34
Arena model for mortgage processing

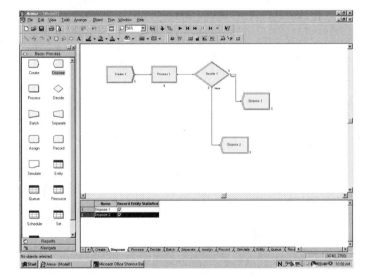

Create module will result in the screen shown in Figure 8.33. To complete the model, we use a *Decide* module to model the *Application Accepted* activity, and *Dispose* modules for both terminating events (Accepted and Declined). The complete model is shown in Figure 8.34.

The next step is to define the data associated with the modules. This is done by double-clicking a module and defining its properties in a dialog box. Clicking on the *Create* module, for instance, brings up the dialog box shown in Figure 8.35. In the *Name* box, enter "Receive Application" (without the quotation marks, this becomes the new name of the module in the *Flowchart View* window). For entity type, enter "Application" (representing the mortgage application—a model may contain several different types of entities, for instance, 30-year mortgage applications and 15-year applications). In the *Time Between Arrivals* section, you may choose several options in the *Type* box. *Random* generates times between arrivals using an exponential distrib-

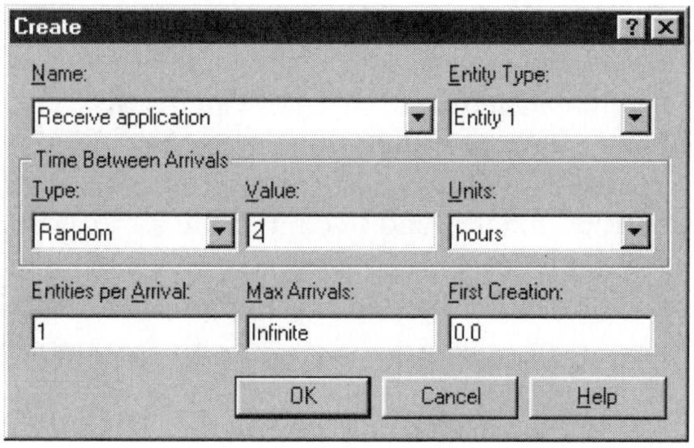

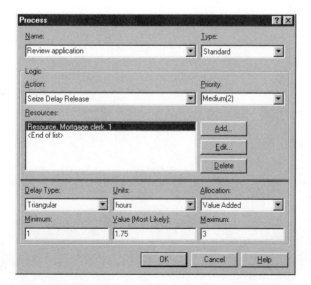

FIGURE 8.35
Create module property dialog box

ution with the mean specified in the *Value* box (2 hours). Choosing *Expression* allows you to select from a variety of standard probability distributions in Arena (normal, uniform, etc.). The *Entities per Arrival* box allows you to create multiple entities to the process at the same time, *Max Arrivals* allows you to limit the number of arrivals during the simulation, and *First Creation* allows you to specify when the first creation occurs. We will leave these at their default values.

Double-clicking on the *Process* module brings up the *Process* property dialog box (Figure 8.36). In the Name box, enter "Review application." From the *Action* list pull-down menu, select *Seize Delay Release*. This causes the entities (applications) to wait for a resource (the mortgage clerk), "seize" the resource, delay for the processing time, then release the resource and continue to the next step in the process. Other options include *Delay*, *Seize Delay*, and *Delay Release*. Explanations of these options can be found in the *Help* topics (*Flowchart modules/Process module/Prompts*). We encourage you to read the help on any module you use for complete information. A list of resources appears in the center of the dialog box. Click on the *Add* button to add a resource for this process. In the dialog box that appears, enter *Mortgage clerk* in the *Resource Name* field and click *OK* to close this box. Finally, we define the process delay distribution as triangular, with parameters of 1, 1.75, and 3.

FIGURE 8.36
Process property dialog box

Next, open the *Decide* module property dialog box (Figure 8.37) and enter *Application Accepted?* in the *Name* field. Since 12 percent of applications are declined, 88 percent are accepted, so enter 88 in the *Percent True* field. Finally, double-click on the first *Dispose* module and name it *Accepted;* then name the second *Dispose* module *Declined*.

Arena allows you to define parameters associated with other model elements, such as resources, entities, queues, and so on. For the mortgage process, we will define the cost rate ($12 per hour) for the mortgage clerk so that the simulation results will report the cost associated with performing the process. Click the *Resource* icon in the *Basic Process* panel (see Figure 8.38) and enter 12 in both the *Busy/Hour* and *Idle/Hour* cells.

To prepare to simulate the model, drag the *Simulate* module to the *Flowchart View* window. Because entities do not flow through this module, it has no connections to the process flowchart. Double-click to open the *Property* dialog box (Figure 8.39), name the project Mortgage review example and define 20 in the *Run Length* field, and select days from the *Run Length Time Units* field. Don't forget to save your work!

To run the simulation, click the *Go* button or choose *Go* from the *Run* menu. Arena will first check to determine if the model is valid, and then will begin. Arena animates the simulation as it is running (see Figure 8.40). You will see small pictures

FIGURE 8.37
Decide module property dialog box

FIGURE 8.38
Resource module for mortgage clerk

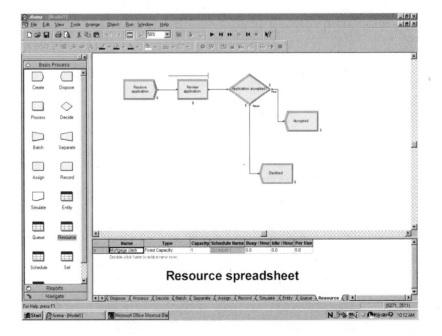

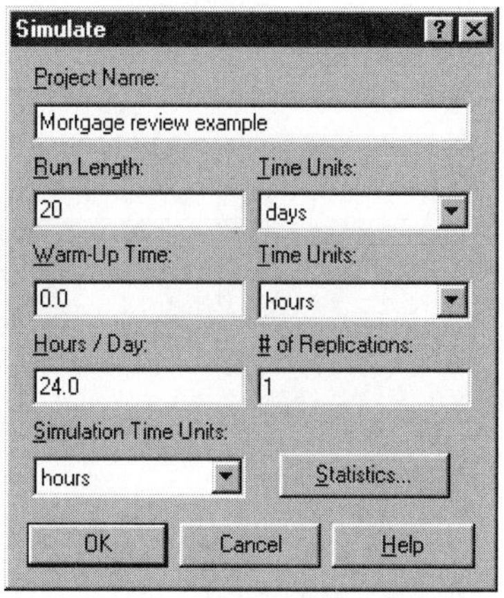

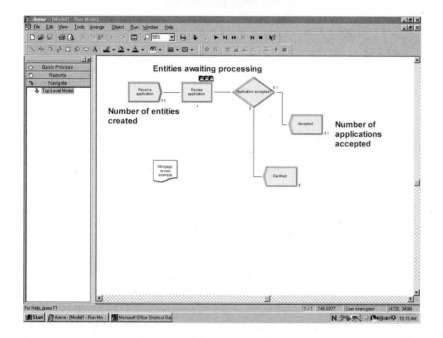

FIGURE 8.39
Simulate module property dialog box

FIGURE 8.40
Animation aspects of mortgage review simulation

resembling pages (the application entities) moving through the model and also waiting for processing by the clerk. Also, a variety of variables change value as entities are created and processed. You can control the simulation speed from the *Run* menu or by pressing the "<" and ">" keys to slow down or speed up the animation.

When the simulation is complete, Arena will ask you if you wish to view the results. In the report window (shown in Figure 8.41) are two panels. The left one is a tree structure listing information available; it can be expanded or collapsed by clicking on the + and − boxes in the same manner as you would in Windows Explorer. Clicking on specific entries will show specific results. For example, the average cycle time is 7.96 hours; this tells how long applications spend in the mortgage review process (question 1 in the problem statement). We also see that the longest time an application spent in review (question 3) was 19.56 hours. You may close the report window by clicking on the window icon to the left of the *File* menu and selecting *Close*. You may also return to the model window by selecting the model file from the *Window* menu.

FIGURE 8.41
Portion of
Arena results
screen

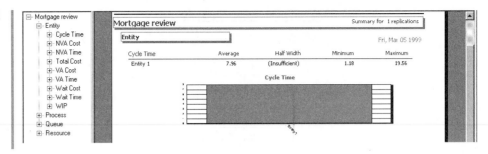

Arena Exercise

Start Arena and open the file *Mortgage Review Example* from the Arena Files folder. Run the simulation to view the animation and find the answers to the remaining questions posed at the beginning of this example.

To illustrate a few other features of Arena, we show how it may be used to simulate the MMI production/inventory model in Example 1 of this chapter. In studying this example, you should compare it to the spreadsheet implementation.

Example 8: An Arena Model of the MMI Production/Inventory Model

Figure 8.42 shows the Arena model and results for the MMI problem. The model consists of two separate process flowcharts. At the top of the figure is an initialization process. The *Create* module creates one entity at time 0.0 and sends it to an *Assign* module. The *Assign* module is used to assign new values to variables, entity types, and other elements of a model. In this case, we define a variable called *Inventory* and set it equal to 250. After this, the entity proceeds to a dispose node and is destroyed.

In the main process flowchart, we begin with a *Create* module that creates 500 entities (representing each week of the simulation), spaced one time unit apart beginning at time 1. (Although Arena only allows you to select time dimensions from seconds to days, the choice in this example is arbitrary.) An *Assign* module sets a variable called *Demand* to a uniform random variate between 120 and 170. Another *Assign* module sets a *Production* variable to the discrete distribution used in the example. At the next *Assign* module, *Inventory* is set to its current value plus *Production* minus *Demand*. We define a new variable, *Ending Inventory*, which will be used to plot the first graph shown in the figure, and set this equal to the current value of *Inventory*. Next, we check to see if the inventory level is less than 200. If so, we increase the inventory level by 100 in the *Add overtime run Assign* module. We use a *Record* module to record the count of the number of overtime runs for the output report (this information is also captured in the counts of the number of entities leaving each branch of the *Decide* module). We then increase the production level by 100 before disposing the entity.

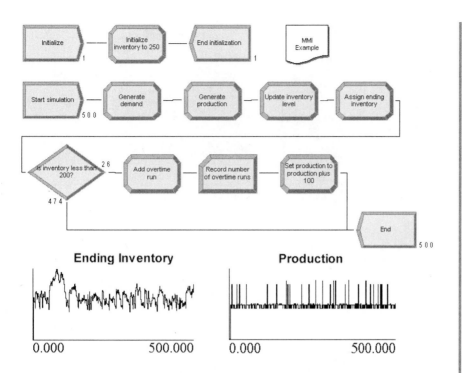

FIGURE 8.42
Arena model
and plots for
MMI example

To create a time plot of the *Ending* inventory and *Production* variables shown at the bottom of Figure 8.42, click the *Plot* button on the toolbar. The *Plot* dialog box (Figure 8.43) appears. Click the *Add* button to add an expression. This opens a new dialog box (Figure 8.44) from which you may select the variable to be plotted, the minimum and maximum values, and the number of history points. *History points* are the most recent values of the variable during the simulation run. If we had plotted the variable *Inventory* instead of *Ending inventory,* we would have included the data points from the *Assign* box *Add overtime run,* which would have given us inappropriate results (i.e., two points whenever an overtime run is made). This is why we needed to define the *Ending* inventory variable. Upon returning to the *Plot*

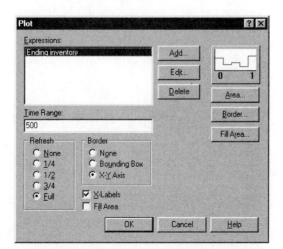

FIGURE 8.43
Arena plot
dialog box

FIGURE 8.44
Plot expression
dialog box

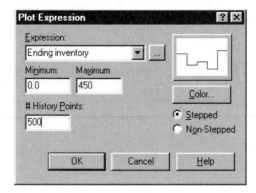

dialog box, change the time range to 500 (length of the horizontal axis), since we are plotting 500 weeks' worth of data.

In viewing the results in Figure 8.42, we see that overtime was required for 26 of the 500 weeks, or about 5 percent of the time. This is similar to the results we found in the spreadsheet simulation.

Arena Exercise

Open the Arena model file *MMI* from the Arena window. Double-click on each module and plot to examine their properties. Create a new plot with *Ending inventory* and *Production* on the same graph (using different colors) and run the model. Find the values of the average production and ending inventory in the output report.

To fully understand the capabilities of Arena and its use in simulation methodology requires more time than can be afforded at an introductory level. Many more-complex simulation models, such as the inventory simulation model discussed earlier in this chapter, can be built using Arena. Arena also has many other options—particularly its graphical animation capabilities—that cannot be fully described within the scope of this book. For example, Arena provides several picture libraries for building visual animations. We encourage you to explore these features through the Help files and the examples provided with the software, and also refer you to Kelton et al. (1998) for additional details.

USING SIMULATION SUCCESSFULLY

The advantages of using simulation for dynamic systems are significant. Through simulation experiments, we can evaluate proposed systems without building them; experiment with existing systems without disturbing them; and test the limits of systems without destroying them. However, as in any modeling approach, there are some drawbacks to using simulation. It requires you to understand fully how a system operates and be able to translate this knowledge into a detailed logical model and computer implementation. A large amount of effort in simulation modeling involves the collection of high-quality data. Therefore, you need a good information system that will capture the required data. You must be able to test the simulation model to see if it works properly and accurately models the actual system in sufficient detail to provide useful answers. These activities all require time and good communication among many individuals. Thus, simulation is often addressed from a project focus. Good management skills are required to coordinate the various activities and people involved in the project.

Verification and Validation

A particularly challenging task is to build an accurate model and convince the end user that the model is a valid representation of the real system under study. Two important considerations when dealing with simulation models are verification and validation. *Verification* refers to the process of determining if a simulation program performs as intended. In other words, have we correctly translated the conceptual model into a working simulation program? Is the computer implementation free from logical errors and bugs? *Validation* is concerned with determining whether the conceptual model is an accurate representation of the real system under investigation.

Several verification techniques exist; many of these are the same used by programmers in writing and debugging computer code. For example, care in developing the logical flowchart will help to minimize logical errors. It is often easier to conceptualize the problem in smaller pieces, or modules, and write the computer code in modules that can be linked together. The code can be checked by someone other than the programmer. The outputs from the model should also be studied for reasonableness. Sometimes some simple analytical results can be used for comparison. If these results are far from the expected results, we might have suspected a programming error. Modern simulation languages have built-in tools to assist in verifying the correctness of a program.

The purpose of simulation modeling is to be able to make decisions about the system that are similar to those that would be made if you were experimenting with the real system. Validation typically consists of three steps:

1. Develop a model with high face validity.
2. Validate model assumptions.
3. Validate model output.

The model should seem reasonable to those who understand the real system. This is called *face validity* and is important in gaining acceptance from the users of the model. Face validity can be strengthened by maintaining good communication with the user during the model development stages. Successful simulation models are usually the result of team efforts involving managers, engineers, and workers involved in the system.

The more complex a system is, the harder it is to validate. Many assumptions are made in simulation models, particularly in regard to the probability distributions used to drive the simulation and behavior of system entities. Thus the probability distributions used to describe probabilistic elements of a simulation should be validated by comparing them with actual data from the system. Similarly, assumptions made regarding the behavior of customers, service personnel, and so on must have a rational basis.

The most objective and scientific means of validation is to compare the output of the model with data from the real system for the same inputs. The idea is to match as closely as possible the inputs used in the real system and the inputs used in the model. The system outputs should closely approximate the model outputs. One should always remember that a real system constantly changes; thus, validation is never really finished. Data need to be periodically updated and the model needs to be retested.

Statistical Issues

Many of the same issues that we addressed in Chapter 7 regarding the analysis of output statistics for Monte Carlo simulation models apply also to systems simulation models. These include replicating model runs to estimate the standard error of the mean and compute confidence intervals on key performance measures, and selecting the proper sample size for replications. Although most systems simulation packages allow you to easily replicate a model, they do not construct output distributions for large numbers of replications as we have seen for risk-analysis spreadsheet simulations using Crystal Ball. Also, large models may consume a significant amount of computer time for just one replication, and it is not practical to replicate the model for 500 or 1,000 trials. Instead, analysts usually use a much smaller number of replications (perhaps as few as 3 to 5) to obtain variance statistics and compute standard errors. Of course, the more replications, the more accurate the results; this is a trade-off that you must make.

Another issue involves the transient period. Whenever we simulate systems that have a transient period for the purpose of estimating steady-state averages, statistical measures that include all the data will be biased. In such cases, we usually do not capture output data for an initial period of time, allowing the simulation to "warm up." This provides more accurate steady-state results. Other techniques are available for handling these statistical issues, but these are beyond the scope of this book.

PROBLEMS

1. A refuse collection company in a small city currently uses trash collection vehicles with a 5-ton capacity. Trucks are dispatched from the landfill at 6 a.m., travel to the landfill, and continue until 2 p.m., at which point they return to the landfill, dump the refuse, and wait until the next morning. Develop a simulation flowchart that describes how to model this system. Management is interested in determining the productivity of the trucks. How might the simulation be used to make a decision on whether to purchase the larger trucks? What other information is needed?

2. A university computer center services students and faculty researchers. During the mornings, demand for student services occurs uniformly every 1.9 to 2.6 minutes. Demand for research support occurs every 10 to 15 minutes. The average service time in minutes for student services is 0.75 minute (distributed exponentially); for faculty research, 3.5 minutes (distributed exponentially). Develop a logical simulation flowchart for this system.

3. The children's board game *Hi-Ho Cherry-O!*® is played as follows: Each child has a tree with 10 cherries. The object of the game is to pick all the cherries into a bucket. A spinner is divided into 7 equal sections:

   1. Pick 1 cherry
   2. Pick 2 cherries

   3. Pick 3 cherries
   4. Pick 4 cherries
   5. Dog eats 2 cherries (place 2 back on the tree)
   6. Bird eats 2 cherries (place 2 back on the tree)
   7. Spill the bucket (place all cherries back on the tree)

a. Develop a logical flowchart for this game.
b. Simulate this game manually five times (until all cherries have been picked)
    to estimate the mean number of moves required to finish the game. (Draw
    random numbers using the Excel function *RAND()* and the techniques de-
    scribed in the previous chapter.)

4. Unstructural Dynamo Research (UDR) is a company that produces software
   designed to enhance company creativity. UDR itself has lately become less than
   creative and has had trouble getting new products to the market. The current
   stock price for the company is $97.50 per share. The daily changes in stock price
   over the last 500 trading days have been analyzed, resulting in the following frequency counts:

Price Change	Frequency
−5	1
−1	20
−1/2	55
−3/8	95
0	212
+3/8	100
+1/2	10
+1	5
+2	2

a. Simulate the stock performance over the next 100 trading days by construct-
    ing a spreadsheet and using the *CB.Custom* function.
b. Replicate the simulation 500 times using Crystal Ball. What is the expected
    ending price of the stock?

5. You have a 3-year-old daughter and are starting to worry about college costs. The
   current cost of a 4-year education at a private school is about $100,000. Experts
   suggest that this will increase somewhere between 1 percent and 5 percent annually over the next 15 years. Develop a simulation model on a spreadsheet to estimate the college costs after 15 years. Assume that the increases will be normally
   distributed with a mean of 3 percent and a standard deviation of 0.8 percent, but
   will not be less than 1 percent or greater than 5 percent. Replicate the model 10
   times and develop a 95 percent confidence interval for the costs.

6. Passengers wait for an airport shuttle service that arrives every 15 minutes. Passengers arrive within each 5-minute period according to the following distribution:

Number of Passengers	Probability
0	0.20
1	0.10
2	0.20
3	0.30
4	0.15
5	0.05

The shuttle holds 10 passengers. Develop a simulation model to find the probability that some passengers will not be able to board the shuttle and must wait for the next one.

7. Water flows into a regional water reservoir in the mountains.[4] The amount of water flowing into the reservoir is a function of rainfall and snow melts. Water leaving the reservoir is a function of evaporation and the controlled flow of stream water through a gate. Too much water flowing through the gate at any one time will result in downstream flooding, but too little water through the gate may cause the reservoir itself to overflow. The decision maker's problem is to determine a constant gate flow rate that maintains a balance between the two extremes. Probability distributions for the input and output processes are given next. The initial water level is 100,000 gallons. Develop a spreadsheet model for this system, assuming that the controlled flow rate is 2,000 gallons per day. Conduct appropriate spreadsheet simulation experiments and analyze the results. What output information would be important to the manager of this system?

Water Input
(thousands of gallons/day)

Amount	Probability
10	0.2
20	0.2
30	0.3
40	0.3
50	0.1

Water Output
(thousands of gallons/day)

Amount	Probability
5	0.15
10	0.10
20	0.25
30	0.25
40	0.25

8. Complete the following table:

Customer	Arrival Time	Processing Time	Start	End	Waiting Time
1	3.2	3.7			
2	10.5	3.5			
3	12.8	4.3			
4	14.5	3.0			
5	17.2	2.8			
6	19.7	4.2			
7	21.1	2.8			
8	26.9	2.3			
9	32.7	2.1			
10	36.9	4.8			

9. A manager is considering changing the replacement policy for three drill bits in a single drill press on the shop floor.[5] The present policy is to replace a drill bit when it breaks or becomes inoperable. One of the supervisors recommended that a new policy be used in which all three drill bits are replaced when any one bit breaks or needs replacement. It costs $100 each time the drill press must be shut down. A drill bit costs $50, and the variable cost of replacing a drill bit is $10 per bit.

   The company that supplies the drill bits has historical evidence that the reliability of a single drill bit is described by an exponential probability distribution with the mean number of failures per hour equal to 0.01. The company would like to compare the cost of the two replacement policies. Use a spreadsheet simulation to arrive at a recommendation.

10. Michael's Tire Company performs free tire balance and rotation for the life of any tires purchased there on a first-come, first-served basis. One mechanic handles this and can usually complete the job in an average of 20 minutes. Customers arrive at an average rate of two per hour for this service.
    a. How long will a customer expect to wait before being served?
    b. How long will a customer expect to spend at the shop?
    c. How many customers will be waiting, on the average?
    d. What is the probability that the mechanic will be idle?

11. For the M/G/1 queueing model, suppose that $\lambda = 6$ and $\mu = 9$. Develop a spreadsheet and graphs for examining the sensitivity of the operating characteristics for the variance Σ^2 as it varies from 0 to 3. What conclusions can you reach?

12. A college office has one photocopier. Faculty and staff arrive according to a Poisson process at a rate of five per hour. Copying times average eight minutes but do not follow an exponential distribution. The standard deviation of copying times is estimated to be two minutes. What are the operating characteristics for this system?

13. Star Savings and Loan is opening up a new branch in Union Township. Market research shows that they can expect an average of 35 customers per hour on Saturdays. Transaction times typically average four minutes.
    a. What is the minimum number of tellers that will be needed?
    b. Compute the operating characteristics for your answer to part a and up to three additional tellers.
    c. How many tellers would you hire if you were the bank manager?

14. Star Savings and Loan is planning to install a drive-through window. Transaction times at the drive-through are expected to average three minutes because customers would use it for simpler transactions. The arrival rate is expected to be 10 customers per hour.
    a. Compute the operating characteristics for the drive-through window queueing system.
    b. The demand for drive-through services is expected to increase over the next several years. If you were the bank manager, how much would you let the average arrival rate increase before considering adding a second drive-through window?

15. Simulate an M/M/1 queueing system with $\lambda = 5$ and $\mu = 7$ for 100, 200, 500, and 1,000 customers (using five replications each) on a spreadsheet. Compute confidence intervals for the average waiting times. How do your results compare with the expected waiting time? What does this say about the use of analytical models in real situations?

16. Customer satisfaction is the most important objective at the Midwestin Hotel. Consumer research has shown that a critical factor in customer satisfaction is check-in time. The hotel currently has five clerks on duty. During peak times, 90 guests arrive each hour and spend three minutes checking in. A quality improvement team is considering two ideas to improve service. One is to have a dedicated clerk serve corporate customers exclusively. Corporate customers account for 30 percent of the hotel's business. Because of prearranged billing information, this will allow the hotel to reduce the registration time for these customers to two minutes on average. A second suggestion is to use a new automated kiosk to allow guests to check themselves in. Approximately 20 percent of the guests might be willing to do this. How would you analyze this information, and what recommendations would you make?

17. A small wholesale distributor sells an automobile after-market item. The distribution of daily demand is as follows:

Daily Demand	Probability
0	.1770
1	.0770
2	.3850
3	.2731
4	.0879

The lead time distribution is as follows (for example, if the lead time is 3, the order arrives on the third day after the order is placed):

Lead Time (days)	Probability
3	.10
4	.75
5	.10
6	.05

Unit cost is $54.90 for order sizes less than 117 units, and $49.40 for order sizes of 117 units or more. The item sells for $61.73. Other information obtained from company records includes the following:

1. Order cost is $25.00 per order.
2. Back order cost is $10.00 per stockout.
3. Unit storage cost is $0.75 per day.

Determine the best reorder point and reorder level for this situation, using simulation analysis and the Crystal Ball *Decision Table* extender.

18. Suppose that you did not know the formula for the return point and upper limit in the Miller-Orr model. Using your spreadsheet modification in Excel Exercise 8.2, further modify the model so that the return point and upper limit are decision variables (recognizin g that the upper limit must be larger than the return point). Design experiments to find the best values (in increments of $500) that minimize total transaction plus opportunity costs using the Crystal Ball *Decision Table* extender with 1,000 trials per simulation. (Warning: this may take a significant amount of computer time—run it overnight!)

19. Use OptQuest (see Chapter 7) to optimize your model in Problem 18.

20. Retailers often observe that daily demand for a product is related to the number of units on the shelf (even beyond the fact that the number of units on the shelf is an upper limit on sales). Customers seem to buy more if the shelf is full. The following table shows the probabilities for different sales levels, given the number of units on the shelf for one product:

Units on the shelf (start of day)	Demand			
	0	10	20	30
0–20	0.1	0.7	0.2	0.0
21–50	0.0	0.1	0.6	0.3
51 or more	0.0	0.0	0.3	0.7

Assume that 100 units is the maximum amount that can be stored on the shelf and that no storage space is available beyond that for this product. The profit per unit sold is $4.75, ordering cost to replenish the shelf is $17 (regardless of the order quantity and whether goods arrive overnight in time to be placed on the shelf for the next day), and demand over available units is simply lost (a stockout), which is estimated to cost $6 per unit.

Evaluate the following two stocking policies with a 20-day simulation.

Policy 1: Restock with 80 more units whenever the number of units on the shelf drops below 21.

Policy 2: Restock every two days with whatever amount is needed to bring the number of units on the shelf up to 100.

Assume that there are currently 63 units on the shelf.

21. Construct an Arena model for the single server queueing system in Figure 8.14. Include a plot of the number in process.

22. Prescriptions at a hospital pharmacy arrive at an average rate of 22 per hour, with exponentially distributed times between arrivals. One pharmacist can fill a prescription at a rate of 12 per hour, or in about 5 minutes, exponentially distributed. Pharmacists are paid $30 per hour, and hospital administrators have attached a cost of $100 per hour to waiting on prescriptions. Develop an Arena model and use it to determine the optimal number of pharmacists to have.

23. A conveyor brings parts to a workstation at a constant rate of one every 10 minutes. Service is random (exponentially distributed). Develop an Arena model to help you find the mean service rate so that the average waiting time does not exceed 5 minutes.

24. A single clerk fills two types of orders. The first type arrive randomly, with a time between arrivals of between 6 and 24 minutes; the second type of order has a time between arrivals of from 45 to 75 minutes. It takes the clerk anywhere between 5 and 15 minutes to fill an order. Develop an Arena model to simulate this system and determine the average number of orders waiting, their average waiting time, and percentage of time the clerk is busy.

25. Computer monitors arrive at a final inspection station with a mean time between arrivals of 9.0 minutes (exponentially distributed). Two inspectors test the units, which take an average of 10 minutes (again exponential). However, only 90 percent of the units pass the test; the remaining 10 percent are routed to an adjustment

station with a single worker. Adjustment takes an average of 30 minutes, after which the units are routed back to the inspection station. Construct an Arena model of this process and determine the average waiting time at each station, number of units in each queue, and percentage of use of the workers.

NOTES

1. Richard Gibson, "Merchants Mull the Long and the Short of Lines," *Wall Street Journal,* September 3, 1998, B1.
2. Systems Modeling Corporation, 504 Beaver Street, Sewickley, PA 15143; (412) 741-3727; **smcorp@sm.com; www.sm.com.**
3. Adapted from *Guide to Arena Business Edition* (Sewickley, Pa.: Systems Modeling Corporation, 1999).
4. Adapted from Mark G. Simkin and Manalur Sandilya, "Simulation on Spreadsheets," undated paper, College of Business, University of Nevada, Reno.
5. Adapted from Larry Cornwell and Doan T. Modianos, "Management Tool: Using Spreadsheets for Simulation Models," *Production and Inventory Management Journal,* Vol. 31, No. 3, Third Quarter 1990, 7–17.

BIBLIOGRAPHY

Bodily, Samuel E. 1986. "Spreadsheet Modeling as a Stepping Stone." *Interfaces,* Vol. 16, No. 5, September–October, pp. 34–52.

Cebry, Michael E., Anura H. deSilva, and Fred J. DeLisio. 1992. "Management Science in Automating Postal Operations: Facility and Equipment Planning in the United States Postal Service." *Interfaces,* Vol. 22, No. 1, January–February, pp. 110–130.

Cornwell, Larry W., and Doan T. Modianos. 1990. "Management Tool: Using Spreadsheets for Simulation Models." *Production and Inventory Management,* Vol. 31, No. 3, Third Quarter, pp. 8–17.

Edie, Leslie C. 1954. "Traffic Delays at Toll Booths." *Journal of the Operations Research Society of America,* Vol. 2, No. 2, pp. 107–138.

Gravel, Marc, and Wilson L. Price. 1991. "Visual Interactive Simulation Shows How to Use the Kanban Method in Small Business." *Interfaces,* Vol. 21, No. 5, September–October, pp. 22–33.

Kelton, W. D., R. P. Sadowski, and D. A. Sadowski. 1998. *Simulation with Arena.* New York: McGraw-Hill.

Landauer, Edwin G., and Linda C. Becker. 1989. "Reducing Waiting Time at Security Checkpoints." *Interfaces,* Vol. 19, No. 5, September–October, pp. 57–65.

Martel, D. L., R. J. Drysdale, G. E. Doan, and D. Boychuk. 1984. "An Evaluation of Forest Fire Initial Attack Resources." *Interfaces,* Vol. 14, No. 5, September–October, pp. 30–42.

Maurer, Ruth A., and Fines H. Munkonze. n.d. "A Simulation Model of an Airport Shuttle Service." Department of Mathematical and Computer Sciences, Colorado School of Mines, Golden, Colo.

Ravindran, A., B. L. Foote, A. B. Badiru, L. M. Leemis, and Larry Williams. 1989. "An Application of Simulation and Network Analysis to Capacity Planning and Material Handling Systems at Tinker Air Force Base." *Interfaces,* Vol. 19, No. 1, January–February, pp. 102–115.

Russell, Robert, and Regina Hickle. 1986. "Simulation of a CD Portfolio." *Interfaces,* Vol. 16, No. 3, May–June, pp. 49–59.

Saunders, Gary. 1987. "How to Use a Microcomputer Simulation to Determine Order Quantity." *Production and Inventory Management Journal,* Vol. 28, No. 4, Fourth Quarter, pp. 20–23.

Thesen, Arne, and Laurel E. Travis. 1992. *Simulation for Decision Making.* St. Paul, Minn.: West Publishing Co.

Willis, David O., Jerald R. Smith, and Peggy Golden. 1997. "A Computerized Business Simulation for Dental Practice Management." *Journal of Dental Education,* Vol. 61, No. 10, October, pp. 821–828.

Appendix A

Normal Curve Areas

The following table gives the standard normal probability in right-hand tail. For negative values of z, areas are found by symmetry. For the probability area to the left of the specified value of z, subtract the area given in the table from 1.

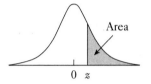

Area

0 z

Second decimal place of z

z	0.00	0.01	0.02	0.03	0.04	0.05	0.06	0.07	0.08	0.09
0.0	0.5000	0.4960	0.4920	0.4880	0.4840	0.4801	0.4761	0.4721	0.4681	0.4641
0.1	0.4602	0.4562	0.4522	0.4483	0.4443	0.4404	0.4364	0.4325	0.4286	0.4247
0.2	0.4207	0.4168	0.4129	0.4090	0.4052	0.4013	0.3974	0.3936	0.3897	0.3859
0.3	0.3821	0.3783	0.3745	0.3707	0.3669	0.3632	0.3594	0.3557	0.3520	0.3483
0.4	0.3446	0.3409	0.3372	0.3336	0.3300	0.3264	0.3228	0.3192	0.3156	0.3121
0.5	0.3085	0.3050	0.3015	0.2981	0.2946	0.2912	0.2877	0.2843	0.2810	0.2776
0.6	0.2743	0.2709	0.2676	0.2643	0.2611	0.2578	0.2546	0.2514	0.2483	0.2451
0.7	0.2420	0.2389	0.2358	0.2327	0.2296	0.2266	0.2236	0.2206	0.2177	0.2148
0.8	0.2119	0.2090	0.2061	0.2033	0.2005	0.1977	0.1949	0.1922	0.1894	0.1867
0.9	0.1841	0.1814	0.1788	0.1762	0.1736	0.1711	0.1685	0.1660	0.1635	0.1611
1.0	0.1587	0.1562	0.1539	0.1515	0.1492	0.1469	0.1446	0.1423	0.1401	0.1379
1.1	0.1357	0.1335	0.1314	0.1292	0.1271	0.1251	0.1230	0.1210	0.1190	0.1170
1.2	0.1151	0.1131	0.1112	0.1093	0.1075	0.1056	0.1038	0.1020	0.1003	0.0985
1.3	0.0968	0.0951	0.0934	0.0918	0.0901	0.0885	0.0869	0.0853	0.0838	0.0823
1.4	0.0808	0.0793	0.0778	0.0764	0.0749	0.0735	0.0721	0.0708	0.0694	0.0681
1.5	0.0668	0.0655	0.0643	0.0630	0.0618	0.0606	0.0594	0.0582	0.0571	0.0559
1.6	0.0548	0.0537	0.0526	0.0516	0.0505	0.0495	0.0485	0.0475	0.0465	0.0455
1.7	0.0446	0.0436	0.0427	0.0418	0.0409	0.0401	0.0392	0.0384	0.0375	0.0367
1.8	0.0359	0.0351	0.0344	0.0336	0.0329	0.0322	0.0314	0.0307	0.0301	0.0294
1.9	0.0287	0.0281	0.0274	0.0268	0.0262	0.0256	0.0250	0.0244	0.0239	0.0233
2.0	0.0228	0.0222	0.0217	0.0212	0.0207	0.0202	0.0197	0.0192	0.0188	0.0183
2.1	0.0179	0.0174	0.0170	0.0166	0.0162	0.0158	0.0154	0.0150	0.0146	0.0143
2.2	0.0139	0.0136	0.0132	0.0129	0.0125	0.0122	0.0119	0.0116	0.0113	0.0110
2.3	0.0107	0.0104	0.0102	0.0099	0.0096	0.0094	0.0091	0.0089	0.0087	0.0084
2.4	0.0082	0.0080	0.0078	0.0075	0.0073	0.0071	0.0069	0.0068	0.0066	0.0064
2.5	0.0062	0.0060	0.0059	0.0057	0.0055	0.0054	0.0052	0.0051	0.0049	0.0048
2.6	0.0047	0.0045	0.0044	0.0043	0.0041	0.0040	0.0039	0.0038	0.0037	0.0036
2.7	0.0035	0.0034	0.0033	0.0032	0.0031	0.0030	0.0029	0.0028	0.0027	0.0026
2.8	0.0026	0.0025	0.0024	0.0023	0.0023	0.0022	0.0021	0.0021	0.0020	0.0019
2.9	0.0019	0.0018	0.0018	0.0017	0.0016	0.0016	0.0015	0.0015	0.0014	0.0014
3.0	0.00135	0.00131	0.00126	0.00122	0.00118	0.00114	0.00111	0.00107	0.00104	0.00100
3.1	0.00097	0.00094	0.00090	0.00087	0.00084	0.00082	0.00079	0.00076	0.00074	0.00071
3.2	0.00069	0.00066	0.00064	0.00062	0.00060	0.00058	0.00056	0.00054	0.00052	0.00050
3.3	0.00048	0.00047	0.00045	0.00043	0.00042	0.00040	0.00039	0.00038	0.00036	0.00035
3.4	0.00034	0.00032	0.00031	0.00030	0.00029	0.00028	0.00027	0.00026	0.00025	0.00024
3.5	0.00023	0.00022	0.00022	0.00021	0.00020	0.00019	0.00019	0.00018	0.00017	0.00017

Index

A

Absolute priorities approach, 180–181
Additive seasonality, 111–113
Add Trendline, forecasting with, 122
Advanced AutoFilter, 58
Aggressive strategy, 255–256
Airline crews, scheduling, 3
Algorithms, 24
Allied-Signal, 97–98
ALPHA/Sim, 361
ALPHATECH, Inc., 361
Alternative decisions, data tables in evaluating, 20–22
Alternative optimal solutions, 169
Alternatives, generating, 249–250
American Airlines, 3, 4
    crew scheduling at, 200–201
    overbooking problem at, 261–262
American option, 258
American Software, 125
AMPL, 184, 233
Analysis Toolpak, 7
Analytical queueing models, 346–354
    and Little's law, 349–350
    M/M/I queueing model, 347–348
    versus simulation, 350–351
Analytical solutions, 24
Anbil, Ranga, 3, 43, 201, 241
Anderson, Anita, 43n
Anderson-Darling test, 73
Andrew, Joseph D., Jr., 44
Andrews, B. H., 98, 139
Annual demand, 30
Answer Report, 170–171

Applied Decision Analysis, 269
Arbitrary service times, 348
Arena, 361
Atre, Shaku, 95
AutoFilter in checking data accuracy, 59
Automatic software, 125
AVERAGE, 45
Average payoff strategy, 255

B

Back orders, 27
Badiru, A. B., 378
Baird, Bruce F., 283n
Balance constraints, 147
Balson, William E., 248, 284
Baumol model for cash management, 37
Bean, L. L. (company)
    call-center demand at, 98
    and competitive advantage, 52–53
    queueing models for telemarketing at,
        353–354
Becker, Linda C., 329, 378
Ben-Dov, Yosi, 4, 43, 201, 241
Binary integer program, 203, 204
Binary variables, 201–202, 203
Binomial distribution, 72
Blackstone, John H., Jr., 321
Blending problem, 152–154
BMPD Statistical Software, 91
Bodily, Samuel E., 378
Boeing, Inc., 287–288
Bohl, Alan H., 201, 241
Bowden, Royce O., 241n
Boychuk, D., 330, 378

Boyd & Fraser Publishing Co., 184, 233
Bradley, Stephen P., 196
Branch and bound, 212
Branching, 212
Break-even probability, 273
Brown, A. A., 8, 43
Brown, Ken, 43*n*
Bryant, Adam, 94*n*
Bubble charts, 62
Buffa, F., 196*n*
Bunn, Derek W., 284
Burke Construction, 205–206
Byrd, Jack, Jr., 284

C

Cafasso, R., 51, 95
Call-center demand at L. L. Bean, 98
Call option, 259
Camm, Jeffrey D., 4, 44, 94*n*, 196, 241
Carlson, Walter, 3, 44
Carraway, R., 241
Carrying charge rate, 30
Carter, Virgil, 249, 284
Cash management, applying inventory theory to, 37–38
Catastrophic risk, assessment of, 248–249
CB Predictor, 7, 125, 126–133
Cebry, Michael E., 329, 378
Central America (ship), 4
Champy, James, 2, 43*n*
Charts, 61–63
Chemco, Inc., 207–208
Chi-square goodness-of-fit test, 72–73
Chorman, Thomas E., 4, 44, 196*n*
Clarke, John R., 284
Clements, D. W., 139
Coefficient
    of determination, 76
    of variance, 258
Cohan, David, 284
Cohen, M., 139
Collegiate athletic board drug testing, 252–253
Colorado Cattle Company, 152–154, 168
Competitive advantage, 52–53
Computerized business simulations, 329
Conditional Formatting, 43*n*
Confidence interval for mean profit, 302–303
Conservative strategy, 256
Constraint coefficients, 141–142
Constraint functions, 142
Constraints, 14
Continuous knapsack problem, 214
Continuous review system, 28–29
Controllable variables, 11

Copeland, Duncan, 3, 44
Cornwell, Larry, 378
Correlated assumptions, 321–323
Correlation, 74–76
Costs, separating fixed and variable, 90
Count, 67
Covariance, 74–76
    applying, in financial portfolio optimization models, 77–78
    and correlation, 66
CPLEX Optimization, Inc., 184
Credit collection, application of linear multiobjective programming model for, 144
Crystal Ball, 291
    building simulation models with, 294–302
    features of, 321–326
    interpreting output, 301
    replication of MMI model, 332–333
    simulation of inventory system, 356
    Tornado Chart Extender, 310
Crystal Ball Pro, 7
Cummins, J., 241
Cunningham, S. M., 98, 139
Custom Autofilter, 57
Customer characteristics in queueing systems, 335–336
Cyclical component, 99

D

Daniel, T. E., 284
Dano, Sven, 196, 241
Dantzig, George B., 196
Darrow, Ross M., 4, 44
Data
    importance of good, 16
    sorting and filtering, 55–59
    storage and retrieval, 51–60
    using *AutoFilter* to check accuracy, 59
    using pivot tables and summarize, 68–71
Data analysis, 50, 66–78
Data base management systems (DBMS), 53
Data management, 50
Data management and analysis
    for health care, 50–51
    for hotel location, 51
    for law enforcement, 50
    for retail operations and strategy, 51
Data mining, 53
Data retrieval methods in optimization modeling, 59–60
Data tables, 17
    in evaluating uncertainty and alternative decisions, 20–22

Data visualization, 60
   charts in, 61–63
   correlation in, 74–76
   covariance in, 74–76
   descriptive statistics in, 67
   geographic maps in, 63–66
   pivot tables in, 68–71
   probability distributions and fitting with Crystal Ball, 71–74
Data warehouse, 51–52
Dearing, P. M., 196
Decision analysis, 247–248
   applications of, 248, 258–260
   of put and call options, 259–260
   technology for, 268–269
Decision Analysis by TreeAge (DATA), 269
Decision branches, 251
Decisioneering, Inc., 125, 287, 288
Decision Focus Inc., 248
Decision nodes, 251
Decision rule, 37
Decisions
   alternatives for, 16, 249–250
   criteria for, 250
   risk in making, 254–260
   structuring problems, 249–254
   understanding risk in making, 254–260
   variables for, 11
Decision strategies, 254
   sensitivity analysis of, 268
Decision technology, 2, 4–8
   spreadsheets in, 5–7
   stand-alone software in, 7–8
Decision trees, 251–254
   analysis of, 7
DeLisio, Fred J., 378
DELPHUS, Inc., 125
Demand Forecasting, 125
Demand forecasts, 204
Dental practice management, dynamic system model for designing, 329
Descriptive models, 8, 11–15
Descriptive statistics, 66, 67
deSilva, Anura H., 329, 378
Detelle, J. D., 8, 43
Deterministic arrivals, 335
Diet problem, 153
Dill, Franz A., 4, 44, 196n
Dimensionality checks, 148
Discrete distribution, simulating outcomes from, 292–294
Distinction Software, Inc., 125

Distribution
   binomial, 72
   exponential, 71
   lognormal, 72
   normal, 71
   Poisson, 71
   triangular, 71
   uniform, 71
Doan, G. E., 330, 378
Dodge, Jeffrey L., 144, 197
Dodge, M., 94n, 95
Double exponential smoothing models, 107, 109
   forecasting with, 109–110
Double moving averages, forecasting with, 107–108
DPL, 269
Drysdale, R. J., 330, 378
Du Pont, 4
DuPont Merck, 287
Durban-Watson statistic, 133
Dynamic inventory systems, modeling and simulating, 354–358
Dynamic production/inventory model, 330–332
Dynamics of waiting lines, 342–346
Dynamic systems, 328–372
   applications of models, 328–330
   modeling and simulating, 330–334
   queueing, 335–338
      analytical models, 346–354
      characteristics in, 337
      customer characteristics in, 335–337
      modeling and simulating, 339–342
      service characteristics in, 336–337
      system configuration, 337–338
   of waiting lines, 342–346

E

Ecker, Joseph E., 144, 197
Economic order quantity (EOQ) model, 34
   validity of, 35–36
Economic order quantity (EOQ) value, 34
Edie, Leslie C., 9, 378
Electronic data interchange (EDI), 53
Emery, Douglas R., 44
Engemann, Kurt J., 284
Erlang, A. K., 335
Evans, James R., 4, 44, 196, 241n, 321
Event branches, 251
Event nodes, 251
Events, 250
Eviews, 125

Excel, 53
    Analysis Toolpak, 67
    AutoFilter, 56–57
    modeling optimization problems in, 145–146
Excel Solver, 7, 34, 166–167
    for integer programs, 212–214
    interpreting sensitivity analysis and reports from, 169–177
    output and linear programming (LP) with lower or upper bounds, 173–174
    in solving nonlinear programming problems, 226–228
Expected value analysis, 263–264
Expected value decision making, 260–265
Expected value of perfect information, 265
ExperCorp, 286–287
Expert Choice, Inc., 269
Expert Choice Professional, 269
Exponential distribution, 71, 294, 336
Exponential smoothing, forecasting with, 106
Exponential utility functions, 275–276
Extend & BPR™, 361

F

Face validity, 371
Facility location model, 223–225
FACTOR/AIM, 361
Farley, A., 241
Feasible solution, 24
Federal Aviation Administration (FAA) work rules, 3
Federal Express, 3
Feinstein, Charles D., 284
Field, Richard C., 197
Financial strategies, 4
Finite calling populations, 349
Finnerty, John D., 44
First-come, first-served (FCFS), 337
Fitzsimmons, James A., 51, 95
Foote, B. L., 378
Forecast chart, 298
Forecaster 2000, 125
Forecasting
    applications of, 97–98
        call-center demand at L. L. Bean, 98
        inventory management at IBM, 98
        to model building, 133–135
        plant closure at Allied-Signal, 97–98
    with CB Predictor, 126–133
    choosing best method, 126
    definition of, 97
    with double moving averages, 107–108
    with exponential smoothing, 106
    with Holt-Winters models, 116–117
    measuring accuracy in, 102–103

    models for, 8, 10
        for supermarket checkout services, 118–120
    models for time series
        with only an irregular component, 102–106
        with seasonal component, 111–115
        with trend and seasonal components, 115–117
        with trend component, 107–110
    with regression models, 117–125
    with single moving averages, 103–104
    software, 125
    time series components, 98–102
Forest fire management
    decision criteria in, 250–251
    dynamic system model for, 330
Freeland, J., 241
Freezing assumptions, 323
Frontline Systems, Inc., 166, 184, 233
Functions, 44–46

G

Gaballa, A., 200, 241
Gale, James R., 284
Gallagher, Timothy J., 44
GAMS, 184, 233
Gass, Saul I., 197
GE Capital, 144
Gelman, Eric, 3, 43, 201, 241
General integer variables, 201
Geographic data maps, 63–66
Geographic information system (GIS) mapping program, 16
Gibson, Richard, 378
GINO, 233
Global optimal solution, 229
Goal-programming approach, 181–184
Golden, Peggy, 378
Goodness-of-fit tests, 72–73, 103
    Chi-square, 72–73
GPSS/H, 361
Granfors, Donna C., 144, 197
Grant, Richard S., 241*n*
Gravel, Marc, 378
GRG2, 233

H

Haas, Stephen M., 284
Hadley, William H., 241*n*
Hahn, Gerald J., 144, 197
Hall, John D., 241*n*
Hall, N., 241
Hammer, Michael, 2, 43*n*
Hanover, Inc., 54, 90, 160, 247
Hax, Arnoldo C., 196
Hayre, Lakhbir, 4, 43, 201, 241

Health care, applications of data management and analysis in, 50–51
Hedgers, 256
Heian, Betty C., 284
Helmer, F., 51, 95
Helmer, M., 145, 196*n*, 197
Hershey, J., 241
Hertz, David B., 321
Heuristics, 25
    as solution to product mix problem, 25–26
Hickle, Regina, 378
HLOOKUP, 46
Holding cost, 30
Holt, Charles, 115
Holt-Winters additive model, 115–116
Holt-Winters multiplicative model, 116
Homer, Jack B., 44
Hosseini, Jinoos, 284
Hotel location, applications of data management and analysis in, 51
Hueter, Jackie, 10*n*
Hulswit, E. L., 8, 43
Hyndman, R. J., 102, 139

I

IF, 45
Imagine That, Inc., 361
Indicator variables
    forecasting with, 123–125
    in regression models, 123
Influence diagram, 8, 11, 30–31
Information, quantitative, 26
Inputs, uncontrollable, 11
Integer, applications of, 200–201
Integer cutoff parameter, 243
Integer knapsack problem, 214
Integer models, 141
Integer optimization, technology for, 233
Integer programming models, 200
    building, 201–212
    solving, 212–218
Integer programs
    Premium Solver in solving, 242–245
    using Solver for, 212–214
Intuition, 8
Inventory Analyst Pro, 125
Inventory holding costs, 27
    per unit, 31–32
Inventory management, 27–38
    at IBM, 98
    setting service level for, 270–272
Inventory operation, 328
Inventory systems
    Crystal Ball simulation of, 356

managing, 27–29
    simulation in optimizing, 357–358
Inventory theory, applying, to cash management, 37–38
Investment allocation, linear programming model for, 148–150
Irregular component, 99

J

Jack's Job Shop, 223
Jackson, W., 196*n*
Janssen, C. T. L., 284
Johnson, R., 196*n*
Just-in-time purchasing, 29, 36

K

Kamesam, P. V., 139
Kanban, 225–226
Kay, E., 94*n*, 95
Keeney, Ralph L., 283*n*
Kelton, W. David, 321, 370, 378
Kendall, D. G., 347
Kessler, L., 241
Kimes, Sheryl E., 51, 95, 95*n*
Kinata, C., 94*n*, 95
Kolmogorov-Smirnov test, 73
Kotler, Philip, 196*n*, 197
Krumm, F. V., 4, 44
Kurtosis, 67
Kwak, N., 145, 196*n*, 197

L

Lagrange multipliers, 228
Landauer, Edwin G., 329, 378
Land O'Lakes, 52
Language multiplier, 227
Lanner Group, Inc., 361
La Quinta Motor Inns, 51
Last-come, first-served (LCFS), 337
Law, Averill M., 319*n*, 321
Law enforcement, applications of data management and analysis in, 50
Lee, H., 139
Leemis, L. M., 378
Leimkuhler, John F., 4, 44
Limits Report, 171
LINDO Systems, 184, 233
Linear integer programs, solving, 242–243
Linear models, 141
    applications of, 143–145
    structure of, 141–143
Linear optimization, technology for, 184
Linear program, spreadsheet model for, 145–146

Linear programming models, 142
 building, 146–163
 dimensionality checks, 148
 examples of, 148–163
 solving, 164–169
Linear programming relaxation, 212
Linear regression, 78
 multiple, 86–90
 simple, 79
 for time series forecasting, 121–123
LINGO, 233
Little, John D. C., 349
Little's law, 349–350
Local optimal solution, 229
Lognormal distribution, 72
Lord Corporation, 144
Ludwig, W., 144, 197
Luenberger, David G., 284n

M

Machol, Robert, 249, 284
Magnanti, Thomas L., 196
Mairose, R., 241n
Makridakis, S., 102, 139
Makuch, William M., 144, 197
Management science, 2, 9–10
 applications and benefits of, 3–4
Managerial uses of sensitivity information, 173
Manual software, 125
Marketing decision variables, 39
Marketing effort, allocation of, 150–152
Markowitz, H. M., 95n, 241n, 242
Markowitz portfolio model, 77, 221–223
Martel, D. L., 330, 378
Martin, R. K., 241n
Mason, Richard O., 3, 44
Mathematical programming, 141
Mathur, K., 242
Maurer, Ruth A., 378
MAX, 45
Maximum, 67
MCI Communications (now MCI Worldcom), 53
McKenney, James L., 3, 44
Mean, 67
Mean absolute deviation (MAD), 103
Mean absolute percentage error (MAPE), 103
Mean interarrival time, 336
Median, 67
Metaheuristics, 231
Microsoft Access, 53
Microsoft Map, 64
Miller, Holmes E., 284
Miller-Orr simulation model, 37, 43
 similarity between MMI and, 334

MIN, 45
Minimum, 67
MINITAB, 91, 125
Minitab, Inc., 91, 125
MINOS, 184, 233
Mixed integer program (MIP), 206, 207
M/M/1 queueing model, 347–348
MMI models
 arena model of, 368–370
 Crystal Ball replication of, 333
 similarity between Miller-Orr simulation models and, 334
Mode, 67
Model analysis, 17–22
 and solution, 32–35
Models, 8
 analyzing and solving, 16–26
 applications of forecasting to building, 133–135
 descriptive, 8, 11–15
 development of, 29–32
 forecasting, 8, 10
 interpreting and using results, 26–27
 involving uncertainty, 16
 optimization, 8, 10, 13–15
 predictive, 8
 product mix, 14–15
 profit, 11–13
 simulation, 8, 10
Modianos, Doan T., 378
Monte Carlo simulation, 286
 applications of, 286–288
 building and implementing models, 288–291
 examples, 303–309
 statistical issues in, 302–303
Moore, L. Ted, 284
Mortgage processing, simulation model of, 363–368
Mortgage valuation models at Prudential Securities, 201
Multi-collinearity, 90
Multidimensional data, 62
Multiobjective decisions, risk trade-offs and, 269–272
Multiobjective linear programs, 142–143
Multiobjective models, 141
 solving, 177–183
Multiobjective optimization models, structure of, 141–143
Multiobjective programming models, applications of, 143–145
Multiperiod production planning, 155–158
Multiple linear regression, 78, 86–90
Multiple servers, 349
Multiplicative forecasting model, 99
 forecasting seasonal time series using, 114–115
Multiplicative seasonality, 113–115

Multiplicative time series model, components of, 99–101
Multistage decision trees, 251
Munkonze, Fines H., 378

N

Nash, Stephen G., 241n
National Cancer Institute, project selection at, 200
National Forest Management Act (1976), 143
National Forest Service, application of linear and multiobjective models for, 143–145
NCR, 53
Nemhauser, G. L., 242
New product introduction, 253
Newsletter problem, 263–264
Newsvendor problem, 304
New venture planning, 286–287
New York City Department of Sanitation, 27
Nichols, Nancy A., 283n
Nodes, 8, 11
Nonlinearity, problems of, 229–231
Nonlinear models, 141
    applications of, 200–201
Nonlinear optimization
    models for, 77, 200
        building, 218–225
    technology for, 233
Nonlinear programming problems, solving, 226–232
Nonlinear programs
    Premium Solver in solving, 242–245
    solving, 243–245
Nonnegativity restrictions, 148
Normal distribution, 71, 293

O

Objective function coefficients, 141
On-line analytical processing (OLAP), 52
Opperman, E., 51, 95
Opportunity loss strategy, 256–257
Optcontrol, Inc., 134
Optimal advertising budget allocation model, 220–221
Optimal Methods Inc., 233
Optimal solution, 24
Optimization and simulation, 312–315
Optimization models, 8, 10, 13–15, 141
    data retrieval methods in, 59–60
    forecasting function in, 134–135
    regression in, 90
    solving, 24–25
Optimization problems, modeling, in Excel, 145–146
OptQuest, 312
Oracle, 53
Ordering cost, 30

Order preparation costs, 27
Outcomes, 250
    defining, 250
    simulating, from discrete distribution, 292–294
Overlay charts, 323

P

Palisade Corporation, 269, 291
Paper production, 201
Parameters, 11
    testing significance of, 83–85
Patty, Bruce, 3, 43, 201, 241
Payoff, 250
Payoff table, 250
Pearce, W., 200, 241
PEER Planning for Windows, 125
Percentiles summary, 298
Perfect information, expected value of, 265
Per-unit contribution margins, 147
Pharmaceutical research, 287
Phillips, Inc., 150–152, 218, 220–221
Pica, Vincent, 4, 43, 201, 241
Pivot tables, 68–71
Plant closure at Allied-Signal, 97–98
Poisson distribution, 71
Poisson process, 335–336
Portfolio optimization model, applying covariance in financial, 77–78
Positive coefficients, 325
Poultry production, 3
Precision Tree, 269
Prediction intervals, 85–86
Predictive models, 8
Premium Solver, 7, 184
    evolutionary algorithm, 231
    software for, 231
    in solving integer and nonlinear programs, 242–245
Premium Solver Plus, 184, 233
Price, Wilson L., 378
Principle of insufficient reason, 255
Probabilistic arrival, 335
Probability distributions
    and data fitting with Crystal Ball, 71–74
    and distribution fitting, 66
    sampling from, 292
Problem Solver, 233
ProcessModel, 361
Process models, 334
Procter & Gamble Company, 4, 64, 162–163
Production/inventory model, 368–370
Production lot size model, 42–43
Product mix model, 14–15
Product mix problem, heuristic solution to, 25–26

Product sourcing model, 163
Profitability simulation, 288–291
Profit models, 11–13
Project management, 287–288
PROMODEL Corporation, 361
Proportional relationships, 147
Prudential Securities, 4
    mortgage valuation models at, 201
Purchase costs, 27
Put and call options
    decision analysis of, 259–260
    evaluating, 258–260
Put option, 258

Q

Qantas, 204
Quantitative information, 26
Quantitative Micro Software, 125
Queen City, Inc., 202–203
Queueing models
    analytical, 346–354
    for telemarketing, 353–354
Queueing systems, 335–338
    characteristics in, 337
    configuration in, 337–338
    customer characteristics in, 335–336
    modeling and simulating, 339–342
    service characteristics in, 336–337
Queueing theory, 335
Queues, 328, 335
    characteristics in queueing systems, 337

R

R^2 value, 219–220
Radar charts, 62
Radloff, David L., 284, 284n
RAND, 45
Random number, 292
Random variates, 293
Range, 67
Rank correlation coefficients, 325
Raturi, R., 241
Ravindran, A., 378
Regression
    in optimization models, 90
    significance of, 83
    strength of relationship, 82–83
Regression analysis, 66, 78–90, 219
    statistical issues in, 82–86
Regression models
    building, 90
    forecasting with, 117–125
    indicator variables in, 123
Reid, R. A., 139

Reorder point, 35
Research and development funding, application of
    linear multiobjective programming model for,
    144
Retail operations and strategy, applications of data
    management and analysis in, 51
Return on investment (ROI), 39
Return-to-risk, 258
Right-hand side values, 142
Risk, 248
    in decision making, 254–260
    quantifying, 257–258
@Risk, 291
Risk assessment for space shuttle, 249
Risk averse, 256
Risk profile, 291
Risk trade-offs and multiobjective decisions, 269–272
Rollback value, 266
Rolle, C. F., 4, 44
Russell, Robert, 378

S

Sabre Decision Technologies, 53
Sadia Concordia SA, 3
Sadowski, D. A., 370, 378
Sadowski, R. P., 370, 378
St. Clair, Christina L., 44
Salkin, H. M., 242
Salvia, A., 144, 197
Sample correlation coefficient, 76
Sample covariance, 74–75
Sample variance, 67
Sampling distribution of means, 301
Sampling from probability distributions, 292–294
Sandilya, Manalur, 378
Santa Cruz MicroProducts, 14–15, 16, 26, 141,
    146–147, 148
SAS Institute, Inc., 91
SAS System, 53, 91
Saunders, Gary, 378
Scenarios, 17
    analysis of, 174–176
Schniederjans, M., 145, 196n, 197
Seasonal component, 99
Seasonality
    additive, 111–113
    multiplicative, 113–115
Seasonal time series, forecasting, using additive
    model, 112–113
Secondary mortgage market, 4
Security checkpoints, dynamic system model for de-
    signing, 329
Semiautomatic software, 125
Sensitivity, 17

Sensitivity analysis, 214–218, 351–352
    of decision strategies, 268
    interpreting Solver reports and, 169–177
    in solving nonlinear program problems, 228–229
Sensitivity charts, 324–326
Sensitivity information, managerial uses of, 173
Sensitivity Report, 171–174
Sensitivity/Supertree, 269
Service characteristics in queueing systems, 336–337
Service level, 270
    setting, for inventory management, 270–272
Service process, 336
Setup cost, 43
Sharma, R., 94n, 95
Sharpe ratio, 258
Shortage costs, 27
Siegel, Matt, 284n
Simkin, Mark G., 378
Simple linear regression, 78, 79
Simulation, 286
    versus analytical models, 350–351
    building, models with Crystal Ball, 294–302
    and optimization, 312–315
    in optimizing inventory systems, 357–358
    successful use of, 370–372
    validation for, 371–372
    verification for, 371–372
Simulation models, 8, 10
    for cash management, 333–334
    for mortgage processing, 363–368
Single exponential smoothing, 105–106
Single moving averages, 102
    forecasting with, 103–104
Single stage decision trees, 251
Site selection, application of linear multiobjective
    programming model for, 144–145
Skewness, 67
Slack variable, 154
SmartForecasts for Windows, 125
Smart Software, 125
Smith, Barry C., 4, 44
Smith, G. N., 43n
Smith, Jerald R., 329, 378
Software
    automatic, 125
    forecasting, 125
    manual, 125
    semiautomatic, 125
    stand-alone, 7–8
Space shuttle, risk assessment for, 249
Speculators, 256
Spider chart, 22
    creating, 48
Spreadsheet model for linear program, 145–146
Spreadsheets, 2, 17

    auditing your, 46
    basic skills in, 44–48
    in decision technology, 5–7
    effective use of, 6–7
SPSS, Inc., 91
SQRT, 45
Stand-alone software in decision technology, 7–8
Standard deviation, 67
Standard error, 67
    of mean, 302
Standard form, 154
Stanford Business Software, 184, 233
Stationary, 335
Statistical analysis, 249
    technology for, 91
Statistical issues in Monte Carlo simulation, 302–303
Statistics summary, 298
STDEV, 45
STDEVP, 45
Steady-state queueing system, 349
Stedman, C., 94n, 95
Steuer, Ralph, 196n
Stinson, C., 94n, 95
Stock options, pricing, 308–309
Stone, John D., 44
Stone, Lawrence D., 4, 44
Stotts, R., 241
Strategic business decisions, 3
Strategic Decisions Group, 269
Strike price, 258
SUM, 45
Sum, 67
Sun Bank, 209, 216–217
Suver, J., 51, 95
Swain, James J., 95
Swart, William, 10n
Sweeney, Dennis J., 4, 44, 196n, 241n
SYMIX/Pritsker Division, 361
System configuration in queueing systems, 337–338
Systems Modeling Corporation, 361
Systems simulation technology, 359–370

T

Taco Bell, 8, 9–10
Tadisina, Suresh K., 196
Tanga, Rajan, 3, 43, 201, 241
Target cell, 34
Taube-Netto, Miguel, 3, 44
Technology evaluation, dynamic system model for
    designing, 329
Technology for statistical analysis, 91
Tekerian, A., 139
Telemarketing, queueing models for, 353–354
Tepe, Jerry, 43n

Terminal nodes, 251
Theil's U statistic, 133
Thesen, Arne, 321, 378
Thommes, M. C., 43n
Time-averaged statistic, 344
Time series components, 98–102
    decomposing, 101–102
Time series forecasting, linear regression for, 121–123
Times series models
    with only irregular components, 102–106
    with seasonal components, 111–115
    with trend and seasonal components, 115–117
    with trend components, 107–110
Tinker (Air Force Base) Integrated Planning Simulation, 360–361
Tool booth improvement, dynamic system model for, 328–330
Tornado chart, 22–23
    creating, 46–48
Total marketing, 39
Transportation models, 59, 162
    for Procter & Gamble's product sourcing decisions, 162–163
Travis, Laurel E., 321, 378
Treasure, search for sunken, 4
TreeAge Software, Inc., 269
TreePlan, 7, 266
Trend, 99
Trend charts, 323–324
Trials, 288
Triangular distribution, 71, 294
Trip Reevaluation and Improvement Program (TRIP), 3
Truck Transport Corporation, 145
Tsubakitani, S., 241
Tufte, E. R., 95
Two-story data table, 21

U

Unbounded problem, 169
Uncertainty, 248
    data tables in evaluating, 20–22
    models involving, 16
Unconstrained optimization problem, 224
Uncontrollable inputs, 11
Uniform distribution, 71, 293
U.S. West, 53
User Solutions, Inc., 125
Utility and decision making, 272–276
Utility theory, 272

V

Validation for simulation, 371–372
Validity, 26
Value decision strategies, optimal expected, 266–268
Variables
    controllable, 11
    decision, 11
    marketing decision, 39
Vennemeyer, Jeanne, 43n
Verification for simulation, 371–372
Vijayan, J., 50, 95
Virgil Center, 249
Visual Basic modules, 7
VLOOKUP, 45–46
Vohio, Inc., 158

W

Waiting lines, 335
    dynamics of, 342–346
Wal-Mart, 52
Watson, Hugh J., 321
Wegryn, Glenn W., 4, 44, 196n
Weighted distances, 223
Weighted objective approach, 177, 179–180
Welsh, Justin L., 248, 284
What-if analysis, 17
What's Best!, 184, 233
Wheelwright, S. C., 102, 139
Whirlpool Corporation, 225–226
Williams, Larry, 378
Willis, David O., 329, 378
Wilson, Donald S., 248, 284
Winters, P. R., 115
WITNESS 9.0, 361
Wolsey, L. A., 242
Wolverine Software Corporation, 361
Woolsey, G., 44
Working capital management, 158–160

Y

Yancik, Richard F., 284

Z

Zicker, J., 94n, 95
Zorn, T., 196n